George A Crofutt

Crofutt's New Overland Tourist and Pacific Coast Guide

George A Crofutt

Crofutt's New Overland Tourist and Pacific Coast Guide

ISBN/EAN: 9783744724456

Printed in Europe, USA, Canada, Australia, Japan

Cover: Foto ©Andreas Hilbeck / pixelio.de

More available books at **www.hansebooks.com**

Episcopal Boarding and Day Schools,
DENVER, COL.

BISHOP SPALDING, Rector and President.

Wolfe Hall exclusively for Girls and Young Ladies.
Jarvis Hall exclusively for Boys and Young Men.

FOUNDED IN 1868–9.—THE OLDEST AND THE BEST IN THE STATE.

Healthy and delightful situation and surroundings. Strict discipline and thorough training in all branches. Large and comfortable buildings, and pleasant home life.
Music a Specialty. Asthmatics cured by the climate.
Christmas term begins first Wednesday in September; Easter term, 1st of February.
For catalogues with terms and other particulars, apply to the Bishop or to the Principal of either school, at Denver, Colorado.

THE ST. JAMES.
D. A. GAGE & CO., Proprietors.

Denver, Colorado.

Fine New Five-Story Hotel in the Centre of Business.

HYDRAULIC ELEVATOR.

GRAND OPERA HOUSE AND POST-OFFICE DIRECTLY OPPOSITE

FINEST VIEW OF THE ROCKY MOUNTAINS
TO BE HAD FROM THREE SIDES OF THE HOTEL.

Rates, $3 and $4 a Day.   Rooms with Baths Extra.

ALBERT C. HALE, E. M., Ph. D.,   MILTON MOSS, Ph. D.,
President of the Faculty,   *Professor of Mineralogy and Metallurgy,*

— OF THE —

COLORADO STATE SCHOOL OF MINES.

HALE & MOSS,
Mining Engineers,
METALLURGISTS AND CHEMISTS,

Examine and report on Mines, Mills and Furnace Property, and give advice on all matters pertaining to Mining, Metallurgy and Technical Chemistry.

P. O. BOX 129,   Golden, Colo.

C. P. HENDRIE, Sec'y and Treas. H. BOLTHOFF, Superintendent.

HENDRIE & BOLTHOFF MFG. CO.

MINING MACHINERY,

HOISTING ENGINES,

Smelting and Concentrating Works.

WESTERN AGENCY FOR

**National Tube Works Company,
Howe's Improved Scales,
Roeblings Sons' Steel Rope,
Knowles' Steam Pumps,
General Mine and Mill Supplies.**

**BOLTHOFF'S PATENT
COMBINATION STAMP MILL,**
All complete wooded.
All sizes from one Pony Mill
of 100 lb. Stamp up to 750 lb.
Stamp.

Office and Warerooms, cor. 17th & Wazee Sts.,
DENVER, COLORADO.

A FULL LINE KEPT IN STOCK.

NEW YORK OFFICE, 115 BROADWAY.

CHAIN & HARDY, Agents for W. H. JACKSON & CO.'S
414 LARIMER STREET, DENVER.

Booksellers
AND
Stationers.
Artists'
Materials,
Pictures
AND
Frames,
Fishing Tackle,
Books,
Maps,
AND
Pictures
OF
Colorado.

Photographs
OF ALL THE
*Principal
Points*
Of interest in
**Colorado,
New Mexico,
Wyoming
and Utah,**
Including
**Yellowstone,
National Park**
And the
Ancient Ruins
of the South West.

DESCRIPTIVE CATALOGUE FREE.

AMERICAN PROG

(See Annex No. 1.)

 *For a condensed description of the*

KANSAS PACIFIC RAILWAY,

—NOW KNOWN AS THE—

"Kansas Division" Union Pacific Railway,

THE COUNTRY THROUGH WHICH IT IS BUILT—SEE PAGE 48

FOR A FULL DESCRIPTION OF

COLORADO,

ITS RAILWAYS, AGRICULTURAL RESOURCES,

Mines of Gold, Silver, and other Precious Metals

ITS WATERING PLACES, GRAND SCENERY, ETC.,

—BUY—

CROFUTT'S GRIP-SACK GUIDE.

THE MAGNIFICENT ENGRAVINGS COST NEAR $10,000.

SOLD ON ALL TRAINS.

UTAH'S BEST CROP.

We were all little jokers once.   Photographed from life, by SAVAGE, Salt Lake City.

CROFUTT'S
NEW OVERLAND
TOURIST,
AND
PACIFIC COAST GUIDE.

CONTAINING A CONDENSED AND AUTHENTIC DESCRIPTION OF OVER

One Thousand Three Hundred Cities, Towns, Villages, Stations, Government Fort and Camps, Mountains, Lakes, Rivers, Sulphur, Soda and Hot Springs, Scenery, Watering Places. and Summer Resorts; WHERE

To look for and hunt the Buffalo, Antelope, Deer and other game; Trout Fishing, etc., etc. In fact, to tell you what is worth seeing—where to see it—where to go—how to go—and whom to stop with while passing over the

UNION, KANSAS, CENTRAL AND SOUTHERN PACIFIC RAILROADS,

Their Branches and Connections, by Rail, Water and Stage,

FROM SUNRISE TO SUNSET, AND PART THE WAY BACK;

Through Nebraska, Wyoming, Utah, Montana, Idaho, Nevada, California, Arizona and New Mexico.

Entered according to Act of Congress in the year 1883, by THE OVERLAND PUBLISHING CO., in the office of the Librarian of Congress, at Washington.

BY GEO. A. CROFUTT,

AUTHOR OF "GREAT TRANS-CONTINENTAL RAILROAD GUIDE," "CROFUTT'S TRANS-CONTINENTAL TOURIST" AND "CROFUTT'S GRIP-SACK GUIDE OF COLORADO."

1883.

OMAHA, NEB. AND DENVER, COL.:

THE OVERLAND PUBLISHING COMPANY

Sold by News Agents on the Railroads, at News Stands and at the Book Stores throughout the United States.

BARKALOW BROS., General News Agents, Union Pacific Railway and Branches; Missouri, Kansas & Texas R. R.; Kansas City, Ft. Scott & Gulf Railway; Kansas City, Lawrence & Southern Railway. RAILROAD NEWS CO., on Atchison, Topeka & Santa Fe; the Denver & Rio Grande, and Burlington & Missouri in Neb.

ELI S. DENISON, General News Agent, Central and Southern Pacific Railroads; General Agent for the Pacific Coast, Sacramento and San Francisco.

15th Volume—1883.

PREFACE.

WITH the world as the book of nature, God as the author, and the Bible as a preface, the precedent for writing a preface is established; and woe be to the Scribe who ignores precedent and custom—he could not live on this planet.

At the present day the preface of a book is read by the public—if at all —in the light of an apology, wherein the author is expected to explain, first, why he did not do better, and, second, why he wrote at all.

FIRST—We have spared neither time, pains nor money to make this a perfect book. Our statements are *concise, plain, unadorned,* and, we believe, *truthful in every particular.* Yet we should shudder at the charge of being *absolutely* perfect.

SECOND—We wrote this book for *Money* and *Love.* For MONEY to help the poor. For LOVE of the far western country—the land of the "Golden Fleece." For love of its broad plains and lofty mountains, its free pure air, healthful climate, magnificent scenery, unrivalled resources, and its unaffected, whole-souled people.

We have taken the traveler with us—in a chatty way—on the longest trip ever attempted by any author in any guide book in the world, and have recorded a telegram of the most important facts and items of information in a trip of over 15,000 miles by rail, steamer and stage coach. We have passed over the longest railroad line in the world, the broadest plains, the loftiest mountains, the finest agricultural and grazing lands, and the most barren deserts; we have climbed from sunrise to eternal snow, only to glide down into perpetual summer, and the orange groves and vineyards of the "Land of the Angels."

We have crossed a level prairie 500 miles in width, then over the most rugged mountains, with frightful chasms almost beneath us, 2,500 feet in depth; and through 100 miles of snow sheds and tunnels. Again, we have stood beneath a dome rising 6,000 feet above our heads, and trees 400 feet in height, and 48 feet in diameter; have strolled amid the redwoods, where they grow so thick that were they felled the ground would be covered to a depth of sixty feet. We have passed through the celebrated Echo, Weber,

Humboldt and Solidad cañons; over the great Colorado desert, around "CAPE HORN" and the "*Dead* Sea," down the *Bitter* and over the *Green* and *Black* waters, *echoing* near the "*Devil's Slide*" and the great "*Sink*" of the Desert; descended into total *darkness*, with jets of boiling *sulphur* on either hand, and finally *through* the DEVIL'S GATE, but *landing safely at the* GOLDEN GATE.

The scenery on this route has been the most varied; we have 8,242 feet *above*, and 366 feet *below* sea level; have taken our breakfast amid the eternal snow, and our supper in a land of perpetual summer, and have glided down from far above "timber line" into a region of continuous bloom, where the luscious fruits ripen each day of the year.

The author first began his explorations of the Trans-Mississippi country in 1860, as a "Pilgrim," and upon the completion of the Pacific railroad line, wrote the *first* descriptive guide of the roads—from actual observation—the "Great Trans-Continental Railroad Guide" of 1869: Soon "Crofutt's Tourist" followed, the publication of which was continued thereafter. The popularity of these books was so great that the sale aggregated, in fourteen years, over 500,000 copies.

The present book describes more than *four times* the extent of country of any book heretofore published, and is profusely illustrated by nearly 100 beautiful engravings, most of which were photographed, designed, drawn and engraved expressly for the author of this work. It also contains the *best* and *most complete* map—in colors—ever published, the plates of which cost over $4,000.

ANNEX—A department in the back part of this book, originated by the author, under which will be found a mass of condensed information, indirectly pertaining to the subject-matter of this work; and under which also contains descriptions of the large, double-page illustrations.

From the first issue of our book, in 1869, imitators have been numerous; no less than thirty-one "Guide-books," "Tourists' Hand-books" and "Books of Travels across the Continent," etc., etc., have been issued, most of which were compiled in the East—without their compilers traveling over one foot of the route, or, at least, not spending more than a few days on the road—while we have spent the best part of every year since 1860 acquiring the information—every item of which we are prepared to verify.

To some "correspondents" across the continent our books have proved an unusual "God-send," enabling them to minutely describe the wonders of the trip passed in the night, while sleeping soundly in a palace car, equally as well as though they were *awake* and in perpetual daylight. Now we do not mean to complain of these experimenters, as they are doubtless "good fellows," but we *do* expect the courtesies usually extended by all honorable writers.

GEO. A. CROFUTT.

DENVER, COLO., MARCH 1883.

INDEX TO ILLUSTRATIONS.

Large Views.

	No.
American Progress	1
Big Trees, Fallen Monarch	5
Cape Horn, Columbia River	10
Castellated Rocks at Green River	2
Fort Point, Golden Gate	4
Falls of the Yellowstone	8
Falls of the Williamette	9
Mirror Lake, Yo-Semite	12
Mt. Shasta, California	15
Nevada Falls, Yo-Semite	13
Sutter's Mill Race	3
Steamboat Rock, Echo	6
Summit Sierras	14
State Capital of California	16
San Francisco and Surroundings	18
The Geysers, California	17
Valley of the Yellowstone	7
Wood Hauling in Nevada	11

Illustrations.

	PAGE.
American River Canon	130
Bee Hive Geyser	24
Burning Rock Cut	75
Big Mule Team	28
Brigham Young	89
Brigham Young's Residence	97
Bloomer Cut	108
Before the Railroad	115
Bird's Eye view of the 'Loop'	214
Bird's Eye View of Plains	13
Crossing the Truckee River	73
Cacti Giganti	235
Crossing the "Loop"	215
Cattle Brands	254
Crossing the "Range" on Snow Skates	60
Devil's Slide, Weber Canon	33
Dale Creek Bridge	37
Down the Weber River	40
Donner Lake Boating Party	98
Devil's Gate, Weber Canon	75
Eagle Rock Bridge	110
Eagle Gate	109
Entering the Palisades	106
Eureka	150
Finger Rock, Weber	42
Forest View, Foot Hill	99
First Steam Train	56
First Mountain Express	152
General Offices U. P.	23
"Giantess," Geyser	113
"Giant" Geyser	146
Hanging Rock, Am. Fork	15
Hanging Rock, Echo Canon	26
High School, Omaha	20
Humbolt House	138
Hydraulic Mining	158
Interior View Snow Shed	72
Interior View Mormon Tabernacle	85
Indians Watching the Pacific Railway	233
James Bridger	77
Leland Standford, C, P. R. R	114
Looking up at Cape Horn	160
Livermore Pass Tunnel	177
Missouri River Bridge	22
Monument Rock, Black Hills	44
Mormon Temple	92
Mormon "Holiness to the Lord"	91
Map of Routes in California	120
Orange Grove and Palms	10
Overland Pony Express	151
One Thousand Mile Tree	84
Profile Map U. P. Ry	35
Packing to Virginia City	115
Palisades of the Humboldt	58
Pricky, the Horned Toad	81
Pulpit Rock, Echo Canon	83
Profile Map of C. P. R. R.	117
Palace Hotel	189
Rounding Cape Horn	159
Sidney Dillon. U. P. Ry	17
Seal of California	150
Steamer "Solano"	186
Summit of the Mountains	55
Seals and Sea Lions	65
Snow Galleries	67
Starvation Camp	71
Snow Sheds	143
Seal Rocks & Pacific Ocean	195
San Pedro's Wife	223
Three Tetons	19
"The Grand," Geyser	104
The Santa Ritas	239
The Last Spike	118
The Maden's Grave	133
Truckee River	136
Utah's Best Crop..Frontispiece.	
Union Depot, Kansas City	48
Union Depot Hotel	20
View of Salt Lake City	69
Yo-Semite Falls	101
Yucca Palm	221

GENERAL INDEX.

Cities, Towns, Villages and Stations.

	PAGE.		PAGE.		PAGE.		PAGE.
Acampo	176	Athlone	210	Benton	31	Bronco	150
Acton	218	Anaheim	224	Benicia	187	Brigham	106-116
Adams	48	Atkins	44	Bennington	108	Bridgeport	185
Abilene	53	Applegate	163	Bernal	204	Biggs	169
Adonde	233	Archer	45	Berenda	211	Bryan	75
Agate	56	Arcade	165	Be-o-wa-we	133	Bridger	78
Alpha	130	Aroyo	56	Bethany	178	Brighton	173
Alameda	181	Argenta	134	Barro	197	Brown's	140-145
Alila	213	Arimo	109	Big Spring	39	Brookville	54
Alpine	218	Armstrong	50	Bingham	93	Brownson	41
Alda	34	Aspen	79	Bitter Creek	70	Brule	39
Alta	95-157	Auburn	104	Bishops	120	Buckeye	170
Alkali	30	Austin	135	Black Buttes	70	Buda	34
Altamont	179	Aurora	64	Black Rock	102	Buck Creek	51
Alder	52	Bantas	178	Blackfoot	111	Bunker Hill	54
Alma	80	Bakersfield	213	Bloomfield	163	Bullion	120
Alvin	27	Battle Creek	109	Blue Creek	116	Burns	44
American Fork	96	Battle Mountain	135	B ue Canon	157	Butte	112
Ames	31	Batavia	185	Bosler	39	Buford	57
Andrews	220	Barton	40	Bowie	241	Bushnell	43
Andersons	144-170	Bavaria	53	Bovine	123	Burlingame	204
Antelope, Neb	43	Baxter	71	Box Springs	130	Buena Vista	162
Antelope, Cal	165	Bealville	216	Box Elder	56	Byers	56
Antioch	172	Belmont	204	Boise	122	Cabazon	229
Antioch Station	179	Bennett	56	Bonneville	116	Carson	147
Anita	170	Beaver Canon	111	Boca	150	Carlyle	55
		Belle Marsh	100	Borden	211	Carquinez	137
		Belvoir	51	Brady Island	37	Cachisa	240
		Benson	230	Brainard	27	Calistoga	197

GENERAL INDEX.—Continued.

Cities, Towns, Villages and Stations CONTINUED.	PAGE		PAGE		PAGE		PAGE
Call's Fork	106	David City	27	Gannett	38	Kansas City	49
Caliente	2,6	Darrance	54	Gardner Pass	131	Kaysville	88
Cana	170	Dana	65	Geyserville	200	Keen	216
Cactus	230	Decota	180	Gerard	217	Kearney Junction	34
Castle	176	Deeth	126	Gibbon	34	Kelton	122
Camas	111	Deming	242	Gila City	232	Kingsburg	212
Cameron	217	Deer Creek	98	Gila Bend	234	Kingsville	52
Caunon	185	Deep Wells	130	Gilmore	25	Kinney's	51
Cascade	156	Denver Junction	40	Gilroy	207	Kit Carson	56
Castle Rock	82	Desert	142	Golconda	137	Knights Landing	171-183
Castroville	208	Deweyville	106	Gold Hill	148	Kress Summit	162
Carlin	128	Detroit	53	Gold Run	158	Lake	122
Carter	77	Devil's Gate	86	Gospel Swamp	225	Latham	70
Carbon	64	Diamond	131	Gorham	54	Laramie	61
Carbondale, Kan	51	Dixon	165	Goshen	212	Lake View	145
Carbondale, Cal	175	Dillon	112	Grainfield	55	Lava Siding	111
Carnadero	207	Dix	43	Granite Point	140	Lawrence, Kan	50
Casa Granda	236	Dexter	39	Grayling	112	Lawrenceburg	52
Camptonville	163	Downeyville	163	Grinnell	55	Lake Point	103
Cedar Point	56	Donahue	198	Grangers	76	Lang	219
Cedar	125-131	Dutch Flat	157	Grantsville	51	Lathrop	178
Central City	32	Dragoon Summit	240	Green River	72	Leavenworth	50
Centerville	89	Dos Palmos	230	Grass Valley	102	Lehi	96
Chappel	41	Downey	224	Gridley	169	Lenape	70
Church Buttes	76	Draper	96	Greenville	07	Lewistown	108
Cheyenne Wells	55	Duncan's Mills	203	Granite Canon	57	Leroy	78
Cheyenne, Wy	45	Duncan	32	Grand Island	32	Lerdo	213
Chico	160	Dunham	169	Gurneyville	200	Lincoln, Neb	27
Chualar	208	Echo	83	Hallville	70	Lincoln	166
Chapman, Kan	53	Edwardsville	50	Half-way House	103	Linwood	50
Chapman, Neb	32	Eagle Rock	111	Hallack	126	Livermore	179
Cicero	175	Ellis, Cal	179	Hamilton	131	Little York	158
Charlestown	210	Egbert	44	Hampton	77	Little Cottonwood	93
Clarkston	107	Edson	65	Hamlet	202	Live Oak	169
Cisco	157	El Casco	229	Harney	59	Lockwood	82
Clay Centre	52	Ellis, Kan	54	Harrisville	105	Lodge Pole	41
Clarks, Neb	32	Elsworth	54	Harper's	63	Lodi	176
Clarks, Nev	144	Elko	7	12	Havens	32	Logan	108
Clear Creek, Cal	170	Elk Grove	175	Hayward's	180	Lomo	189
Clear Creek, Neb	27	Elm Creek, Neb	35	Hay Ranch	129	Lookout	63-147
Clipper Gap	163	Flin Creek	54	Hays	54	Loray	124
Clifton	52-107-241	Elmira	185	Healdsburg	200	Lordsburg	241
Cloverdale	200	Elkhorn	26	Hendrey	37	Lorenzo	180
Clyde	52	Emigrant Gap	157	Hillsdale	44	Loring	50
Cluro	132	Empire, Nev	147	Hilliard	79	Los Angeles	220
Collingston	107	Emory	82	Hoge	50	Lovelocks	140
Colyer	55	Essex	150	Hollester	207	Lucin	123
C. H. Mills	150	Evanston	80	Honeyville	106	Maracopa	234
Colfax	161	Evans, Nev	129	Hooker	170	Madera	211
Colorado Junction	56	Eureka	131-148	Hot Springs	142	Madrone	205
Coin	137	Ewing	107	Howard	203	Malad	127
Coyote	36	Fairfield	185	Howells	63	Mammoth Tank	230
Contention	240	Fermont	50	Huffakers	144	Manhattan	52
Como	64	Farmington	88	Humboldt	139	Market Lane	111
Concordia	52	Flowing Wells	230	Hugo	56	Martinez	179
Colusa	171	Florin	175	Huron	212	Mariposa	210
Columbus	31	Fink's Springs	230	Hutton's	63	Marston	76
Colton, Neb	41	Fillmore	68	Hyde Park	108	Marysville	167
Colton, Cal	228	Folsom	173	Hyrum	107	Matlin	122
Croydon	85	Fort Harker	54	Idaho, Idaho	122	Maxwell	37
Cooper Lake	63	Fort Riley	52	Independence	124	Mayfield	205
Corinne	116	Fort Fred Steele	66	Indio	229	McConnells	175
Cornwall	179	Fort Saunders	61	Iron Point	137	McPherson, Kan	53
Cottonwood	170	Fowler	212	Ione	175	McPherson, Neb	37
Council Bluffs	16	Forest City	163	Illinoistown	161	Mead	27
Cozad	36	Franktown	145	Iowa Hill	161	Medicine Bow	64
Cheston, Cal	185	Franklin	106	Josseyn	36	Medina	51
Creston, Wy	68	Fremont, Neb	27	Jordan	93	Melrose, Cal	181
Curtis	183	Frisco	102	Junb	101	Melrose, Mon	113
Cucamonga	228	Freeport	172	Julesburg	40	Menlo Park	205
Duvisville	183	Fresno	212	Junction, Roseville	164	Me-no-kew	52
		Fulton	200	Junction City, Kan	52	Mendon	107
		Galt	175	Junction, Cal	201	Mercede	210
		Garfield	139	Junction, Utah	93	Mercer	27

GENERAL INDEX.—Continued.

Cities, Towns, Villages and Stations. CONTINUED.	PAGE		PAGE		PAGE		PAGE
Mescal	230	Oreana	140	Ross Fork	109	St. George	51
Mesquite	230	Oroville	107	Roscoe	39	St. Helena	197
Michigan Bar	175	Osino	127	Rossville	52	Stevenson	35
Midway	109	Otto	50	Rozel	121	Stein's Pass	241
Milbra	204	Otego	124	Rye Patch	139	Steamboat Springs	145
Millard	26	Overton	35	Rutherford	197	Storms	162
Milford, Kan	52	Oxford	107-109	Russel	54	Stormsburg	27
Milford, Utah	101	Painted Rocks	234	Savanna	227	Stockton, Cal	176
Mill City	139	Palisade	129	Salt Lake	90	Stockton, Utah	104
Mill Station	145	Pantano	239	Sacramento	165	Stone House	137
Millis	80	Pajaro	207	Santa Monica	221	Stranger	50
Millville	107	Paradise	107	Santa Ana	224	Strong's Canon	155
Mineral	130	Paris	108	Santa Clara, Cal	205	Spadra	227
Mirage, Colo	56	Payson	100	Santa Cruz	205	Stanwix	234
Mirage, Nev	142	Paddock	32	Santa Rosa	199	Suisun	185
Miser	63	Plum	36	San Leandro	180	Solomon	53
Mississippi Bend	172	Papillion	25	San Pablo	188	Solano	187
Modesto	210	Peru	75	San Simon	241	Summit, C. P.	155
Monell	70	Petaluma	199	San Bruno	204	Summit, U. N	107
Mouida	111	Petersons	86	San Miguel	204	Sumner	213
Mojava	217	Percy	64	San Gabriel	226	Summit, E. & P.	131
Mono	100	Pequop	124	San Fernando	220	Summit Siding, Kan	54
Montello	123	Perryville	51	San Francisco	190	Summit Siding, Neb	25
Monterey	208	Peko	125	San Juan, S	207	Sunol	180
Montpelier	108	Pinole	188	San Juan, N	163	Swan Lake	109
Monument	122	Port Neuff	109	San Rafael	201	Table Rock	70
Moore's	124	Pasadena	225	San Quintin	201	Tamarack	157
Moore's Summit	50	Piedmont	79	San Jose	206	Tamalpais	201
Moute	227	Picacho	236	San Joaquin	227	Taylorsville	202
Moleen	128	Pilot Knob	230	San Mateo	204	Tecoma	123
Mokelumne	175	Pine Station	130	San Diego	224	Tehamma	170
Morano	210	Pino	164	San Juan Capistrano	225	Tennants	206
Morganville	52	Pine Bluffs	43	San Bernardino	228	Terra Cotta	54
Mound House	148	Puente	227	San Gorgonio	229	Terrace	123
Murphys	195	Piute	137	Salida	210	Tehachapi	217
Mystic	150	Placerville	174	Salt Wells	71	Texas Hill	234
Natividad	224	Pleasanton	180	Salina, Kan	53	Thompson	196
Nadeau	217	Pleasant Grove	99	Salinas	208	Thayer	71
Napa	196	Plum Creek	36	Sandy	94	Thumel	32
Napa Junction	185-196	Pomona	227	Salvia	144	Tie Sning	59
N. E. Mills	103	Point Rocks	70	Sargents	207	Tipton, U. P	70
Newton	107	Potter	42	Santaquin	100	Tipton, C. P	213
Newhall	219	Proctors	150	Sand Creek	218	Tiblow	50
Newport	224	Prosser Creek	150	Soco	122	Tooele City	103
Nowman	51	Provo	99	Sosma	170	Tomales	202
New Castle	164	Providence	107	Separation	68	Tombstone	240
New Cambria	53	Promontory	117	Sepulveda	220	Toano	124
Nelson	100	Pyramid	241	Sentinel	231	Tocolumna	202
Nephi	101	Quarry	116	Seven Palms	220	Toltec	236
Nichols	38	Raspberry	139	Schuyler, Colo	56	Topeka	51
Niles Junction	64	Ravena	219	Schuyler, Neb	31	Tortuga	230
Niles	180	Rawling	67	Sheridan, Cal	107	Tonganoxie	50
North Platte	38	Reeds	137	Sheridan, Kan	55	Truckee	150
North Bend	31	Red Buttes	60	Sherman	57	Tracy, U. P	44
Nord	170	Redwood City	204	Shoshone	134	Tracy, C. P	178
North San Juan	163	Redding	170	Shady Run	157	Transfer Grounds	20
Norwalk	224	Red Bluffs	170	Silver Lake	52	Tres Pinos	207
Oakland, East	182	Red Desert	76	Shelton	34	Tremont	184
Oakland, Wharf	180	Red Dog	158	Silver Creek	32	Tryone Mills	203
Oakland	182	Red Rock	112-236	Silver City, Nev	146	Tucson	236
Oak Knoll	197	Reno, Kan	50	Silver City, Idaho	122	Tulasco	126
Oakville	197	Reno, Nev	144	Siegel	51	Tulare	213
Odessa	35	Rio Vista	172	Sidney	41	Tule	137
O'Fallons	30	Riverside	27-111-228	Silver Station	148	Uintah	86
Ogalalla	30	Richmond	108-240	Simpson	64	Yuma City	246
Ogden, Kan	52	Richland, Kan	51	Soledad	200	Yuba Station	167
Ogden, Utah	87	Rillito	236	Soto	170	Yuba City	168
Olema	202	Ripon	210	Solon	68	You Bet	158-162
Omaha	23	Rose Creek	139	Sonoma	198	Youtsville	197
Ombey	122	Rock Creek	63	Springfield	108	York	100
		Rock Springs, Kan	54	Springville	100	Valparaso	27
		Rock Springs, Wy	71	Spring Hill	111	Valley Ford	202
		Rocklin	164	Spanish Fork	100	Valona	187
		Rogers	31	St. Mary's	52	Vallejo Junction	187

GENERAL INDEX.—Continued.

Cities, Towns, Villages and Stations, CONTINUED.

	PAGE.
Valley	27
Vallejo	185
Verdi	150
Vina	170
Virginia Dale	80
Visalia	212
Vista	144
Virginia City, M	113
Virginia City, N	148
Victoria	54
Wa-Keeney	53
Walker	54
Wakefield	52
Wallace	54
Wamego	52
Wadsworth	142
Wasatch, Wy	81
Wasatch, Utah	94
Wash-a-kie	70
Washoe	145
Wahoo	27
Waterloo	27
Watsonville	207
Warren	36
Washington	183
Weber Quarry	107
Weber	86
Webster	183
Walters	229
Wells	125
Wellsville	107
Weston	107
White Plains	142
Wheatland	167
Whitney	166
Wild Horse	56
Winnamucca	137
Wilmington	223
Windsor	200
Willard	105
Wilkins	71
Willards	129
Wilcox	64-240
Williams, Mon.	111-184
Willow Island	86
Wood's Crossing	89
Wier	40
Willson's	54
Williamson	51
Wolcotts	66
Woodland	183
Wood River	34
Wyoming	58
Wyandotte	50
Yuma	230

U. S. Forts and Camps

	PAGE.
Omaha Barracks	24
Camp at Sidney	42
Camp Lowell	238
Camp Bowi	240
Ft. Kearny	34
Ft. McPherson	37
Ft. Douglas	92
Ft. Sedgwick	40
Ft. Morgan	43
Ft. D. A. Russell	46
Ft. Larimie	46
Ft. Fetterman	46
Ft. Casper	46
Ft. Reno	46
Ft. Phil. Kearney	46
Ft. C. F. Smith	46
Ft. Saunders	61
Ft. Yuma	231
Ft. Fred Steele	66
Ft. Bridger	77
Ft. Halleck	125
Ft. Riley	52
Ft. Harker	54
Ft. Wallace	55
Ft. Hall	111
Mare Island	185

Mineral and Medical Springs, HOT AND COLD.

Pages ...... 67, 71, 76, 79, 89, 90, 96, 101, 102, 105, 106, 112, 113, 127, 133, 134, 137, 142, 145, 156, 184, 196, 198, 199, 200, 207, 209, 230, 238.

Railroads.

	PAGE.
American Fork	96
Amadoro Branch	175
Atchison, Topeka & Santa Fe	240-242
Bingham Canon	93
Burlington & Missouri	32-34
Black Hills "	32
Carbondale Branch	51
Central Pacific	115
Carson & Colorado	148
California Pacific	183
California Northern	167
Denver & Rio Grande Western	67
Echo & Park City	83
Eureka & Palisade	120
Fremont & Elkhorn	29
Grand Island & St. Paul Br.	33
Junction City & Ft. Kearny	52
Kansas Pacific	49
Los Angeles & Independence	221
Leavenworth, Lawrence & Galveston	51
Leavenworth Branch	50
North Pacific Coast	200
Northern	184
Nevada Central	135
New Railways	104
Nevada County	161
Oregon Short Line	76
Republican Valley	27
Sioux City & Pacific	27
Santa Cruz	207
San Pablo & Tulare	176
San Francisco & North Pacific	198
Sacramento Valley	173
Southern Pacific	203
Stockton & Visalia	176
Stockton & Copperopolis	176
San Diego Railroad	224
Salina & Southwestern	53
South Pacific Coast	205
St. Joseph & Denver	33
Solomon Railroad	53
Union Pacific	18
Utah Central	88
Utah Southern	92
Utah Western	102
Utah & Northern	105
Virginia & Truckee	144
Wasatch & Jordan Valley	94
Western	100-104
Wilmington Div	223

Annex Index.

No.		Page.
1	American Progress	243
2	Passage Ticket Men	243
3	Baggage Check "	243
4	Rates of Fare	244
5	Our Western Country	244
6	High School	248
7	First Steam Train	56
8	The Madrone Tree	164
9	The Manzanita	164
10	Jack Slade	248
13	Snow Difficulties	249
15	State Capital of Cal.	173
16	Castellated Rocks	72
17	Memories of Ft. Brid'r	249
18	Hauging Rock, Utah	97
19	Steamboat Rock	250
20	Paddy Miles' Ride	250
21	Salt Lake	251
22	" "	251
23	Discov'y of Califor'a.	226
23	The Coast Range	226
23	The Rainy Season	226
24	Hauling Ore in Hides	146
25	Life of Bigham Young	251
26	National Park	252
27	Ocean Steamships	253
28	Col. Hudnut's Survey	253
29	Western Stock Rais'g	254
30	The Great Cave	255
31	Nevada Falls	209
32	Pioneer Mail	218
33	The Donner Party	256
34	Roll 'Em Through	256
35	Val. of the Yollows'e.	257
36	Falls " "	257
37	" " Willi'etto.	215
38	Cape Horn	232
39	Wood Hauling	232
40	Mirror Lake	209
41	Pony Express	151
42	Sierra Nevada Moun's	138
43	Mt. Shasta	214
44	Woodward Gardens	257
45	The Geysers	184
46	Bird's Eye View	214
47	Ancient Ruins	257
48	Painted Rocks	258
49	Viewing Progress	258
50	Palace Hotel	258
52	"Prickey"	126
53	Route to Yo-Semite	184
55	The "Boss Cactus"	184
58	Sacramento Depot	173
63	Mammoth Snow Plow	126
64	Arizona	125
65	Emigrant Sleeping Cars	260
66	Black Hills R. R.	261

Miscellaneous.

	PAGE.
Calaveras Big Trees	176
California Windmills	175
Down the Sacramento	172
Excursions No. 1	193
" " 2	196
" " 3	198
" " 4	200
" " 5	203
First Gold Discovery	174
Hints	14
Humboldt Well	125
Laramie Plains	61
Yo-Semite Valley and Big Trees	212
Montana	112
New Alemaden Quick-Silver Mines	206
Sutro Tunnel	148

ORANGE GROVE, WITH FAN PALM TREES IN THE FOREGROUND. (See page 226.)

Ocean to Ocean, Overland.

Around the Circle.

SUNRISE—As the city of Halifax, in the Province of Nova Scotia and Dominion of Canada, is the extreme eastern terminus of the grand system of North American railways, which extend from its Atlantic portal across the continent 3,646 miles to San Francisco, its Pacific brother, it would seem to be the most proper point in the East from which we should *first* start on our journey with the tourist or emigrant for the same destination.

At Halifax, the morning sun, as it rises from its apparent cold water bath in the broad ocean on the east, casts its golden rays down upon the *first* rail-track that spans a continent, and from the moment the light strikes these iron bands of civilization and progress, it seems to follow them up, step by step, through populous cities, over mighty rivers, across broad, treeless plains, and towering snow-capped mountains, on, on! toward the tropical regions of the Orient. Every foot of the route, every object of interest or being is minutely inspected, while rolling over to its daily bath, in the mighty Pacific Ocean of the West. Our course is in the same general direction, but our *time* will be slower; as we shall linger by the way, and shall, after noting the principal routes east of the Missouri River, take the traveler with us over the Union, Kansas, and Central Pacific railroads to the Pacific coast—sunset; thence eastward over the Southern Pacific via Los Angeles, Yuma, Tucson, through Arizona and New Mexico, back toward sunrise, making a grand circle, one continuous run by rail of over 4,000 miles, which, with the numerous side tours by rail, steamer and stage, will comprise full 15,000 miles of travel.

Come along with us! but first discard the dress-coat of style, and put on the wrapper of simplicity and ease; fill your purse with coin, open your eyes and let us learn something of the extent, riches, varied resources, grandeur and wonders of what was a few years ago known only as the "Great American Desert."

We shall first take a run over the various Eastern branches of the Union Pacific, the "Denver Short Line" included, interview the noted gold fields of the Black Hills of Dakota and climb to the summit of their namesakes of Wyoming, the highest station on the road, where we can look off into the great State of Colorado, with her magnificent mountain ranges, peaks, parks and mineral wonders. Then, after whirling over the broad plains of Laramie, we shall mount to the summit of the "Rockies," and maybe, amid the clouds, stand astride the great Continental Divide.

Descending, we rattle through the cañons of Echo, Weber and the Devil's Gate to the land of Zion, where mothers-in-law and white-haired babies are in the majority.

We shall glance at the resources of the territory and spin all over Utah by the various railroads, visit the noted mines, temples, tabernacles and Mormon wonders, ascend to the "Mount of Prophecy," bathe in the "Dead Sea," interview the famous soda and hot springs, snatch a rose from the garden of the Prophet, then bound away to the northward over the Utah & Northern to Montana; again scale the great mountain divide and take a peep at the Yellowstone National Park. the lakes, springs, geysers, waterfalls, etc. —the " wonderland."

We will stand by the spot where the "last spike" was driven, which united the East and the West by iron bands, and over where the "ten miles of track was laid in one day;" we shall run along beside Salt Lake,— the great dead sea,— down the Humboldt, and over the Eureka & Palisade railroad to the Eureka and White Pine country. The "Palisades," as well as the "Lake" and the "Sink" of the Humboldt will be visited, as also the "Reese River Country," Nevada Desert, and the hot, spurting springs of Nevada. We shall visit the most noted silver country in the world — Virginia City, Gold Hill and Carson, via the Virginia & Truckee railroad; we will take a trip over Lake Tahoe and fish in Donner; ascend the Sierras and roll through more than fifty miles of snow sheds and tunnels, one continuing for 28 miles. Then around " Cape Horn " and to the old mining towns of Grass Valley and Nevada, over the Nevada County Narrow Gauge railroad, one of the finest in the world. We will take a run all over California, visit the " Big Trees," Yo-Semite Valley, the " Geysers," " Redwood Forests," " Seal Rocks," "Quicksilver Mines," Santa Cruz, Santa Barbara, Monterey, Calestoga, and the grape vineyards and wine cellars of Sonoma and Napa counties. We will visit Mount Shasta and the Upper Sacramento Valley; Coloma, where gold was first discovered; Mt. Diablo, the lofty peak of the Contra Costa, and Mt. Tamalpais, the huge sentinel of the Coast Range, at the Golden Gate, where we are apparently near sunset. After taking a hasty glance at Oregon and the Columbia River, we shall direct our course south and eastward, *toward Sunrise*, up the great San Joaquin Valley, over the " Loop " of the "Tehachapie Pass," and out on the great "Mojave Desert," rolling down the *infamous* Soledad Cañon,— the "Robbers' Roost,"— and through the San Fernando Mountains, out into the valley, and to the " city of the angels," Los Angeles, with its tropical fruits, orange orchards, and eternal summer.

From Los Angeles we shall "take in" Santa Monica—a charming watering place — and around to Wilmington Harbor, where we can interview the "Woman of the Period." We will visit Santa Ana, San Barnerdino, Anaheine, Riverside, and not forgot the "Gospel Swamp." At San Gabriel we will find the oldest Mission building on the coast in ruins; where orange trees are over 100 years old and loaded down with the golden fruit. We will have a run through the great vineyards and fruit orchards of this tropical region, inspect the mammoth cactus pads and the huge palm trees. "Progress" from this point turns more to the *Eastward*. We will follow its track and pass over the San Barnerdino mountains, and descend into the " Great Colorado Desert,"— rolling down, down to the *sea level*, where one would suppose " Progress " would naturally stop, unless she had a boat or a diving suit; but *no*, our train starts again downward; ye gods! down, down we go, *under the sea level two hundred and sixty-six feet*, where sulphur springs, mud geysers, salt and many other kinds of springs—both *hot* and *cold*—are very numerous, forcibly reminding one of the " old version;" but, as our modern teachers have done away with that old "bugaboo," we suppose they would not hesitate to visit with us this remarkable and very interesting region, and—gather a *speciman*.

Let us see, we are living in a fast age; the sun makes very good time, but "Old Sol" is aged, has run in the same old groove for too many years to retain much of the spirit of Progress. It is within the memory of many, how Morse, with his lightning, beat the old luminary, and we are now "talking all around him." Steam on the rail is next in speed; "one mile a minute" is not uncommon. The trip from New York to San Francisco, a distance of 3,296 miles, was commenced June 1, 1877, by Jarret & Palmer, on a special train, and the run made in 83 hours, 53 minutes, and 45 seconds, an average of

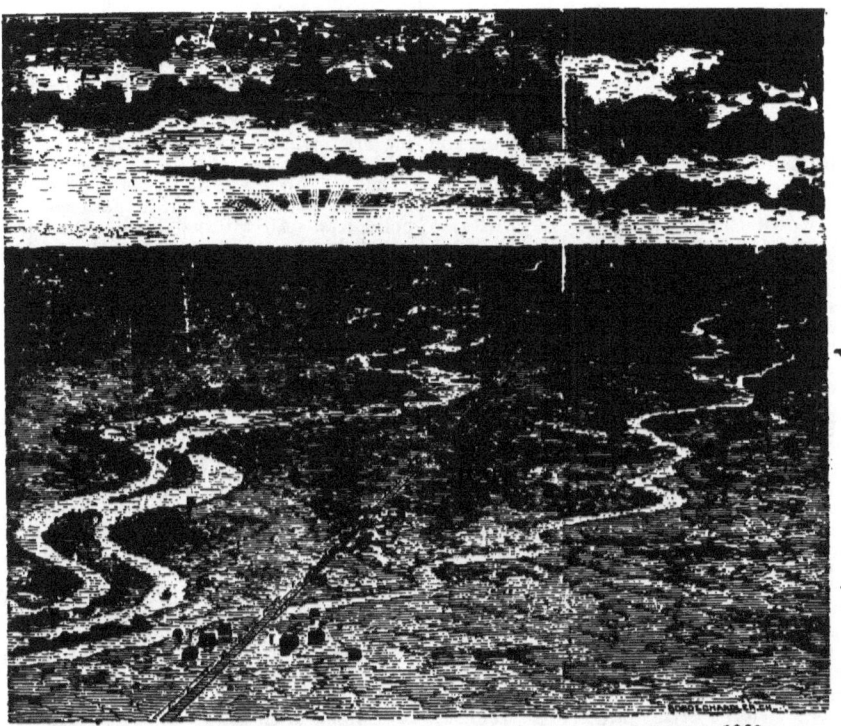

BIRD'S EYE VIEW OF THE PLAINS, FROM LOUP FORK RIVER. 1869.

39 miles an hour including stops. Sol must look sharp, or steam will also beat him in the race.

WEST TO THE MISSOURI RIVER—We shall not attempt a minute description of the various railroad and steamboat routes, east of the Missouri River. Each possesses its own peculiar attractions, a few of which will be briefly noted hereafter.

Passengers from the Eastern Atlantic sea-board, contemplating a trip to the Pacific coast, or the trans-Missouri country bordering the great Pacific railroad, can have their choice of five through "Trunk Lines," four American and one Canadian, which find their way by different routes, to a connection with the Union Pacific railroad, on the east bank of the Missouri River, midway between Council Bluffs and Omaha.

These five lines are the New York Central and Hudson River railroad, the Erie railway line, the Pennsylvania Central, the Baltimore & Ohio railroad, and the Grand Trunk, of Canada.

The railroad connections by these lines are almost innumerable, extending to almost every city, town, and village in nearly every State and Territory in the United States and Dominion of Canada; the regular through trains of either line make close and sure connections with the Pacific road, while the fares are the same. Sleeping cars are run on all through trains —most luxuriant palaces. The charges are *extra*, or about $3 per day—24 hours.

Only first-class passengers can procure berths in the sleeping cars.

HINTS BEFORE WE START.

1. Provide yourself with Crofutt's New Overland Tourist, and then be particular to choose such routes as will enable you to visit the cities, towns, and objects of interest that you desire to see, without annoyance or needless expense.

2. Greenbacks are good everywhere, so there is no longer any necessity of changing them for gold.

3. Never purchase your tickets from a stranger in the street, but over the counter of some responsible company. When purchasing tickets, look well to the date, and notice that each ticket is stamped at the time you receive it. Then make a memorandum on the blank in the ANNEX No. 2, of your Guide Book, of the name of the road issuing the ticket, destination of ticket, form, number of ticket, consecutive number, class and date. In case you lose your ticket, make known the fact *at once* at the office of the company, showing the memorandum as above described, and steps can be taken immediately to recover the ticket, if lost or stolen, or to prevent its being used by any one else. By attention to such slight and apparently unimportant matters as these, travelers may recover their loss and save themselves much inconvenience.

4. Before starting out, provide yourself with at least one-third more money than your most liberal estimate would seem to require, and do not lend to strangers or be induced to play at their games, *if you do, you will surely be robbed.*

5. Endeavor to be at the depot at least fifteen minutes before the train leaves, thereby avoiding a crowd and securing a good seat.

6. You will need to show your ticket to the baggage-man when you ask him to check your baggage; then see that it is properly checked, and make a memorandum of the number of the check in the blank of the ANNEX No. 3; this done, you will need to give it no further attention until you get to the place to which it is checked.

7. Persons who accompany the conductor through the cars, calling for baggage to be delivered at the hotels or other places, are generally reliable, but the passenger, if in doubt, should inquire of the conductor, and then be careful to compare the number of the ticket received from the agent in exchange for your check, to be sure that they are the same.

8. Don't grumble at everything and everybody or seek to attract attention; remember only boors and uneducated people are intrusive and boisterous.

9. Remember this: "Please" and "Thanks" are towers of strength. Do not let the servants excel you in patience and politeness. All railroad employes are instructed to be gentlemanly and obliging at all times.

10. *And finally*—Do not judge of the people you meet by their clothes, or think you are going west to find foo's; as a millionaire may be in greasy buckskin, a college graduate in rags, and a genius with little of either, while in the breast of *each* beats an *honest heart.*

---o---

For Rates of Fare, see ANNEX No. 4.

---o---

ROUTE 1.—From HALIFAX take the Inter-Colonial and Grand Trunk railways, through the Province of New Brunswick via Quebec, Montreal, Victoria Bridge, along the shore of the St. Lawrence River, Thousand Islands, and La Chine Rapids, Toronto, Hamilton, Niagara and Detroit, where connections are made with routes 2 and 3. Another route is by Maine Central via Portland, and *then* the Grand Trunk, or, via Boston and the Central Vermont and the Grand Trunk.

From BOSTON there are quite a number of lines. One, as above described, is through Vermont and Canada; another is by the Boston & Albany railroad to Albany, where connections are made with route 2; another is the new "Hoosac Tunnel" route, through the mountain and tunnel of that name,—25,081 feet in length, double track; cost $16,000,000—cut through a mountain which rises 1,900 feet above the track. This route is a very desirable one, passes through the entire length of the State of Massachusetts, and connects with route 2, at Troy, New York. Another line is via Springfield, or Shore Line, to New York city; or, you can take part "rail" and the steamships on Long Island Sound, of which there are three first-class lines, comprising some of the finest boats in the world.

From NEW YORK city, passengers who desire to visit NIAGARA—whose thundering cataracts, in volume of waters, far surpass all other waterfalls in the known world—

HANGING ROCK, AMERICAN FORK R.R., UTAH.
See Annex No. 13.

may also view the great Suspension Bridge over Niagara River, which, undoubtedly, is one of the finest structures of its kind in this country. They can have choice of two trunk lines.

ROUTE 2.—The New York Central & Hudson River line, passes up the glorious old Hudson, the magnificent river upon the bosom of which Fulton launched his "experiment," the *first* steamboat *ever* constructed. This road is built almost on the river brink, upon the eastern bank, which slopes back in irregular terraces, presenting from the car window one of the finest, if not *the* finest, panoramic view in the world. On the right are many small cities, towns and villages, with groves, parks, gardens, orchards, and alternate rich fields, with here and there, peeping out from beneath the trees, the magnificent country villa of the nabob, the substantial residence of the wealthy merchant, or the neat and tasteful cottage of the well-to-do farmer. Then come the "Palisades of the Hudson," and then again a repetition of the beauties above described, while to the west of our train rolls the river, with numberless steamboats tugs, barges, small boats, and sailing vessels of all kinds and classes, while beyond, on the west bank, is spread out a succession of scenery not much unlike that seen on the eastern side. This line passes through Central New York, the "Garden Spot of the State," via Albany,— the Capital of the State,—Troy, Utica, Rochester, to Suspension Bridge, Niagara, and Buffalo.

The direct western connections of this route are at Suspension Bridge, with the Great Western and Michigan Central and at Buffalo with the Canada Southern and the Lake Shore & Michigan Southern, via Dunkirk and Cleveland.

ROUTE 3.—The Erie railway line traverses the southern portion of the State of New York, via Binghampton, Corning, and Buffalo. The track of the Erie is the *broad* gauge; the cars are very wide and commodious. This route affords the traveler a view, while crossing and recrossing the Delaware, of scenery and engineering skill, at once grand, majestic, and wonderful. The direct western connection of the Erie is the Lake Shore & Michigan Southern, at Dunkirk and Buffalo; and the Canada Southern, at Buffalo—with the Great Western and Michigan Central, at Suspension Bridge; and the Atlantic & Great Western, at Corry, Penn.

ROUTE 4.—The Pennsylvania Central line receives passengers in New York and PHILADELPHIA, and conveys them the entire length of the State of Pennsylvania, via Harrisburg—the capital of the State, —to Pittsburgh, the most extensive iron manufacturing city in the United States. The landscape on this line, and especially while passing along the Susquehanna River, and the charming "Blue Juniatta," and over the Alleghanies, presents scenery most grand; while the fearful chasms and wonderful engineering skill displayed at the "Great Horse-shoe Bend," and at other points, are second only to that displayed at "Cape Horn" on the Sierra Nevada mountains. At Pittsburgh, the Central connects with the Pittsburgh, Fort Wayne & Chicago, —*one* of the *best* roads in this country— and also with lines, via Columbus and Indianapolis, and St. Louis, or Cincinnati, Indianapolis and St. Louis.

ROUTE 5.—Is via the cities of Philadelphia and BALTIMORE, by the Baltimore & Ohio. By this line, passengers are afforded an opportunity of visiting the capitol at Washington, and thence, via Harper's Ferry, "over the mountains" to Wheeling. It is said by some travelers that the scenery by this line is unsurpassed by any on the continent. The western connections are at Chicago, Cincinnati, and St. Louis.

From CINCINNATI passengers can have choice of several first-class competing lines, via either Chicago or St. Louis, or via the Burlington route—direct, via Burlington, Iowa, where connection is made with the Burlington & Missouri, for Council Bluffs.

From ST. LOUIS passengers can take the "Wabash Line" direct, with elegant sleeping and dining cars, or the Missouri Pacific, via Kansas City, and the Kansas City, St. Joseph & Council Bluffs, via St. Joseph, Mo., and arrive at Council Bluffs.

From CHICAGO there are four first-class roads. The Chicago & Northwestern was the first road built to the Missouri River, where the first train arrived Jan. 17, 1867, which route is via Clinton and Cedar Rapids. The Chicago, Rock Island & Pacific road, which passes through the cities of Rock Island, Davenport and Des Moines: the Chicago, Burlington & Quincy, via Galesburg and Burlington, form the "Burlington Route," and the Chicago, Milwaukee & St. Paul. These four roads are known as the "Iowa Pool Lines," and are equipped with all the modern improvements. Magnificent drawing room sleeping cars run with all through trains; also, dining cars, in which meals are served for 75 cents.

All trains from the East and South stop a few moments at Council Bluffs before proceeding to the Transfer Grounds, two miles further west. Let us take a look at

Council Bluffs--This city is in the western portion of the State of Iowa, about three miles from the Missouri River, at the foot of the bluffs. It is the county seat of Pottawattomie county, and contains a population of about 18,400. It is four miles distant from Omaha, Neb., with which city it is connected by hourly steam cars. The explorers, Lewis and Clark, held council with the Indians here in 1804, and named it Council Bluffs. It is one of the oldest towns in Western Iowa. As early as 1846, it was known as a Mormon settlement, by the name of Kanesville, which it retained until 1853, when the legislature granted a charter designating the place as the City of Council Bluffs.

The surrounding country is rich in the chief wealth of the nation—agriculture.

Council Bluffs includes within her corporate limits 24 square miles. The buildings are good; the town presents a neat, tasty, and, withal, a *lively* appearance; street-cars traverse the principal streets; churches and schools are numerous. The State Institute for the Deaf and Dumb is located near the city, to the southeast. The Ogden, is the principal hotel, and the *Daily Nonpareil*, and the *Daily Globe*, are the principal newspapers.

By a decision of the United States Supreme Court, the *eastern* bank of the Missouri River is the terminus of the Union Pacific railroad. The terminus is now known as the Transfer Grounds.

CASTELLATED ROCKS, GREEN RI

WYOMING, (See Annex No. 16.) (2.)

SIDNEY DILLON.

Among the men of progress in America there will be found no name more distinctly representative or more thoroughly in unison with the spirit of the age, than that of Mr. Sidney Dillon, President of the Union Pacific railroad. Born in Northampton, Montgomery county, New York, on the 7th of May, 1812 at which place his father was a well-to-do farmer, he came of sterling stock—his grandfather having been a Revolutionary soldier.

From early childhood his life has been an active one, given almost wholly to the advancement of the internal improvements of his country. When a mere lad, he commenced his railroad life as an errand boy, on the Mohawk & Hudson railroad— the *first* railroad built in his native State—running from Albany to Schenectady. (ANNEX No. 7, page 56.) He next entered the service of the Rensselaer & Saratoga—then we hear of him as overseer of a contract on the Boston & Providence, and several other roads. In 1838, he took his first contract, and completed it with profit in 1840, from which time his contracts have been very numerous. Among these was "Clay Hill," two miles from West Troy, on the Troy & Schenectady railroad. Mr. Dillon next built twenty-six miles of the Hartford & Springfield, six miles of the Cheshier, and ten miles on the Vermont & Massachusetts. Besides the above, he has been engaged in the construction of the Rutland & Burlington; Central, of New Jersey; the Morris canal; the Boston & New York Central; the Philadelphia & Erie; the Erie & Cleveland; the Morris & Essex; the Boston, Hartford & Erie; the Iowa; the New Orleans, Mobile & Chattanooga, the Canada Southern; the Union Pacific, and many others. The last great work upon which Mr. Dillon has been engaged is the "Fourth Avenue improvement," New York. The contract involves $7,000,000, and is a work of great magnitude. Suffice it to say, that he has been engaged in over forty of the leading public works of America, and that the contracts with which he has been engaged have amounted to over $100,000,000. The career of Mr. Dillon teaches the lesson, that, at the hands of a man thoroughly conversant with his business, persevering, energetic, faithful to trust, upright in his relations with his fellow-men, *success is sure.*

In person, Mr. Dillon is tall, exceedingly well built, and combines suavity of manner with great promptness of decision in action. He was married in 1841, residence in New York City, and devotes his whole time to directing the interests of the

Union Pacific Railroad.

Official Headquarters, R. R. Building, Omaha, Neb., and 44 Equitable Building, Boston, Mass.

SIDNEY DILLON,......*President*.......*New York.*
ELISHA ATKINS,......*Vice President*,......*Boston.*
H. MCFARLAND,......*Sec. and Treas.*,...... "
S. H. H. CLARK,......*Gen'l Manager*,...*Omaha.*
THOS. L. KIMBALL,....*Ass't Gen. Manag'r,* "
E. P. VINING,.........*Freight Traffic Mg'r* "
J. W. GANNETT,......*Auditor*,........... "
J. O. BRINKERHOFF,..*Chief Engineer*,.... "
LEAVITT BURNHAM,...*Land Comm'r*,..... "
J. J. DICKEY, ...*Sup't Telegraph*,.. "
P. P. SHELBY ....*Gen'l Freight Agent,* "
J. W. MORSE,........*G, P. Agent*,........ "
S. B. JONES.........*Ass't* "
C. S. STEBBINS......*Ger' Ticket Agent,* "
F. KNOWLAND, *General Eastern Agent,*
287 BROADWAY, NEW YORK.
M. T. DENNIS, *Gen'l Agent for New England,*
BOSTON, MASS.

Though but little faith was at first felt in the successful completion of this great railway, no one, at the present day, can fail to appreciate the enterprise which characterized the progress and final completion of this road, its immense value to the Government, our own people, and the world at large.

By the act of 1862, the time for the completion of the road was specified. The utmost limit was July 1, 1876.

The first contract for construction was made in August, 1863, but various conflicting interests connected with the location of the line delayed its progress, and it was not until the 5th day of November, 1865, that the ceremony of breaking ground was enacted at a point on the Missouri River, near Omaha, Neb.

The enthusiast, Mr. Train, in his speech on the occasion of breaking ground, said the road would be completed in five years. Old Fogy could not yet understand Young America, and, as usual, he was ridiculed for the remark, classed as a dreamer and visionary enthusiast; the greater portion of the people believing that the limited time would find the road unfinished. But it was completed in *three years, six months, and ten days.*

Most Americans are familiar with the history of the road, yet but few are aware of the vast amount of labor performed in obtaining the material with which to construct the first portion. There was no railroad nearer Omaha than 150 miles eastward, and over this space all the material purchased in the Eastern cities had to be transported by freight-teams at ruinous prices. The laborers were, in most cases, transported to the railroad by the same route and means. Even the engine, of 70 horse power, which drives the machinery at the company's works at Omaha, was conveyed in wagons from Des Moines, Iowa, that being the only available means of transportation at the time.

For five hundred miles west of Omaha the country was bare of lumber save a limited supply of cottonwood on the islands in and along the Platte River, wholly unfit for railroad purposes. East of the river, the same aspect was presented, so that the company were compelled to purchase ties cut in Michigan, Pennsylvania, and New York, which cost, delivered at Omaha, $2.50 per tie.

Omaha, at that time, 1863, contained less than 3,000 population, mostly a trading people, and the railroad company were compelled to create, as it were, almost everything. Shops must be built, forges erected, all the machinery for successful work must be placed in position, before much progress could be made with the work. This was accomplished as speedily as circumstances would permit, and by January, 1866, 40 miles of road had been constructed, which increased to 265 miles during the year; and in 1867, 285 miles more were added, making a total of 550 miles on January 1, 1868. From that time forward the work was prosecuted with greatly increased energy, and on May 10, 1869, the road met the Central Pacific railroad at Promontory Point, Utah Territory—the last 534 miles having been built in a little more than fifteen months; being an average of nearly one and one-fifth mile per day.

By arrangements with the Central Pacific Railroad Company, the Union in 1870 relinquished to the Central 46 miles of road, and again in 1875, another strip of 6 miles, leaving the entire length of the Union, 1,032 miles, and its junction with the Central at Ogden, Utah.

———o———

For SNOW DIFFICULTIES, see ANNEX No. 13

The Three Tetons. By Thos. Moran.

THE UNION PACIFIC DEPOT.

The Transfer Grounds—are about two miles west of Council Bluffs, and about half a mile east of the Missouri River Bridge. Here, all passengers, baggage, express, and mails on arriving from the eastward, change to cars of the Union Pacific. A large fine building affords ample accommodation for passengers, and for the transaction of all kinds of business connected with the transfer. The tracks of the eastern roads terminate at the eastern front. Between these tracks are long wide covered platforms along which passengers, mails, baggage, etc., reach the depot, and after passing through the building, find the Union Pacific trains waiting on the west side.

The Union Depot is a model of convenience, built of brick with stone trimmings,—two stories. On the first floor are two large waiting rooms for ladies and gentlemen, ticket, telegraph and express offices, baggage and news rooms, restaurant, lunch counter, barber, etc. The second story of the building is divided into rooms for hotel accommodations, the parlors are elegant, the rooms,—40 in number—are very large, furnished in the best manner, with hot and cold water, gas, annunciators, etc.

A "Dummy" train leaves the depot for Omaha every hour through the day, and horse cars from the south side of the depot for Council Bluffs, regularly.

The Emigrant House—so-called, is situated a short distance west of the Depot on the north side of the track. It was built by the Union Pacific Co., and is run by their direction for the accommodation and protection of their emigrant passengers. The building is of wood, has 70 rooms comfortably furnished, accommodating 200 persons, and charges only sufficient to cover cost; plain, substantial meals, 25 cents each.

The handling of the baggage at this depot is no small item. The baggage room is very large, as well as the number of "smashers," but, as the latter are always under the eye of Mr. Traynor, General Baggage Agent of the road, the baggage is not only handled quickly but with a *fatherly* care.

Passengers will here re-check their baggage, and secure tickets in one of the PALACE SLEEPING CARS that accompanies all through trains, and thereby insure an opportunity for a refreshing sleep, as well as a palace by night and day. This, however, costs an extra fee. The charges are over the Union Pacific from Omaha, $8.00; from Ogden, over the Central Pacific, to the Pacific Coast cities, $6.00. But as all cannot *afford* to ride in palace cars, secure—pre-empt, if you please—the best seat you can, and prepare to be *happy*.

☞ Sleeping car and stop-over privileges are not allowed on second and third-class tickets. [See Note, p. 149.] Baggage can be checked only to the destination of second and third-class tickets—100 lbs. allowed free on each full, and 50 lbs. on each half-ticket of all classes. Extra baggage is $10 to $15 per 100 lbs. according to class. Passengers holding first-class tickets to San Francisco, with pre-paid orders for steamer passage to trans-Pacific ports, will be allowed 250 lbs. baggage, free, on presentation of such orders to the baggage agent at the Transfer, or Omaha; on second-class tickets, 150 lbs., free. Orders for steamship passage can be purchased at the Transfer, or Omaha depot ticket offices.

For rates of Fare see ANNEX No. 4.

LUNCH BASKETS—With only two or three exceptions, all the eating-houses on this line are good. The *good* ones have our approval, but the others—Oh! well—should you provide yourself with a small basket of provisions, and use it accordingly, it would tend to preserve your temper. The accommodations at all the principal stations for those who wish to "stop over" a day or two, are ample; charges, from $3.00 to $4.00 per day.

FOR A BRIEF SKETCH OF OUR WESTERN COUNTRY — THE FAR WEST — CONDENSED HISTORY — ORGANIZATION OF THE PACIFIC RAILROAD—LAND GRANT —COST OF CONSTRUCTION—MATERIAL USED— IMPORTANCE OF THE ROAD — FACTS IN BRIEF—GRUMBLERS – See ANNEX No. 5.

ONE WORD MORE—As you are about to launch out upon the broad, sweeping plains, the barren desert, and the grand old mountains—for all these varied features of the earth's surface will be encountered before we reach the Pacific Coast—lay aside *all* city prejudices and ways for the time; leave them *here*, and for once be *natural* while among nature's loveliest and grandest creations. Having done this, you will be prepared to enjoy the trip—to appreciate the scenes which will rise successively before you. But, *above all* forget everything but the journey; and in this consists the *great secret* of having a good time generally. Are you ready?—The bell rings, "All aboard" is sounded, and our train leaves the "Transfer Grounds," and directs its course due west towards the

MISSOURI RIVER BRIDGE—The construction of this bridge was first authorized by Congress on the 25th of July, 1866, but very little was done until March, 1868, when work commenced, and was continued from that time until July 26, 1869, when it was suspended. Nothing more was done until April, 1870, when a second contract was made with the American Bridge Company of Chicago, and work again commenced. On the 24th of February, 1871, Congress passed a special act authorizing the Union Pacific Railroad Company to construct this bridge across the Missouri River, and to issue bonds to the amount of $2,500,000.

The county of Douglas, Nebraska, voted, under certain conditions, aid in county bonds to the amount of $250,000. Also, Pottawattomie county, Iowa, voted, under certain conditions, aid to the amount of $205,000.

This bridge is a notable structure (see illustration), one-half mile in length, with the approaches over one mile.

It is located below the old depot, and opposite that part of the city of Omaha known as "Train-Town," and has a single track.

The bridge is known as a "Post's Patent." The hollow iron columns are 22 in number, two forming a pier. These columns are made of cast iron one-and-three-fourths inches in thickness, 8½ feet in diameter, 10 feet long, and weigh 8 tons each. They are bolted together air-tight, and sunk to the bed-rock of the river, in one case, 82 feet below low-water.

After these columns are seated on the rock foundation, they are filled up twenty feet with stone concrete, and from the concrete to the bridge "seat," they are filled with regular masonry. From high-water mark to the bridge "seat," these columns measure 50 feet. The eleven spans are 250 feet in length, making

MISSOURI RIVER BRIDGE—OMAHA IN THE DISTANCE.

the iron part, between abutments, 2,750 feet.

These columns were cast in Chicago, and delivered in the shape of enormous rings, 10 feet in length. When they were being placed in position the workmen would take two or more rings, join them together, place the co'umn where it was to be sunk, cover the top with an air-lock, then force the water from the column by pneumatic pressure, ranging from 10 to 35 pounds per square inch. The workmen descend the columns by means of rope-ladders, and fill sand-buckets, which are hoisted through the air-lock by a pony-engine. The sand is then excavated about two feet below the bottom of the column, the men come out through the air-ock, a leverage, from 100 to 300 tons, is applied, the pneumatic pressure is removed, and the column sinks, from three inches to two and one-ha'f f et—in one ins'ance, the co'umn s'eadily sank down 17 feet. When-ever the column sinks, the sand fills in from 10 to 30 feet—in one instance, 40 feet. This has to be excavated before another sinking of a few inches can take place, making altogether a slow and tedious process.

Soon after crossing the bridge, our train stops in the Omaha depot—a large building with one enormous span overhead, built in the most substantial manner, of iron and glass, with six tracks running through it from end to end. On the south side are ample waiting and dining-rooms, express, telegraph, baggage, ticket, and other offices. Passengers who wish to stop over, will find o mnibuses at the depot to take them and their baggage to the hotels, or any point in the city; fare, 50 cen:s; or, they will find street cars on the north side of the depot, that leave every five minutes, passing the princ'pal hotels, and running the whole length of the city; fare, ͜ cents.

GENERAL OFFICES, UNION PACIFIC RAILWAY CO.

Omaha—This is one of the most progressive cities in the West. It is the county seat of Douglas county, situated on the western bank of the Mo. River, on a slope about 50 feet above high-water mark, with an altitude of 966 feet above sea level. The first "claim cabin" was built here in 1854, and the place named *Omaha*, after the Omaha Indians.

It is related that the first postmaster of Omaha used his hat for a postoffice, and many times, when the postmaster was on the prairie, some anxious individual, would chase him for miles until he overtook the traveling postoffice and received his letter. "Large oaks from little acorns grow," says the old rhyme, 'tis illustrated in this case. The battered-hat postoffice has given place to a first-class postoffice, commensurate with the future growth of the city, the last census giving a population of 30,642.

In 1875, the Government completed a large court-house and postoffice building, using a very fine quality of Cincinnati free-stone. It is 122 feet in length by 66 feet in width—four stories high—cost $350,000, and is one of the most attractive buildings in the city.

The State capitol was first located here, but was removed to Lincoln in 1868. Omaha though the first settlement made in Nebraska, is a young city. The town improved steadily until 1859, when it commenced to gain very rapidly. The inaugurating of the Union Pacific railroad gave it another onward impetus, and since then the growth of the city has been very rapid. There are many evidences of continued prosperity and future greatness, one of which, is the fact that there are *no* dwelling houses in the city "To Let;" yet the records show there were over $2,000,000 expended during 1881 in new buildings and city improvements.

Omaha is the headquarters for half dozen railroads, has three daily papers, the *Herald*, *Bee* and *Republican*, besides several periodicals; a grand opera house and several large first-class hotels building (with 36 small ones in operation). The amount of jobbing business, banking, packing, manufacturing, grain and produce handling, etc., done in Omaha is *immense* and rapidly increasing. There are twenty-four churches in the city and numerous public and private schools. The Omaha people *are proud of their public schools*. The city has near $500,000 invested in *free* school property.

The High School is the finest building of its kind in the Western country, and stands on Capitol Hill, on the site of the old State House, the highest point in the city, and is the first object which attracts the attention of travelers approaching from the East, North, or South. Its elevation and commanding position stand forth as a fitting monument to attest a people's intelligence and worth. [See ANNEX No. 6.]

To the north of the High School building is the Creighton College, just completed at a cost of $55,000, with a further endowment of $100,000 the gift of Mrs. Edward Creighton. The building is 54 by 126 feet—three stories and a basement—capable of accommodating 480 pupils. It is a *free* school and conducted by the Jesuit Fathers.

Omaha has a beautiful driving park, known as Hanscom Park, the gift of Mr. A. J. Hanscom and Mr. Jas. G. McGath. It is situated about one mile southwest from the center of the city, contains 80 acres, is in as romantic and picturesque a locality as one could wish, with a perfect forest of shade trees and beautiful walks, fountains, etc., and, in fact, is "a thing of beauty," and to the citizens of Omaha, we believe, will be a joy "forever."

Omaha is headquarters of the Department of the Platte. Fort Omaha was established in 1868; the barracks are 8 in number capable of accommodating 1,000 men. They are situated 3 miles north and in full view of the city. Latitude, 40 deg. 20 min.; longitude, 96 deg. from Greenwich. Eighty acres of land are held as reserved, though no reservation has yet been declared at this post. There is an excellent carriage-road to the barracks and a fine drive around them, which affords pleasure parties an excellent opportunity to witness the dress-parades of the "boys in blue." It is a favorite resort, the parade, the fine drive and improvements around the place calling out many of the fashionable pleasure-seekers. The grounds are planted with shade trees, and in a few years will become one of the many pleasant places around Omaha.

The post is the main distributing point for all troops and stores destined for the West.

These barracks were erected for the purpose of quartering the troops during

BEE HIVE, OR FAN GEYSER, YELLOWSTONE PARK.

the winter season when their services were not required on the plains, and as a general rendezvous for all troops destined for that quarter.

Besides the Union Pacific, there are two other railroads that branch off from Omaha. The Omaha & Northern Nebraska, and the Burlington and Missouri in Nebraska. These roads were chartered under the general railway act which gave two thousand acres of land for every mile of road completed before a specified time. The route of the Northern is five miles up the Missouri River Valley, then northwest to the valley of the Papillion, thence to the Elkhorn River, and up the Elkhorn Valley to the mouth of the Niobrara. It is now completed, and cars are running to Tekamah, Burt county—about 80 miles from Omaha. At Blair connections are made with the Sioux City & Pacific railroad.

The route of the B. & M. R. R. in Neb. Branch, is down the Missouri River Valley, where it crosses the Platte and runs to Lincoln. This road is under the management of the Burlington & Missouri River Railroad Co., and makes close connections at Orepolis with the main line of the B. & M. for the East and West.

At Omaha, are located the general offices of the Union Pacific Railroad Company, in a fine, large building just completed and fitted up in the most complete and convenient manner. This company employs about 9,000 men; this includes the men employed in the Laramie rolling mill and in the coal mines. There are about 5,000 employed on the road proper. there are 2,050 employed in the shops at Omaha, and about 800 more in the various offices, at stations, on the track, and at the depots at Omaha, making near 3,000 in all on the Omaha pay rolls.

About one mile above the bridge, on the low lands fronting the river, the railroad company have located their principal shops and store-houses. They are built of brick, in the most substantial form, and with the outbuildings, lumber yard, tracks, etc, cover about 30 acres of ground.

The machine shop is furnished with all the new and most improved machinery, which is necessary for the successful working at all the branches of car and locomotive repairs or car construction. The round-house contains 20 stalls; the foundry, blacksmith shop, car and paint shops, are constructed and furnished in the best manner. The company manufacture all of their own cars. The passenger cars, in point of neatness, finish, strength of build and size, are unsurpassed by any and rivaled by few manufactured elsewhere. It is the expressed determination of the Union Pacific Company to provide as good cars and coaches for the traveling public in style and finish as those of any Eastern road. They reason, that as the great trans-continental railroad is the longest and grandest on the continent, its rolling-stock should be equally grand and magnificent. From the appearance of the cars already manufactured, they will achieve their desires. On the same principle, we proposed to make our BOOK superior to any other. *Haven't we done so?*

☞ [For time, distances, altitudes, etc., see Table in back of book.]

Our train runs along through the southern suburbs of the city, on an ascending grade, 3 2-10 miles to

Summit Siding—a flag station, where trains seldom stop. Elevation, 1,142 feet, 176 higher than the Omaha depot. But our route is now downward for 6 3-10 miles to

Gilmore—The country around this station is rich prairie land, well cultivated. A small cluster of buildings stands near the road; the station is of little importance, merely for local accommodation.

Continuing our descent 5 miles, we reach

Papillion (Pap-e-o)—Here we are at the lowest elevation on the whole line, excepting Omaha, which is 6 feet lower. From this station to Sherman, on the Black Hills, 585 miles, it will be a gradual up-grade, rising in the distance 7,270 feet.

Papillion is the county seat of Sarpy county and has the usual county buildings, some of which are fine structures. The *Times*, a weekly paper, is published here. The station is on the east side of Papillion River, a narrow stream of some 50 miles in length, which, running southward, empties into Elkhorn River, a few miles below the station. The bridge over the stream is a very substantial wooden structure. The country about the station has been improved very much within the last few years; it has fully doubled its population, which is now about 1,000, and evidences of thrift appear on every hand.

Soon after leaving the station, we cross the Papillion River, and 6 4-10 miles brings us to

HANGING ROCK, ECHO CANYON, UTAH.

Millard—an unimportant station to the tourist, is situated in the midst of a fine agricultural section—two stores, a flouring mill and hotel; population about 300. Eight miles further, we come to

Elkhorn—which is on the east bank of Elkhorn River, and of considerable importance as a point for freight traffic—it being the outlet of Elkhorn River valley.

ELKHORN RIVER—is a stream of about 300 miles in length. It rises among the hills of the Divide, near where the headwaters of the Niobrara River rise and wend their way toward their final destination, the Missouri. The course of Elkhorn creek, or river, is east of south. It is one of the few streams in this part suitable for mill purposes, and possesses many excellent mill sites along its course. The valley of this stream averages about eight miles in width, and is of the best quality of farming land. It is thickly settled by Germans for over 200 miles of its length from its junction with the Platte River. The stream abounds in native fish, as well as a great variety of "fancy brands" from the East—a car load of which were accidently emptied into the water at the bridge, while en route to be placed in the lakes and streams of California, during the spring of 1873.

Wild turkeys on the plains, and among the low hills, along with deer and antelope, afford sport and excitement for the hunter. The river swarms with ducks and geese at certain seasons of the year, that come here to nest and feed. The natural thrift of the Settlers is manifested in his well-conducted farms, comfortable houses, surrounded by growing orchards, and well-tilled gardens. There is no pleasanter valley in Nebraska than this, or one where the traveler will find a better field for observing the rapid growth and great natural resources of the Northwest; and should he choose to pass a week or more in hunting and fishing, he will find ample sport and a home with almost any of the settlers.

Waterloo—two miles from Elkhorn, is a small side-track station where passenger trains seldom stop. It has a flouring mill, store, school house, and some neat little cottages of well-to-do farmers.

Valley—is 4 3-10 miles further, and shows a marked improvement within the last few years; there are a score of new buildings in sight ; elevation, 1,-147 feet. The curious who wish to note the elevation—station by station—are referred to the "Time Tables" at back of the book, where the figures will be found for each station on the whole line of road.

Omaha & Republican Valley Branch-leads off southwesterly from Valley, and is completed to Stromsburgh, 90 miles distant.

The road crosses the Platte River on a pile bridge 2,200 feet long, enters Saunders county, crosses the river bottom, and reaches CLEAR CREEK, the first station from Valley, seven miles distant.

Passing on over a rolling prairie, five miles further, we come to

MEAD—This is a small place, composed of thrifty farmers.

WAHOO—seven miles further, is the county seat of Saunders county. It is situated on a broad plateau, in the midst of a farming country, where they raise 60 bushels of corn, 50 bushels of oats, and 25 bushels of wheat to the acre, and other crops in proportion. There are three flouring mills in the town. The court house, school and other buildings are very good. Near the town is located one of the finest fair grounds and race tracks in the State. The Wahoo house is the principal hotel.

Leaving Wahoo, the face of the country becomes more rolling, and after crossing Wahoo Creek and making a run of eight miles, we reach WESTON, still young, but a prosperous little place, from which it is eleven miles to

VALPARAISO—another small place of four stores, a hotel and school house, situated on Oak Creek, on which is some very fine land, as well as large groves of oak.

BRAINARD—comes next, thirteen miles from Valparaiso, after rolling over numerous cuts and rough country. Ten miles more, and we arrive at

DAVID CITY—county seat of Butler county. Population about 300. There are four churches and several schools. The Saunders house is the principal hotel.

From David City it is ten miles to RISINGS, seven to SHELBY, seven to OSCEOLA and five to Stromsburg. The **Lincoln Branch**—leads off from VALPARAISO, nine miles to RAYMOND and ten miles to LINCOLN, the Capital of the State.

Mr. S. H. H. Clark, Gen. Manager of the Union Pacific, is President of both of the above named roads, and it is understood they are owned by the Union Pacific parties, under whose management they are conducted.

We will now return to Valley, and proceed westward.

From Valley the Bluffs on the south side of the Platte River can be seen in the distance, but a few miles away, in a southwesterly direction. Soon after leaving the station we catch the first glimpse of the Platte River, on our left. Six miles further over the broad plain brings us to

Mercer—an unimportant station, from which it is five miles to

Fremont—the county seat of Dodge county, situated about three miles north of the Platte River, and contains a population of about 3,500. The regular passenger trains on the "Overland" route stop here 20 minutes for dinner, both from the east and west.

The public buildings include a jail and court-house, seven churches, and some fine school houses. Also a fine opera house. Ten years ago we said: "It was a thriving place in the midst of a beautiful country." Now it is a *city* of no mean pretentions. Within the past ten years there has been built nearly 1,000 dwelling houses, with stores of all kinds in proportion.

Fremont supports several newspapers and hotels, and is the shipping point for a large amount of grain, hay and live-stock raised in the country to the northward.

The Sioux City & Pacific Railroad connects here with the Union Pacific and runs through to Missouri Valley Junction, Iowa, where it connects with the Chicago & Northwestern Railway.

Mule Team, in 1869, loaded with Boilers and Machinery, weighing 54,000 pounds, en route from Elko to White Pine

It is claimed this route is 33 miles shorter to Chicago than via Omaha, but we do not know of any through travel ever going by this line, and judge the local travel to be its sole support.

The Fremont & Elkhorn Valley railroad to the northward is completed to Norfolk, 80 miles, and trains are running regularly. This road runs through a very rich and well cultivated country, where wheat yields as high as 30 bushels to the acre.

Fremont is connected with the south side of the Platte by a wagon bridge that cost over $50,000.

THE PLATTE RIVER.—We are now going up the Platte, and for many miles we shall pass closely along the north bank; at other times, the course of the river can only be traced by the timber growing on its banks. Broad plains are the principal features, skirted in places with low abrupt hills, which here, in this level country, rise to the dignity of "bluffs."

It would never do to omit a description of this famous stream, up the banks of which so many emigrants toiled in the "Whoa, haw" times, from 1850 to the time when the railroad superseded the "prairie schooner." How many blows from the ox-whip have fallen on the sides of the patient oxen as they toiled along, hauling the ponderous wagons of the freighters, or the lighter vehicles of the emigrant! How often the sharp ring of the "popper" aroused the timid hare or graceful antelope, and frightened them away from their meal of waving grass! How many tremendous, jaw-breaking oaths fell from the lips of the "bull-whackers" during that period, we will not even guess at; but pious divines tell us that there is a Statistician who has kept a record of all such expletives; to that authority we refer our readers who are fond of figures. Once in a while, too, the traveler will catch a glimpse of a lone grave, marked by a rude head-board, on these plains; and with the time and skill to decipher the old and time-stained hieroglyphics with which it is decorated, will learn that it marks the last resting-place of some emigrant or freighter, who, overcome by sickness, laid down here and gave up the fainting spirit to the care of Him who gave it; or, perchance, will learn that the tenant of this rentless house fell while defending his wife and children from the savage Indians, who attacked the train in the gray dawn of darker night. There is a sad, brief history connected with each told to the passer-by, mayhap in rude lines, possibly by the broken arrow or bow, rudely drawn on the mouldering head-board. However rude or rough the early emigrants may have been, it can never be charged to them that they ever neglected a comrade. The sick were tenderly nursed, the dead decently buried, and their graves marked by men who had shared with them the perils of the trip. Those were *days*, and these plains the *place* that tried men's mettle; and here the Western frontiersman shone superior to all

HIGH SCHOOL, OMAHA.

others who ventured to cross the "vast desert," which stretched its unknown breadth between him and the land of his desires. *Brave, cool and wary as the savage, with his unerring rifle on his arm, he was more than a match for any red devil he might encounter. Patient under adversity, fertile in resources, he was an incalutable aid at all times; a true friend, and bitter foe.* This type of people is fast passing away.

The change wrought within the last few years has robbed the plains of its most attractive feature, to those who are far away from the scene—the emigrant train. Once, the south bank of the Platte was one broad thoroughfare, whereon the long trains of the emigrants, with their white-covered wagons, could be seen stretching away for many miles in an almost unbroken chain. Now, on the north side of the same river, in almost full view of the "old emigrant road," the cars are bearing the freight and passengers rapidly westward, while the oxen that used to toil so wearily along this route, have been transformed into "western veal" to tickle the palates of those passengers, or else, like Tiny Tim, they have been compelled to "move on" to some new fields of labor.

To give some idea of the great amount of freighting done on these plains we present a few figures, which were taken from the books of freighting firms in Atchison, Kansas. In 1865, this place was the principal point on the Missouri River, from which freight was forwarded to the Great West, including Colorado, Utah, Montana, &c. There were loaded at this place, 4,480 wagons, drawn by 7,310 mules, and 29,720 oxen. To control and drive these trains, an army of 5,010 men was employed. The freight taken by these trains amounted to 27,000 tons. Add to these authenticated accounts, the estimated business of the other shipping points, and the amount is somewhat astounding. Competent authority estimated the amount of freights shipped during that season from Kansas City, Leavenworth, St. Joe, Omaha and Plattsmouth, as being fully equal, if not more than was shipped from Atchison, with a corresponding number of men, wagons, mules and oxen. Assuming these estimates to be correct, we have this result: During 1865, there were employed in this business, 8,960 wagons, 14,620 mules, 59,440 cattle, and 11,220 men, who moved to its destination, 54,000 tons of freight. To accomplish this, the enormous sum of $7,289,300 was invested in teams and wagons, alone.

But to return to the river, and leave facts and figures for something more interesting. "But," says the reader, "Ain't the Platte River a fact?" Not much, for at times, after you pass above Julesburg, there is more fancy than fact in the streams. In 1863, teamsters were obliged to excavate pits in the sand of the river-bed before they could find water enough to water their stock. Again, although the main stream looks like a mighty river, broad and majestic, it is as deceiving as the "make up" of a fashionable woman of to-day. Many places it looks broad and deep; try it, and you will find that your feet touch the treacherous sand ere your instep is under water; another place, the water appears to be rippling along over a smooth bottom, close to the surface; try that, and in you go, over your head in water, thick with yellowish sand. You don't like the Platte when you examine it in this manner. The channel is continually shifting, caused by the vast quantities of sand which are continually floating down its muddy tide. The sand is very treacherous, too, and woe to the unlucky wight who attempts to cross this stream before he has become acquainted with the fords. Indeed, he ought to be introduced to the river and all its branches before he undertakes the perilous task. In crossing the river in early times, should the wagons come to a stop, down they sank in the yielding quicksand, until they were so firmly imbedded that it required more than double the original force to pull them out; and often they must be unloaded, to prevent the united teams from pulling them to pieces, while trying to lift the load and wagon from the sandy bed. The stream is generally very shallow during the fall and winter; in many places no more than six or eight inches in depth, over the whole width of the stream. Numerous small islands, and some quite large, are seen while passing along, which will be noticed in their proper place.

The Platte River has not done much for navigation, neither will it, yet it drains the waters of a vast scope of country, thereby rendering the immense valleys fertile; many thousand acres of which, during the past few years, have been taken up and successfully cultivated.

The average width of the river, from where it empties into the Missouri to the junction of the North and South Forks, is not far from three-fourths of a mile; its average depth is *six inches*. In the months of September and October the river is at its lowest stage.

The lands lying along this river are a portion of the land granted to the Union Pacific railroad, and the company are offering liberal terms and great inducements to settlers. Much of the land is as fine agricultural and grazing land as can be found in any section of the Northwest. Should it be deemed necessary to irrigate these plains, as some are inclined to think is the case, there is plenty of fall in either fork, or in the main river, for the purpose, and during the months when irrigation is required, there is plenty of water for that purpose, coming from the melting snow on the mountains. Ditches could be led from either stream and over the plains at little expense. Many, however, claim that in ordinary seasons, irrigation is unnecessary.

From Omaha to the Platte River, the course of the road is southerly, until it nears the river, when it turns to the west, forming, as it were, an immense elbow. Thence along the valley, following the river, it runs to Kearny, with a slight southerly depression of its westerly course; but from thence to the North Platte it recovers the lost ground, and at this point is nearly due west from Fremont, the first point where the road reaches the river That is as far as we will trace the course of the road at present.

The first view of the Platte Valley is impressive, and should the traveler chance to behold it for the first time in the spring or early summer, it is then very beautiful; should he behold it for the first time, when the heat of the summer's sun has parched the plains, it may not seem inviting; its beauty may be gone, but its majestic grandeur still remains. The eye almost tires in searching for the boundary of this vast expanse, and longs to behold some rude mountain peak in the distance, as proof that the horizon is not the girdle that encircles this valley.

When one gazes on mountain peaks and dismal gorges, on foaming cataracts and mountain torrents, the mind is filled with awe and wonder, perhaps fear of Him who hath created these grand and sublime wonders. On the other hand, these lovely plains and smiling valleys—clothed in verdure and decked with flowers—fill the mind with love and veneration for their Creator, leaving on the heart the impression of a joy and beauty which shall last forever.

Returning to Fremont—and the railroad—we proceed seven miles to

Ames—formerly called Ketchum—only a side track. Near this station, and at other places along the road, the traveler will notice fields fenced with a fine willow hedge, which appears to thrive wonderfully. Eight miles further we reach

North Bend—which is situated near the river bank, and surrounded by a fine agricultural country, where luxuriant crops of corn give evidence of the fertility of the soil. The place has materially improved within the last few years and now has some fine stores, two hotels, a grain elevator, and about 75 dwellings and places of business, and a population of about 350. Young cottonwood groves have been set out in many places—good fences built, and altogether the town has a progressive appearance.

Leaving the station, for a few miles the railroad track is laid nearer the river's bank than at any point between Fremont and North Platte. Seven miles from here we arrive at

Rogers—a new station, and apparently one of promise—7 4-10 miles further is

Schuyler—the county seat of Colfax county, containing 1,000 inhabitants, and rapidly improving. It has five churches, two very good hotels, with courthouse, jail, school-houses, many stores, a grain elevator, and several small manufactories. The bridge over the Platte River, two miles south, centres at this town a large amount of business from the south side of the river.

From Schuyler it is 7 8-10 miles to

Benton—formerly called Cooper; later, Richland—a small side-track station, from which it is eight miles to

Columbus—the county seat of Platte county, a substantial growing city, which contains about 2,500 inhabitants, has two banks, six churches, several schools, good hotels, and two weekly newspapers,—the *Platte Journal* and the *Era*. The Hammond is the principal hotel.

Columbus—from its location in the midst of the finest agricultural lands in the Platte Valley, with the rich valley of the Loup on the north—has advantages that will, at no distant day, make it a city of many thousand inhabitants.

George Francis Train called Columbus the geographical center of the United States, and advocated the removal of the National Capitol to this place. We have very little doubt, should George ever be elected President, he will carry out the idea, when we shall behold the Capital of the Union located on these broad plains—but we shall not buy corner lots on the strength of the removal.

In July and August, 1867, Columbus was a busy place, and the end of the track. Over 10,000,000 lbs. of Government corn and other freight was re-shipped from here to Fort Laramie, and the military camps in the Powder River country. The Burlington and Missouri River Railroad reaches this place from the southward, and the Omaha, Niobrara & Black Hills Railroad leads off to the northward. See ANNEX No. 66.

Soon after leaving Columbus we cross Loup Fork on a fine bridge, constructed in the most substantial manner. This stream rises 75 miles northeast of North Platte City, and runs through a fine farming country until it unites with the Platte. Plenty of fish of various kinds are found in the stream, and its almost innumerable tributaries. These little streams water a section of country unsurpassed in fertility and agricultural resources. Game in abundance is found in the valley of the Loupe, consisting of deer, antelope, turkeys and prairie chickens, while the streams abound in ducks and geese.

From Columbus it is 7 6-10 miles to

Duncan—formerly called Jackson —surrounded by well cultivated fields.

Passing along, and just before reaching the next station, we cross a small stream called Silver Creek. From Duncan it is 10 1-10 miles to the station of

Silver Creek—This section of country has improved very rapidly during the last few years, and we notice many substantial evidences of thrift in every direction—many new buildings.

To the northeast of this station is the old Pawnee Indian Reservation, but not visible from the cars. It covered a tract of country 15x30 miles in area, most of which is the best of land. About 2,000 acres are under cultivation. The tribe, numbering about 2,000, were removed to the Indian Territory in 1878 by the Government and the lands sold at auction.

Again we speed westward, six miles to

Havens—from which it is 5 3-10 miles to

Clark's—a small station named in honor of the Gen. Manager of the road. The surrounding country is remarkably rich in the chief wealth of a nation —agriculture, and has made rapid progress. Of late years, several new stores, a church, school house and many dwellings have been erected, indicating permanent prosperity. From Clark's it is 5 6-10 miles to

Thummel—and 5 6-10 miles more to

Central City—formerly Lone Tree, the county seat of Merrick county. It contains a population of about 900, and is surrounded by thrifty farmers. The "old emigrant road" from Omaha to Colorado crosses the river opposite this point, at the old "Shinn's Ferry."

The more recent settlers of Lone Tree, call the place "Central City," in anticipation of the early completion of the Nebraska Central railroad to this place. Cottonwood trees have been planted by many of the settlers about their homes, which present a cheerful and homelike appearance.

Passengers should notice the railroad track—for 40 miles it is constructed as *straight as it is possible to build a road.* When the sun is low in the horizon, at certain seasons, the view is very beautiful. Rolling along 3 4-10 miles, and we arrive at

Paddock—seven miles more, to

Chapman's—a small place, comprising a few buildings, near the station, but the country around about is a broad prairie, and nearly all improved and settled by thrifty farmers.

From Chapman, we continue west 5 5-10 miles to

Lockwood—a small side-track station. Six miles further and we are at our supper station,

Grand Island—the county seat of Hall county, which contains a population of about 1,500. It is provided with the usual county buildings, several banks, churches of various denominations, good schools, several hotels, many stores, some very pretty private residences, and two weekly newspapers, grain elevators and one of the largest steam flouring mills in the State.

Grand Island is a regular eating station, where trains going west stop 30 minutes for supper, and those for the East have the

SUTTER'S MILL-RACE—WHERE GOLD W

2

DEVIL'S SLIDE, WEBER CANYON, UTAH.

same length of time for breakfast. The eating-house is on the right or north side of the track, in a large, new building, and the meals served are very good. It is claimed that this town will become a great railroad center—in proof of which we notice the completion of the St. Joseph & Denver City Railroad to this place— from the south—in 1879, and the completion of the Grand Island & St. Paul branch of the Union Pacific to St. Paul —22 miles northward—up the Loupe Fork River. The Union Pacific Ry. Co. have located here machine and repair shops, round-house, etc., being the end of the first power division west of Omaha.

This station was named after Grand Island in the Platte River, two miles distant, one of the largest in the river, being about 50 miles in length by four in width. The Island is well wooded—cottonwood principally, and, some years after completion of the railroad was a government reservation.

When the road was first built to Grand Island, buffalo were quite numerous, their range extending over 200 miles to the westward. In the spring, these animals were wont to cross the Platte, from the Arkansas and Republican valleys, where they had wintered, to the northern country, returning again, sleek and fat, late in the fall; but since the country has become settled, few, if any, have been seen. In 1860, immense numbers were on these plains on the south side of the Platte, near Fort Kearny, the herds being so large that often emigrant teams had to stop while they were crossing the road. At

Chinese Cheap Labor—"Work for nothing and board yourself"—from the inhabitants of your neighbor's hen-roost.

It is said that in San Francisco the people can drink, and carry more without staggering, than in any city of the world.

Fort Kearny, in 1859 and 1860, an order was issued forbidding the soldiers to shoot the buffalo on the parade ground.

Proceeding westward 7 7-10 miles, we reach

Alda—a small station just east of Wood River.

After crossing the river, the road follows along near the west bank for many miles, through a thickly settled country, the farms in summer being covered with luxuriant crops of wheat, oats and corn. Wood River rises in the bluffs, and runs southeast until its waters unite with those of the Platte. Along the whole length of the stream and its many tributaries, the land for agricultural purposes is surpassed by none in the Northwest, and we might say in the world. The banks of the river and tributaries are well woo led, the streams abound in fish and wild-fowl, and the country adjacent is well supplied with game, deer, antelope, turkeys, chickens, rabbits, etc., forming a fine field for the sportsman.

This valley was one of the earliest settled in Central Nebraska, the hardy pioneers taking up their lands when the savage Indians held possession of this, their favorite hunting-ground. Many times the settlers were driven from their homes by the Indians, suffering fearfully in loss of life and property, but as often returned again, and again, until they succeeded in securing a firm foothold. To-day the evidences of the struggle can be seen in the low, strong cabins, covered on top with turf, and the walls loop holed, and enclosed with the same material, which guards the roofs from the fire-brands, bullets and arrows of the warriors.

From Alda, it is 8 1 10 miles to

Wood River—a small stat on. Here can be seen one of the old-fashioned specimens of plains station-men, in the person of Charley Davis. He keeps an eating-house and saloon, where freight and emigrant trains often stop for meals. Charley's specialty is the "Jerusalem Pickle." A good "square meal" is served for 50 cents.

Passing on 7 5-10 miles, we reach

Shelton—a side-track, where a flouring mill, store, and a few dwelling houses constitute the place. To the westward 5 8-10 miles, is

Gibbon—It is situa'ed in the midst of a fine farming country, was once the county seat of Buffalo county, and is a thriving place, with a population of about 100.

Proceeding, it is 8 4-10 miles to

Buda—formerly Kearny—later, Shelby—a station of little account.

Westward again four miles, and we reach a place of some importance,

Kearny Junction—the county seat of Buffalo county—named for the old fort of that name on the south side of the river, nearly opposite.

In 1873, the first few buildings were erected here, since which time the place has improved wonderfully. It now contains a population of over 1,200, with two weekly papers.

The citizens, as a class, are enterprising, law-abiding representatives from nearly every State in America, with a few from foreign countries.

Here the B. & M. R. R. in Neb. comes in from the south—crossing the Platte River—two miles distant—and forms a junction with the Union Pacific. This road runs through a rich, well-settled agricultural country.

The local business coming in on the B. & M. and the Union Pacific makes this place one of unusual activity and business promise. The town contains the usual county buildings, which are built of brick, has two banks, six fine churches, two schools, many stores of all kinds, several hotels—the Atkins and the Grand Central are the principal—and some fine private residences.

The country around the town is not as good agricultural land as we have seen further to the eastward, yet some good crops of grain are raised, and large quantities are hauled here, to be shipped to the East, West and South.

From this point *west*, the country is occupied principally by the stock men.

Stages leave here daily, except Sunday, for the Republican Valley, and all intermediate points, carrying the U. S. mail to Franklin, Bloomington, Republican City, Orleans and Melrose, where connections are made with stages for every town in the Upper Republican Valley and Northern Kansas.

Let us take a look at the grounds on which stood old

FORT KEARNY—This post was first established at Fort Childs, Indian Territory, in 1848, by volunteers of the Mexican war—changed to Fort Kearny in March, 1849. In 1858 the post was re-built by the late Brevet-Colonel Charles May, 3d Dragoons. It is situated five miles south of Kearny station, and nine miles via Burlington & Missouri railroad from Kearny junction, on the south bank of the Platte, which is at this point three miles wide, and full of small islands. The fort is in latitude 40 deg. 33 min., longitude 99 deg. 6 min.

In the fall of 1872, all the Government buildings, worth moving, were removed to North Platte and Sidney, on the Union Pacific Railway, 291 and 412 miles, respectively, west from Omaha, and the post abandoned. The remains of the dead bodies of soldiers, buried at Kearny, were taken up and reintered in the National Cemetery, at Fort McPherson.

Two miles above the Fort, on the south bank, is Kearny City, in the early days more commonly called "Dobey Town." This was once a great point with the old Overland Stage Company, and at that time contained about five hundred inhabitants, the greater portion of which left upon the abandonment of the line and the south-side route of travel. But we are told that settlers are coming in fast, and it will soon regain its "old time" figures.

Returning to Kearny Junction, 5 9-10 miles brings us to

Stevenson—a side-track,—unimportant. Again, four miles west is

Odessa—another small station,—from which it is 6 3-10 miles to

Elm Creek Station—a small place of several stores and a few dwellings.

Soon after leaving the station, we cross Elm Creek, a small, deep, and quite lengthy stream. It was well wooded before the advent of the railroad, the timber consisting almost entirely of red elm, rarely found elsewhere in this part of the country.

From Elm Creek station it is nine miles to

Overton—This is another small station of a few buildings. It is situated on a branch of Elm Creek.

The Platte Valley along here, and for the fifty miles over which we have just passed, is very broad; nearly all the best land has been taken up, or pur-

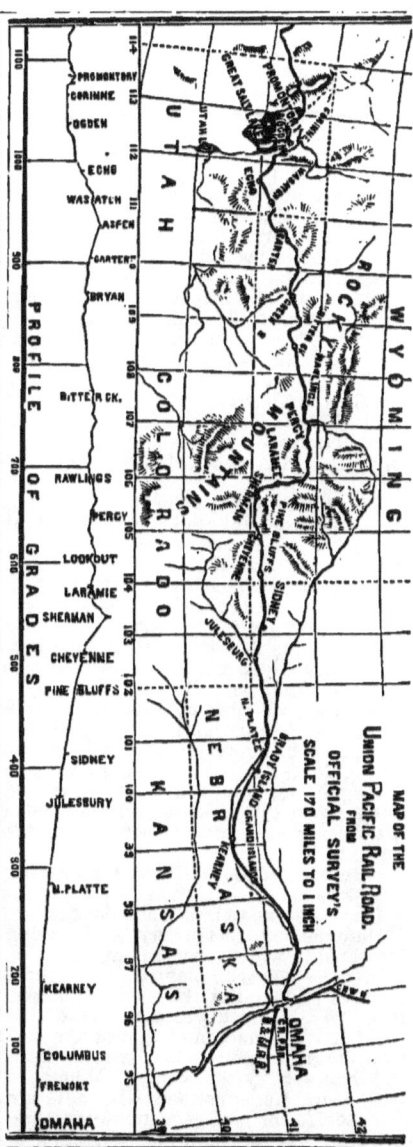

To be well armed and rea'y for a fight, is "to be heeled."

The Indians on the plains call the locomotives and cars "bad medicine wagons."

chased, but only a small portion is under cultivation.

Passing on, 4 miles brings us to

Josselyn—a side-track station, named after the paymaster of the road, a much more important person to the employes than the station, as trains do not always stop here, but roll on five miles further to

Plum Creek—the county seat of Dawson county. It contains a population of about 800, has a fine, brick court-house, two churches, a school-house, several hotels, four stores, a bridge across the Platte, to the south, and a weekly newspaper—the *Pioneer*. The town was named after an old stage station and military camp, situated on the south side of the river, on Plum Creek, a small stream which heads in very rugged bluffs southwest of the old station, and empties its waters into the Platte—opposite Plum Creek station on the railroad.

This old station was the nearest point on the "old emigrant road" to the Republican River, the heart of the great Indian rendezvous, and their supposed secure stronghold, being but about 18 miles away. Around the old Plum Creek station many of the most fearful massacres which occurred during the earliest emigration were perpetrated by the Sioux, Cheyenne, and Arapahoe Indians. The bluffs here come very close to the river, affording the savages an excellent opportunity for surprising a train, and, being very abrupt and cut up with gulches and cañons, affording them hiding-places, from which they swooped down upon the luckless emigrant, often massacring the larger portion of the party.

Returning to the railroad, 7 8-10 miles brings us to

Coyote—an unimportant station. Here the bottoms are very wide, having increased in width for many miles.

In early days, all along the river, for a distance of 50 miles, the islands and lowlands were covered with cottonwood timber, but since the completion of the railroad, the greater portion have been cut down and removed by the settlers. Where, in 1860, were huge cottonwoods, now are wheat-fields, or young cottonwoods and willows. We are now in a section of country where large quantities of hay are put up annually for shipment,

Passing on 6 miles, we reach

Cozad—About one-fourth mile before reaching this station, on the right, we cross the 100th meridian, marked by a sign, which reads, in large letters "100th Meridian."

This place was named by a gentleman from the East, who purchased 40,000 acres of land from the railroad company here, and laid out a town. It has not been a "huge success" as a speculation, so far, but by a thorough system of irrigation could be made very productive. There are a few good buildings at and near the station, and some herds of cattle and sheep range near by; in fact, this section of country is more adapted to stock-raising than it is for agricultural purposes.

The high bluffs to the south and west—our road here runs nearly north—looming up in the distance, are on the south side of the Platte River, 25 miles distant.

From Cozad, it is five miles to

Willow Island—population 100—named from an island in the Platte River, near by, the second in size in that river. For some distance before reaching this station, large herds of cattle and sheep can be seen, particularly on the opposite side of the river, where can also be seen some of the old adobe ranches of the days when the "overland stage" was the fastest method of crossing these plains.

We are now beyond the agricultural section, and are entering the great grazing region of the West.

For some years after the completion of the road the traveler could see, near this place, and in fact for many miles beyond North Platte, some of the old log houses of the early settlers, with their sides pierced with loop-holes and walled up with turf, the roofs being covered with the same material, which reminds one of the savage against whom these precautions were taken. In fact, from here up the river, the traveler will doubtless observe many of the rude forts along the roadside as well as at the stations. The deserted ranches to be met with along the "old emigrant road," on the south side of the river, are fortified in the same manner. The fort was generally built of logs, covered on top and walled on the side in the manner described. They are pierced with loop-holes on all sides, and afforded a safe protection against the Indians. They generally stood about fifty yards from the dwelling, from which an underground passage led to the fort. When attacked, the settlers would retreat to their fortification where they would fight it out; and until the Indians got "educated," many a "red brother" would get a shot—to him unawares—which would send him to his "Happy Hunting-ground."

As we pass along to the next station, 10 3-10 miles, the passenger will note that our direction is nearly north, with the bottom lands getting narrower as we proceed.

Warren—This is simply a side-track, where trains seldom stop. The grass here is short and thick on the upland, and coarse and tall on the bottoms.

Sand-hills close in on the right, and the river on our left, as we proceed eight miles further to

Brady Island—This station derives its name from an island in the Platte River, which is of considerable size. In early times many wandering bands of Indians were wont to cross the river at this point, and for months at a time camp on the Island or on the river banks.

Hendrey—is a side-track 4 9-10 miles further, and 4 2-10 miles from

Maxwell—formerly McPherson Station. It is five miles from the Platte River and seven miles from old "Cottonwood Springs" on the opposite side of the river, with which it is connected by a bridge, a great improvement on the old ford.

The country round about is fertile, with some timber on the river bottoms. A large amount of fine meadow land adjoins the station, from which are cut thousands of tons of hay.

FORT McPHERSON—is situated on the south side of the Platte River, near Cottonwood Springs. The post was established February 20, 1866, by Major S. W. O'Brien, of the 7th Iowa Cavalry. It was originally known as "Cantonment McKeon," and also as "Cottonwood Springs." At the close of the war, when the regular army gradually took the place of the vol-

DALE CREEK BRIDGE, NEAR SHERMAN, ON THE BLACK HILLS OF WYOMING.

unteers who had been stationed on the frontier during the rebellion, the names of many of the forts were changed, and they were re-named in memory of those gallant officers who gave their lives in defense of their country. Fort McPherson was named after Major-General James B. McPherson, who was killed in the battle before Atlanta, Georgia, July, 22d, 1864. Supplies are received via McPherson Station. Located in latitude 41 deg., longitude 100 deg. 30 min.

The next station is 7 7-10 miles further, named

Gannett—a side-track—nearly five miles from where the trains cross the long trestle bridge over the

,NORTH PLATTE RIVER—This river rises in the mountains of Colorado, in the North Park. Its course is to the northeast from its source for several hundred miles, when it bends around to the southeast. We shall cross it again at Fort Steele, 402 miles further west. The general characteristics of the stream are similar to tho e of the South Platte.

For 100 miles up this river the "bottom lands" are from 2 to 15 miles wide, very rich, and susceptible of cultivation, though perhaps requiring irrigation. Game in abundance is found in this valley, and bands of wild horses at one time were numerous.

Fort Laramie is about 150 miles from the junction—near where the Laramie River unites with this s'ream.

On the west bank of the river, 80 miles north, is Ash Hollow, rendered famous by General Harney, who gained a decisive victory over the Sioux Indians here, many years ago.

About one mile beyond the bridge and 5 8-10 miles from Gannett is situated

North Platte City—the county seat of Lincoln county. Elevation, 2,789 feet; distance from Omaha, 291 miles. Here is the end of the Eastern and the commencement of the Mountain Division. For altitude of each station see "Time Table" No. 2 at the end of the Book. This is a regular eating station for the trains on the "Denver Short Line." Breakfast going west; supper coming east.

The road was finished to this place, November, 1866. Here the company have a round-house of 20 stalls, a blacksmith and repair shop, all of stone. In these shops are employed—regularly—76 men, besides those engaged in the offices and yard. The Railroad House is the principal hotel.

North Platte has improved very rapidly during the last three years, and contains about 2,000 population. Churches, hotels, county buildings, and scores of dwellings have been built, or are in course of erection. A new bridge has been completed across the South Platte River. Two weekly papers are published here, and several others projected. Settlers' houses, and tens of thousands of cattle, sheep and horses are to be seen in every direction. The advantages of this place, as a stock range and shipping point, exceed all others on the line of road.

Messrs. Keith, Barton, and Dillon, citizens of North Platte City, have a herd of 15,000 head of cattle—on the North Platte above the City—and there are many other parties living at or near this city, who own herds of from 500 to 5,000 head. In this country a man that only owns 500 head, is counted a "poor shoat"—one to be pitied.

North Platte, in its palmiest days, boasted a population of over 3,000, which was reduced in a few months after the road extended, to as many hundreds. Until the road was finished to Julesburg, which was accomplished in June, 1867, all freight for the West was shipped from this point; then the town was in the height of its prosperity; then the gamblers, the roughs and scallawags, who afterward rendered the road accursed by their presence, lived in clover—for there were hard-working, foolish men enough in the town to afford them an easy living. When the town began to decline, these leaches followed up the road, cursing with their upas blight every camp and town, until an enraged and long-suffering community arose in their own defense, binding themselves together, *a la vigilantes*, and, for want of a legal tribunal, took the law into their own hands, and hung them to the first projection high and strong enough to sustain their worthless carcasses. But many "moved on," and we shall hear of them again many times before we are through.

From North Platte our route is due west. It is 8 4-10 miles to

Nichol—an unimportant side-track. NorthPlatte city is in plain sight—as is also the North and South Platte Rivers—and the Valleys of the same.

From Nichols it is 8 5-10 miles to

O'Fallon's Station—situated in the sand hills, where the bluffs on the right come close to the river. On the south side of the river are the famous O'Fallon's Bluffs, a series of sand hills interspersed with ravines and gulches which come close to the river's bank, forming abrupt bluffs, which turned the emigrants back from the river, forcing them to cross these sand hills, a distance of eight miles, thro' loose yielding sand, devoid of vegetation. Here as well as at all points where the bluffs come near the river, the emigrants suffered severely, at times, from the attacks of the Indians. Opposite, and extending above this point is a large island in the river, once a noted camping ground of the Indians. O'Fallon's Bluffs are the first of a series of sand hills, which extend north and south for several hundred miles. At this point the valley is much narrower than that thro' which we have just passed. Here we first enter the "alkali belt," which extends from this point to Julesburg—about 70 miles. The soil and water are strongly impregnated with alkaline substances. The country on both sides of the river is occupied exclusively for grazing purposes. The first volume of this book instructed passengers to keep their "eye peeled" for buffalo, as we are now getting into the buffalo range. During the spring of 1873-74 immense numbers roamed over this country, along the road for 100 miles westward, but few, if any, have been seen since that time. Passing along up the narrow bottom, with the bluffs along our right, 7 3-10 miles brings us to a side-track, called

Dexter—Trains seldom stop here, and 7 2-10 miles further we reach

Alkali—on an alkali bottom. This station is directly opposite the old stage station of that name on the south side of the river. After leaving the station the road passes thro' the sand-bluffs, which here run close to the river's bank. A series of cuts and fills, extending for many miles, brings us to the bottom land again. From Alkali, it is 9 6-10 miles to

Roscoe—another side-track station. Passing along over a narrow bottom, with sand bluff cuttings, at intervals, 9 6-10 miles we come to

Ogalalla—the county seat of Keith county. The settlers here are all more or less engaged in stock-raising. It is the river crossing for large droves of cattle en route for the Indian reservation, Fort Laramie and the Black Hills country, to the northward. Near this station, several years ago, at a point where the road makes a short curve and crosses the mouth of a ravine, the Indians attempted to wreck a passenger train, by suddenly massing their ponies on the track ahead of the locomotive. The result was, some score or more of ponies were killed, without damaging the train, while the men used their "pistols" and guns pretty freely on the Indians, who were apparently greatly *surprised*, and who now called the locomotive "Smoke wagon—big chief! Ugh!! no good!"

Passing on 1 6-10 miles we pass

Bosler—a side-track from which it is 8 miles to

Brule— near is the old California Crossing, where the emigrants crossed the river when striking for the North Platte River and Fort Laramie, to take the South Pass route overland. On the south side of the river, opposite in plain view, is the old ranche and trading post of the noted Indian trader and Peace Commissioner—Beauve—now deserted.

Passing along over cuts and fills, 9 7-10 miles, we reach

Big Springs—The station derives its name from a large spring, the first found on the road, which makes out of the bluffs, opposite the station, on the right hand side of the road, and in plain view from the cars. The water is excellent, and will be found the best along this road. It was at this station where the "Blue Spring's robbery" took place, Sept. 18th 1877. A party of twelve masked men took possession of the station, bound and gagged the men, cut the telegraph wires, when the western train arrived took possession of it with guns and revolvers, in the name of "hands up". The robbers secured $65,000 from the express car, $1,300 and four gold watches from passengers, then mounted their horses and allowed the train to proceed. No person was killed or injured, but all were very badly frightened. Immediately after the robbery, a reward of

$10,000 was offered for the arrest of the perpetrators, and several have been caught and have paid the penalty of the crime with their lives. About one-half of the money has been recovered.

After leaving this station, we pass by a series of cuts and fills, and another range of bluffs, cut up by narrow ravines and gorges. At points, the road runs so near the river bank, that the water seems to be right under the cars. But we emerge again after 7 8-10 miles and come to

Barton—a small signal station of very little importance, from which it is 2 7-10 miles to

Denver Junction—Here the new "Omaha & Denver Short Line" branches off to the left.

In 1873-4, a railroad bed was graded up the north side of the Platte river, in the interest of the U. P. Ry. Co., but for some reason the ties and iron were not laid until the summer of 1881. On November 6th of that year the first through passenger trains commenced making regular trips. The stations and distances are as follows; (See time table in back of book. MILES.

Denver Junc. to Sedgwick	14.8
Sedgewick to Crook	15.6
Crook to Iliff	15.5
Iliff to Sterling, (Dinner Station.)	11.8
Sterling to Buffalo	12.7
Buffalo to Snyder	16.9
Snyder to Deuel	12.9
Deuel to Orchard	17.6
Orchard to Hardin	17.6
Hardin to LaSalle	15.4

Fom Denver Junction to LaSalle, to connect with Cheyenne Div. U. P. Ry.,	150.8
From LaSalle to Denver	46.4

From Denver Junction to Denver	197.2
" Omaha to North Platte	291.
" North Platte to Denver Junction	80.4

Omaha to Denver, via "Short Line,"	568.6
Omaha to Denver, via Cheyenne	622.

Difference in favor of "Short Line,"...... 53.

THE PLATTE RIVER, west of North Platte city, is called the South Fork of the Platte. We have ascended it almost on its banks, over 350 miles, and shall now leave it, as the "Overland Route" turns to the right, and northwest, and follow up the narrow valley of Lodge Pole Creek, to Egbert, about 100 miles distant. The South Fork of the Platte, up which the "Short Line" is built, rises in the South Park of the Rocky Mountains of Colorado, about 280 miles distant. The valley extends from the Junction up the river about 217 miles, to where the river emerges from the mountains. The average width of the valley is about three miles, the soil of which, in places, is very rich, producing good crops with irrigation, large quantities of hay, and most excellent grazing. It now supports, with the adjoining uplands, vast herds of cattle, sheep and horses.

We refer the reader, for full information in regard to Colorado, her mineral, stock-raising, and varied resources, watering places, and scenic attractions, to *Crofutt's Grip-Sack Guide of Colorado*. Sold on all trains.

From Denver Junct'n it is 6 miles to

Wier—formerly Julesburgh, station. Elevation 3,394 feet. Until 1868, this was an important military, freight, and passenger station, since when it declined to a simple way station. The Union Pacific was completed to this place the last of June, 1867, and all Government freight for the season was shipped to this point, to be reshipped on wagons to the north and west. At that time Julesburgh had a population of 4,000; now the town is almost deserted. During the "lively times," Julesburgh was the roughest of all towns along the Union Pacific line. The roughs congregated there, and a day seldom passed but what they "had a man for breakfast." Gambling and dance houses constituted the greater portion of the town; and it is said that morality and honesty clasped hands and departed. We have not learned whether they have returned; and really we have our doubts about their ever having been there.

Before the railroad, the last of Utah and California emigration that came up the Platte crossed opposite the station, and followed up the valley of Lodge Pole Creek to Cheyenne Pass.

The old, old, town of Julesburgh, was situated on the south side of the Platte river nearly opposite this station and was named for Jules Burgh who was brutally assasinated as will be related in ANNEX No. 10.

Near this old town was the site of

FORT SEDGWICK—this post was established May 19, 1864, by the Third U. S. Volunteers, and named after Maj. Gen'l John Sedgwick., Col. 4th Cavalry, U. S. A., who was killed in battle at Spottsylvania C. H., Va., May 9th, 1864. It is located in the northeast corner of Colorado, on the south side of the Platte river, four miles distant, on the old emigrant and stage road to Cllorado, in plain view from the cars. Latitude 31 deg., longitude 102 deg. 30 min.—now abandoned. During the winter of '65-'66, most of the wood used at Julesburg and Fort Sedgwick, was hauled on wagons from Denver, at an expense of from $60 to $75 per cord, for transportation alone, and was sold to Government, by contract, at $105 per cord. The wood cost in Denver about $20. Besides this the contractors were allowed by Government to put in what hard wood they could get at double price, or $210 a cord, many thought this to be a "pretty soft snap." The "hard wood" was obtained in the scrub-oak bluffs of Colorado, 30 miles south of Denver, and cost no more for transportation than did the soft.

From Wier it is 10 miles to

Chappell—a small side-track where passenger trains seldom stop, and 9 1-10 further to

Lodge Pole.—another side-track. This valley is narrow, but with bluffs, and a great open prairie country to the northward, extending to the North Platte river, a distance of 30 miles, affords the finest grazing range, and large herds of cattle, and numerous bands of antelope can be seen while passing on up the valley.

Colton—is a small station, 10 miles from Lodge Pole. It was named in honor of Francis Colton, Esq., a former general passenger agent of the road.

From Colton it is 7 7-10 miles to

Sidney—named after the president of the road. This is a regular eating station, where trains stop 30 minutes, those from the East for breakfast, and from the West for supper. Sidney is the county seat of Cheyenne county, Neb., and within the last few years has improved in buildings, and increased in population, until it now contains about 1,500 people. The "Lockwood" the largest hotel, is situated a little to the west of the station from which start the daily stages for Deadwood in the Bl'ck Hills of Dakota. Distance 267 miles.

BLACK HILLS GOLD MINES—For many years anterior to the building of the Pacific Railway vague reports were circulating among old plainsmen and miners, of rich gold deposits in the Black Hills and Big Horn country, but until Gen. Custer. with a military expedition, penetrated to, and explored the region about Harney's Peak in '74, and reported gold abundant, the soil rich, the country well timbered, and most desirable, nothing definite was known. In '75 the gold-seekers began their pilgrimage to the "Hills," in '76 the numbers were greatly increase, but in '77 the great rush was at its height. These Hills lie between the 43d and 45th degrees of latitude, and the 103d and 105th parallels of longitude; are about 100 miles long and 60 miles wide. Besides extensive and rich veins of gold and silver yielding quartz, there are found to be vast beds of coal, iron, copper, lead and mica. Placer mines are also numerous, many of which are worked with profit. The country is well watered, the mountains covered with timber, while the valleys are very rich and productive agricultural lands. For grazing purposes the country about and adjacent to the "Hills" is unequaled, and stock thrives the year around upon the native grasses. The population of this region, at present, is not far from 20,000; the greater portion are engaged in quartz mining. The ores are worked principally by the stamp process, some of the largest mills in this country being located here. The mills now in operation aggregate 1,192—stamps, thundering away night and day, the yield of which, including the placer mines, for 1881 exceeded $4,500,000. Deadwood is the principal city, out of a half a hundred cities, towns, villages and prosperous mining settlements. Sidney is the chief out-fitting point for the "Hills," and freight in large quantities is shipped from here on wagons, and it is claimed this is the shortest and most comfortable route. Sidney has some good business blocks and private residences. The railroad company have a 10-stall round house, machine shop, a large freight warehouse and depot building.

To learn all about Colorado, buy "Crofutt's Grip-Sack Guide." It is a complete Encyclopedia of the State.—Sold on the trains.

The principal outfitting store at Sidney is owned by Mr. Chas. Moore, the pioneer ranchman of the old South Platte route; but 'Charley' talks *poor*. Besides his big stock of goods, he has *only* about 5,000 head of cattle and sheep,—and by the way, SIDNEY is not much behind in the number of prosperous stock-men. Scores of her citizens own from 500 to 5,000 head, within range of the late "Cattle King," Iliff, to the south, on which graze 30,000 head.

The Government has established a military post at this station, and erected extensive barracks and warehouses. The post is on the south side of the track, a little to the east of the station. The old "PostTrader" at this place, Mr. James A. Moore, recently deceased, was an old pioneer, and the hero of the "Pony Express." June 8th, 1860, he made the most remarkable ride on record. Mr. Moore was at Midway stage station on the south side of the Platte, when a very important Government despatch arrived for the Pacific Coast. Mounting his pony, he left for Julesburg, 140-miles distant, where, on arriving, he met an important despatch *from* the Pacific; resting *only seven minutes*, and, *without eating*, returned to Midway, making the "round trip"—280 miles—in fourteen hours and forty-six minutes. The despatch reached Sacramento from St. Joseph, Mo., in eight days, nine hours and forty minutes.

From Sidney it is 9 miles to

Brownson—Passenger trains do not stop. The station was named after Col. Brownson, who was with the Union Pacific from the first, and a long time their general freight agent. The valley along here is very narrow, with high rocky bluffs on each side. It is 9 9-10 miles further to

Potter—Large quantities of wood and ties are usually stored here, which are obtained about 20 miles north of this point, on Lawrence Fork and Spring Canyon, tributaries of the North Platte River. Potter, although not a large place, is situated

FINGER ROCK, WEBER CANYON, UTAH.

near a very large city, called
Prairie Dog City—one of the largest cities on the whole line of the road. At this point, and for several miles up and down the valley, the dwellings of the prairie dogs frequently occur, but three miles west of the station they are found in large numbers, and there the great prairie dog city is situated. It occupies several hundred acres on each side of the road, where these sagacious little animals have taken land and established their dwellings without buying lots of the company. (We do not know whether Mr. Land-Commissioner, intends to eject them or not.) Their dwellings consist of a little mound, with a hole in the top, from a foot to a foot and a half high, raised by the dirt excavated from their burrows. On the approach of a train, these animals can be seen scampering for their houses; arrived there, they squat on their hams or stand on their hind feet, barking at the train as it passes. Should any one venture too near, down they go into their holes, and the city is silent as the city of the dead.

It is said that the opening in the top leads to a subterranean chamber, connecting with the next dwelling, and so on through the settlement; but this is a mistake, as in most cases a few buckets of water will drown out any one of them. The animal is of a sandy-brown color, and about the size of a large gray squirrel. In their nest, living with the dog, may be found the owl and rattlesnake, though whether they are welcome visitors is quite uncertain. The prairie dog lives on grasses and roots, and is generally fat; and by many, especially the Mexicans, considered good eating, the meat being sweet and tender, but rather greasy, unless thoroughly par-boiled. Wolves prey on the little fellows, and they may often be seen sneaking and crawling near a town, where they may, by chance, pick up an unwary straggler. But the dogs are not easily caught, for some one is always looking out for danger, and on the first intimation of trouble, the alarm is given, and away they all scamper for their holes.

Court-House Rock—About 40 miles due north from this station is the noted Court-House Rock, on the North Platte River. It is plainly visible for 50 miles up and down that stream. It has the appearance of a tremendous capitol building, seated on the apex of a pyramid. From the base of the spur of the bluffs on which the white Court-House Rock is seated, to the top of the rock, must be nearly 2,000 feet. Court-House Rock to its top is about 200 feet. Old California emigrants will remember the place and the many names, carved by ambitious climbers, in the soft sand-stone of which it is composed.

Chimney Rock—is about 25 miles up the river from Court-House Rock. It is about 500 feet high and has the appearance of a tremendous, cone-shaped sandstone column, rising directly from the plain. The elements have worn away the bluffs, leaving this harder portion standing.

The next station is nine miles distant, called

Dix—formerly Bennett—a side-track for the accommodation of stockmen residing near. The name of the station is in honor of Gen. Dix, of New York. Passenger trains seldom stop, but roll on 9 2-10 miles further where they *do stop*, at

Antelope—It is situated at the lower end of the Pine Bluffs, which at this point is near the station, on the left.

This station is in the center of what the plains-men call "the *best* grass country in the world," as well as one of the best points for antelope on the route. For article on stock-raising, see Annex No. 29.

Six miles further and we come to

Adams—an unimportant side-track, from which it is 5 9-10 miles to

Bushnell—This is another unimportant side-track, near the boundary line between Nebraska and Wyoming Territory. Passenger trains do not stop, but pass on ten miles further to

Pine Bluffs—where cattle-shipping is the principal business transacted at the station.

During the building of the road, this place was known as "Rock Ranche"—and a tough ranche it was. Considerable pitch pine wood was cut for the railroad in the bluffs, a few miles to the southward, from which the station derives its name. The bluffs are on the left hand side of the road, and at this point are quite high and rocky, extending very near the track.

Fort Morgan—was established in May, 1865, abandoned in May, 1868, and its garrison transferred to Laramie. It is about 60 miles north of this station, on the North Platte River, at the western base of what is known as Scott's Bluffs. Latitude 40 deg. 30 min.; longitude 27 deg.

Our course from this station is more to

the westward, for 5 6-10 miles to

Tracy—a small side-track, where passenger trains seldom stop. It is 5 6-10 miles further to

Egbert—another unimportant side-track.

Near this point we leave Lodge Pole Creek, from which to the source of the stream in the Black Hills, about 40 miles away, the valley presents the same general appearance until it reaches the base of the mountains. Bears, deer and wolves abound in the country around the source of the stream, and herds of antelope are scattered over the valley. At one time beavers were plenty in the creek, and a few of these interesting animals are still to be found in the lower waters of the stream, near to its junction with the Platte. This valley was once a favorite hunting-ground of the Sioux and Cheyennes, who long resisted the attempts of the Government to remove them to a reservation to the northward.

Passing on up a dry ravine 6 3-10 miles, we come to

Burns—another small side-track—and nothing else—which is 5 7-10 miles from

Hillsdale—When the road was being constructed from this place to Cheyenne, a large amount of freight was re-shipped from here on wagons. *Then*, it was a busy place, *now*, only a water-tank and side-track. The station was named after a Mr. Hill, one of the engineering party who was killed near this place by the Indians while he was engaged in locating the present site of the road.

About 50 miles to the south is "Fremont's Orchard," on the South Platte River, about 60 miles below Denver City, Colorado, and in that State. It was named after Col. Fremont, who discovered the point in his exploring expedition. It consists of a large grove of cottonwood trees, mostly on the south side of the river.

MONUMENT ROCK, BLACK HILLS, U.P.R.R.

The river here makes an abrupt bend to the north, then another to the south, cutting its way through a high range of sand-hills—the third range from the Missouri River. Where the river forces its way through the bluffs, they are very high and abrupt on the south side. The two bends leave a long promontory of sand hills, the end of which is washed by the waters. At a distance, this grove of cottonwoods on the bottom land reminds one of an old orchard, such as is often seen in the Eastern States.

Near Fremont's Orchard is located the Green Colony, at Green City, which numbers about 100.

Passing on from Hillsdale up a ravine, which gradually becomes narrower as we ascend, with bluffs on either hand, 6 2-10 miles, we come to

Atkins—a side-track. Passing on, our train gradually rises on to the table-land, and *th-n*, if the day be a fair one, the traveler can catch the first glimpse of the Rocky Mountains, directly ahead. On the right he can catch glimpses of the Black Hills of Wyoming, stretching their cold, dark ruggedness far away to the right, as far as the eye can see; but the bold, black line—the dark shadow on the horizon, which will soon take tangible shape and reality, but which now seems to bar our way as with a gloomy impenetrable barrier, is the "Great Rocky Mountain Chain," the back-bone of the American continent, though bearing different names in the Southern hemisphere. The highest peak which can be seen rising far above that dark line, its white sides gleaming above the general darkness, is Long's Peak, one of the highest peaks of the continent. Away to the left rises Pike's Peak, its towering crest robed in snow. It is one of those mountains which rank among the loftiest. It is one of Colorado's noted mountains, and on a fair day is plainly visible from this point, 175 miles distant.

From Atkins it is 5 4-10 miles to

Archer—situated on the high tableland, where the cars seldom stop—is eleven miles from Hillsdale; and a little farther on, the cars pass through the *first* snowshed on the Union Pacific road, emerging with Crow Creek Valley on the left.

After passing through a series of cuts and fills, the track of the Denver Pacific railroad can be seen on the left side, where it passes over the bluffs to the southeast. Directly ahead can be seen, for several miles, the far-famed "Magic City of the Plains," 8 4-10 miles from the last station—

Cheyenne—which is the capital of Wyoming, the largest town between Omaha and Ogden. Passenger trains from the East and West stop here 30 minutes, for dinner—and no better meals can be had on the road than at the Railroad House. Distance from Omaha, 516 miles; from Ogden 516 miles—just *half* the length of the Union Pacific road; distance to Denver, Colorado, 106 miles.

Cheyenne is the county seat of Laramie county. Population about 6,000. Elevation 6,041 feet. It is situated on a broad plain, with Crow Creek, a small stream, winding around two sides of the town. The land rises slightly to the westward. To the east it is apparently level, though our table of elevations shows to the contrary. The soil is composed of a gravelly formation, with an average loam deposit. The sub-soil shows volcanic matter, mixed with marine fossils in large quantities. The streets of the town are broad and laid out at right angles with the railroad.

Schools and churches are as numerous as required, and society is more orderly and well regulated than in many western places of even older establishment. The church edifices are the Presbyterian, Congregational, Episcopal, Methodist, Catholic, and several of other denominations. The city boasts of a $40,000 court-house, a $70,000 hotel—the Inter-Ocean—many new blocks of buildings, among which are, an opera house, banks, and stores of all kinds, besides many fine private residences, also a grand lake or reservoir for supplying the city with pure water, conducted by canal from Crow Creek, from whence smaller branches run along the sidewalks for the irrigation of gardens, trees and shrubbery, which will soon make the city a place of surpassing beauty. It also boasts of a race-course and some good "steppers." It has two daily newspapers, the *Leader* and the *Sun*, both of which issue weeklies.

Cheyenne has the usual small manufactories, among which the item of saddles is an important one, as the saddle of the plains and most Spanish countries, is a different article altogether from the Eastern "hogskin." When seated in his saddle, the rider fears neither fatigue nor injury to his animal. They are made for use—to save the animal's strength, as well as to give ease and security of seat to the rider. The best now in use is made with what is known as the "California tree." The old firm of E. L. Gallatin & Co., make these saddles a specialty, and fill orders from all over the western portion of the United States, Mexico and South America.

The railroad company's buildings are of stone, brought from Granite Canyon, 19 miles west. They consist of a round-house of 20 stalls, and machine and repair shop, in which are employed 50 men. The freight office and depot buildings are of wood. The freight office was opened for business during the first part of November, 1867, at which time the road was completed to this station.

No land is cultivated around Cheyenne, except a few small gardens around Crow Creek. The soil is good, and the hardiest kinds of vegetables and grains could be raised successfully with irrigation. Grazing is the main feature of the country.

The Railroad House, before which all passenger trains stop, is one of the finest on the road, and has ample accommodations for 60 guests. The dining-room, which everybody patronizes, as it is celebrated for its good fare, is tastefully ornamented with the heads and horns of the buffalo, deer, elk, antelope, mountain sheep, and other game, all preserved and looking as natural as life; here, too, is a great variety of other interesting specimens.

The other hotels are the Inter-Ocean, Delmonico, on the European plan, Dyer's, Simmon's, and Metropolitan.

EARLY TIMES—On the fourth day of July, 1867, there was *one house* in Cheyenne —no more. The first Mayor of Cheyenne was H. M. Hook, an old pioneer, elected August 10, 1867, who was afterwards drowned in Green River, while prospecting for new silver mines.

In the spring of 1869, there were 6,000 inhabitants in the place and about the vicinity; but as the road extended westward, the floating, tide-serving portion followed the road, leaving the more permanent settlers, who have put up substantial buildings of brick and stone, which mark a thriving and steadily growing city.

Cheyenne, at one time, had her share of the "roughs" and gambling hells, dance-houses, and wild orgies; murders by night and day were rather the rule instead of the exception. This lasted until the business men and quiet citizens, tired of such doings, and suddenly an impromptu vigilance committee appeared on the scene, and several of the most desperate characters were found swinging from the end of a rope, from some convenient elevation. Others, taking the hint, which indicated they would take a rope unless they mended their ways, quietly left the city. At present Cheyenne is orderly and well-governed.

In the fall of 1869, Cheyenne suffered severely by a large conflagration, which destroyed a considerable portion of the business part of the town, involving a loss of The inhabitants, w rebuilt, in many durable material tl

GOVERNMENT

FORT D. A. Russ tablished July 31, and intended to acc panies. It is three on Crow Creek, w of the enclosure. L longitude 10 deg 4 side-track with the at Cheyenne. The ment—12 store-hou the fort and the tow Several million stores are gathered forts to the northw The reservation or ated was declared 28th, 1869, and con

FORT LARAMIE lished August 12tl Sanderson, Mount once a trading po Fur Company, was ernment, through I

DOWN THE WEBER RIVER, NEA

pany's agent, for the site of a military post. It was at one time the winter quarters of many trappers and hunters. It is also noted as being the place where several treaties have been made between the savages and whites—many of the former living around the fort, fed by Government, and stealing its stock in return. The reservation, declared by the President on the 28th of June, 1869, consists of 54 square miles. It is situated 89 miles from Cheyenne—the nearest railroad station—on the left bank of the Laramie, about two miles from its junction with the North Platte, and on the Overland road to Oregon and California. Latitude 42 deg. 12 min. 38 sec.; longitude 104 deg. 31 min. 26 sec.

FORT FETTERMAN—This post was named in honor of Brevet Lieutenant. Col. Wm J Fetterman, Captain 18th Infantry, killed at the Fort Phil. Kearny massacre, December 21st, 1866, established July 19th, 1864, by four companies of the Fourth Infantry, under command of Brevet Colonel William McE. Dey, Major Fourth Infantry It is situated at the mouth of La Poele Creek, on the south side of the North Platte River, 135 miles from Cheyenne, 90 miles south of Fort Reno, and 70 miles northwesterly from Fort Laramie; latitude 42 deg. 49 min. 08 sec., longitude 105 deg. 27 min. 03 sec. The reservation of sixty square miles was declared June 28th, 1869. Cheyenne is the nearest railroad station. The regular conveyance from Cheyenne to the Fort is by Government mail ambulance and Black Hills stages.

FORT CASPER—was situated on the North Platte River, at what was known as "Old Platte Bridge," on the Overland road to California and Oregon, 55 miles north of Fort Fetterman; was built during the late war; re-built by the 18th Infantry in 1866, and abandoned in 1867. Its garrison, munitions of war, etc., were transferred to Fort Fetterman. The bridge across the Platte at this place cost $65,000—a wooden structure, which was destroyed by the Indians shortly after the abandonment of the post.

FORT RENO—was established during the war by General E. P. Connor, for the protection of the Powder River country It was situated on the Powder River, 225 miles from Cheyenne, 90 miles from Fort Fetterman, and 65 miles from Fort Phil Kearny. It was re-built in 1866 by the 18th Infantry, and abandoned in July, 1868.

FORT PHIL. KEARNY—was established July, 1866, by four companies of the 18th Infantry, under command of Colonel H. B. Carrington, 18th Infantry. This post was situated 290 miles north of Cheyenne, in the very heart of the hunting grounds of the northern Indians, and hence the trouble the troops had with the Indians in establishing it. Near this post is where the great massacre took place in 1866. It was abandoned in July, 1868.

FORT C. F SMITH—was established in 1866, by Brevet Lieutenant-Colonel N C Kinney, Captain 18th Infantry, and two companies of that regiment. It was at the foot of the Big Horn Mountain, on the Big Horn River, 90 miles from Fort Phil. Kearny, and 380 from Cheyenne. It was abandoned in July. 1868

Here the thoughtful will note, that the Government established four forts in this northern Powder River country, for the protection of the white man as against the Indian To the occupancy of the country the Indians protested, and the Government acceded, and made a treaty yielding up possession of the whole country north of the North Platte River—the Black Hills *included*—and abandoned the posts and the country to the Indians. When gold was discovered in this—*acknowledged*—Indian country, and the white man commenced to invade it—in search of gold—the Government attempted to prevent their trespassing, and to keep faith with the Indians and Gen. Sheridan issued his orders against this invasion, and sent soldiers to arrest all parties in the "Hills," and prevent others from going to them. Finally, the Government "winked" at emigration which it could not, or would not prevent. What see we now? The white man has *taken* the Indian's country, that our Government has acknowledged belonged to the latter, has *driven* the Indians out, beggars as they are, with only the bread that the Government chooses to toss to them. We are no " Indian lover " but, if the Government had a right to build these posts, they should never have abandoned them; having abandoned them, and treated with the Indian, as an equal, where is our boasted "civilization," when, though the lands do contain gold, we *take* them without a "thank you," as the elephant would crush a toad, Does *might* make *right?*

Plains teamsters call a meal a "grub-pile"

Yes! it is the finest depot building in the western country! And do you know that loss than a quarter of a century ago the beautiful site it now occupies was covered with a thicket and immense cottonwood and sycamore trees? In fact, it was at that time the Indians' home. Ho is gone—his "pony lodge" is now the palace car. In place of the battle-ax and scalping knife of this and the adjacent country twenty-five years ago, is now the plow, cultivator and reaping machine.

Yes! and the blood-stained soil of even a later date, now is occupied by a class of the most law-abiding and prosperous people. Where once lurked the cunning *red* savage and *white* ruffianism was rampant,

UNION PACIFIC RAILWAY DEPOT, KANSAS CITY, MO.

now are *schools*, and the twin children of ignorance and scoundrelism have "moved on." God bless the common schools! they are the germs of true civilization.

In this grand depot connections are made with the passenger trains on all the roads entering Kansas City.

The principal railroad connections at Kansas City are—The Union Pacific (Kansas Pacific Division); Chicago, Rock Island & Pacific; Chicago, Burlington & Quincy; Missouri Pacific; Hannibal & St. Joseph; Kansas City, St. Joseph & Council Bluffs; Wabash, St. Louis & Pacific; Chicago & Alton; Kansas City, Fort Scott & Gulf, and Atchison, Topeka & Santa Fe.

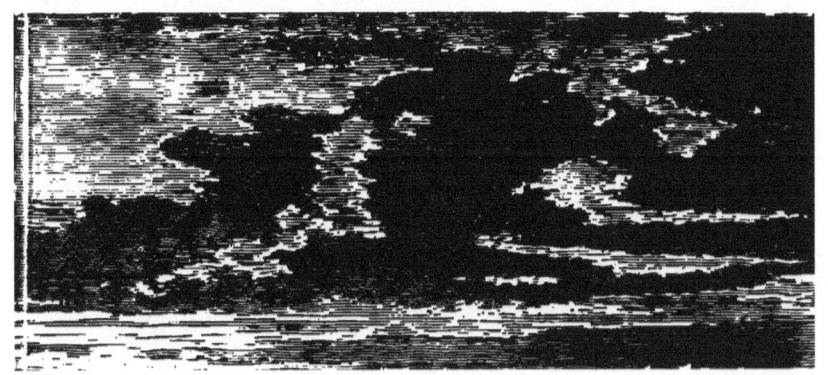

UNION PACIFIC RAILWAY.
KANSAS PACIFIC DIVISION.
D. E. CORNELL,..........................GENERAL AGENT, KANSAS CITY.

Passengers at Kansas City for the "Overland Route," *via* Colorado, Utah and Nevada, will step into the Palace Cars and superb coaches of the Kansas Pacific Division of the U. P. Ry.; pass through Denver and connect at Cheyenne, Wyo., with the "Overland" train from Omaha. See "Time Table."

To write the history and record the prosperity of the Kansas Pacific and the country tributary, in the brief space allotted for that purpose in the "Overland," it will be impossible to more than register a "telegram" of the most important matters, those of the greatest interest to the tourist or emigrant.

The Kansas Pacific Railway Company, formerly the "Leavenworth, Pawnee and Western," was incorporated by Act of Congress July 1, 1862, to construct a railroad and telegraph line from the Missouri River, at the mouth of the Kansas, to connect with the Pacific Railroad of Missouri, to the 100th meridian of longitude, upon the same terms and conditions as provided for the construction of the Pacific Railroad across the continent, and to meet and connect at the meridian above named.

The route proposed was from the mouth of the Kansas River to the junction of the Republican Fork, at Fort Riley; thence up the Republican, over the "divide" and Platte River and connect with the Union Pacific near Kearny Station.

Work commenced on the "K. P." at Wyandotte, Kansas, September 1, 1863.

By Act of Congress of July 2, 1864, the company acquired additional rights; and again, by amendment of the original Act, approved July 3, 1866, was authorized to change their route and build westward—on the 39th parallel—from Fort Riley up the Smoky Hill River to Denver, in Colorado; thence to a junction with the Union Pacific at or near Cheyenne, Wyoming.

The road was completed to Denver in 1870, and in 1872, by the purchase of a controlling interest in the Denver Pacific, reached Cheyenne; and again, by a sale in 1879, passed to the management of Union Pacific parties, where it still remains. Distance from Kansas City to Denver, 639 miles; from Denver to Cheyenne, 106 miles. The branch lines, six in number, make a mileage, respectively: 32 miles, 46 miles, 57 miles, 70 miles, 21 miles and 23 miles; total, branches, 249 miles; total, whole line, 994 miles.

Kansas City is the eastern terminus of the main line of the Kansas Pacific Railway. Prior to 1839 the place was known as "Westport Landing," but in that year was changed to Kansas City, with a population—mostly traders, hunters and trappers—of 300. *Now* it claims 61,000. The city is built on a high bluff on the south bank of the great bend of the Missouri River, just below the mouth of the "Kaw"—or Kansas River. Its central location has, from the first, enabled it to control a large trade with the country to the west and southward, which, since the advent of railroads, has grown to an enormous business. The Missouri Pacific was the *first* railroad completed to Kansas City from the eastward, where it arrived October 1, 1865, since which time *nine* have arrived to bid for and share the business which often taxes their entire combined capacity. As a live stock center—cattle, sheep and hogs—and for slaughtering, packing or shipping, Kansas City has no equal in the western country. The stock yards, beef and pork packing establishments are *immense*—are situated on the bottom lands in the western part of the city, south of the Union Depot, and are well worthy a visit by the traveller.

The *first* bridge over the Missouri was commenced at Kansas City, and its completion celebrated July 4, 1869.

Kansas City possesses all the modern improvements—horse railroads, gas, water works, etc.; churches and schools in great numbers, opera house, theatre, daily papers, and of hotels, a few dozen, chief of which are the Coates, St. James and Pacific.

WYANDOTTE, Kansas, is about two miles west, across the Kansas River, and might well be called a suburb of Kansas City—it is connected by horse cars—has a population of about 6,000, many of whom do business in Kansas City and reside in Wyandotte. The town is on a portion of the lands once owned by the Delaware Indians, who sold them in 1842 to the Wyandotte Indians, the remnants of a tribe from the State of Ohio. The lands are in a high state of cultivation, and large orchards of fruit are numerous.

The Kansas State Institution for the Blind is located at Wyandotte.

Leaving the Union Depot—which is used by all the railroads in common that enter Kansas City—we soon cross the Kansas River and the STATE LINE, pass ARMSTRONG at the end of one mile—where are located the machine shops of the Kansas Pacific—and following along on the west bank of the river one-half mile further to MUNCY SIDING. The river in places is close on the left; the bottoms are wide and covered with trees, with here and there a clearing. On the right the view is obstructed by high bluffs covered with brush or small trees. Continuing on 4.4 miles we come to EDWARDSVILLE; 3.6 miles more to TIBLOW, and 3.2 miles to LORING, from which it is three miles to LENAPE, and 4.4 miles more to LINWOOD, where Stranger Creek is crossed.

The timber on the Kansas River bottoms consists of red and burr oak, hackberry, ash, hickory, cottonwood and sycamore. The acreage under cultivation is increasing, and after a run of five miles from Linwood we pass FALL LEAF, an unimportant station, beyond which the country fairly "spreads out," and we get a *first view* of the great rolling prairies of Kansas.

LEAVENWORTH JUNCTION is the next station; distance 4.5 miles from Fall Leaf, 86 miles from Kansas City and 32 miles from Leavenworth. Let us take a run over the

Leavenworth Branch.—This road runs through a section of country the greater portion of which is under cultivation. The stations are RENO, five miles; TONGANOXIE, two miles; MOORE'S SUMMIT, two miles; BIG STRANGER, three miles; HOGE, four miles, and five more to FAIRMONT, the most important station on the line. It is situated on a portion of the Delaware Indian Reservation, first purchased from the Indians by the Kansas Pacific Railway Company, and by them re-sold to a class of farmers and stock raisers who have become prosperous.

After leaving Fairmont several small stations are passed—the first, PENITENTIARY, where the State institution of that name is located—and a run of ten miles brings us to

Leavenworth, situated on the west bank of the Missouri River, and contains a population of 18,000. It was settled in 1854, and is surrounded by a section of country of unsurpassing fertility. Leavenworth has all the metropolitan features of a big city—horse railroads, gas, water works, a big railroad bridge over the Missouri, twenty-six churches, exclusive of a Catholic cathedral that cost $130,000, nine banks, six daily papers and a score or more of hotels, besides quite a number of manufactories.

Fort Leavenworth is two miles north of the city—established in 1827—and is now the headquarters' Department of the Missouri.

Returning to the Junction, about one mile and we are at

BISMARK GROVE—On the right of the road. It contains about 40 acres heavily timbered with oaks and elms, in the center of which is a beautiful lake. This grove has become widely known of late as the place where the first National Temperance Camp Meeting was held.

From the Grove it is one mile to

Lawrence.—So named for the millionaire Lawrences, of Boston, Mass. The city proper is opposite the depot, on the south side of the Kansas River, about one mile distant, reached by several fine bridges. It is situated in the midst of the richest and most fertile section of Kansas, as well as being the most beautiful city in the State. Here, too, are street railroads, gas, water works, and, in fact, all the improvements and conveniences found in the large cities east. Population, about 15,000. Settled in 1854. Raided by Quantrell's band, August 21, 1863, who burned the town and murdered upwards of 100 unarmed citizens.

In the southern portion of the city is located the State University, on the summit of Mount Oread; from which point you look u,.on a very beautiful landscape, dotted, in all directions, with hundreds of farm houses.

The CARBONDALE BRANCH of the "K. P." leads off from this place—32 miles to Carbondale. The stations and distances between, are: Siegel, 7.5 miles; Belvoir, 5 miles; Richland, 6.3 miles; Kinney's 8.2 miles; Summit, 4 miles; Carbondale, 1 mile.

Th Leavenworth, Lawrence and Galveston Railroad, coming in from the South, is another element of prosperity for the city.

From Lawrence it is 6.4 miles to BUCK CREEK, a small station, then three more to WILLIAMSTON, and 3.2 miles to PERRYVILLE. These are all small stations, surrounded by a thrifty farming community, and are growing in importance. About one mile beyond Perryville, we cross Grasshopper River, upon which are located several flouring mills, and small manufactories. The river is well timbered,—oak, hickory, elm, ash, cottonwood and soft maple, principally.

The Grasshopper unites with the Kansas River, opposite the old town of Lecompton, of "Lecompton constitution" notoriety. The soil is a black loam, and very productive. The lands were once a portion of the Delaware Indian Reservation. From Perryville it is 3.2 miles to

M dina.—The town was laid out in 1860, and with the near surroundings, has a population of about 1,500, mostly engaged in agricultural pursuits.

Two miles north of the station is located the old "Indian Mill Farm," which has been under cultivation for over 35 years. From Medina it is 2.5 miles to NEWMAN from which it is 5.2 miles to GRANTSVILLE, a small station of the west bank of Muddy Creek. This section is noted, if at all, for its "Osage Orange" hedges, some of which are very fine. Six miles further, and our road crosses the track of the Atchison, Topeka and Santa Fe Railroad, which is completed from Atchison and Kansas City to Deming, in New Mexico, with the Pacific Coast for an objection point. The crossing is only a few hundred yards from the depot at

Topeka—The capital of the State. Population, 15,433. Here passenger trains stop 20 minutes for meals.

Topeka is in Shawnee County, situated on the north bank of the Kansas River, and surrounded by a very rich and fertile country; was located in 1854. The river is crossed at Topeka on one of the "King Iron Tubular Bridges," a solid structure 900 feet in length, composed of six spans, resting on stone piers, built from the "bed rock" in the river.

The Capitol is a fine building, built of what is called in this country "Junction City Marble," a white magnesian limestone, found in many places in the State. It can be quarried in blocks from - to 10 tons in weight, and when fresh from the quarry is very easy to saw. The Government buildings at Fort Riley are built of this kind of stone, and has proved satisfactory. The Capitol cost $400,000.

While we are here at the Seat of Government, we will note a few items in regard to the State. Kansas has an area of 52,053,520 acres, of which 40,000,000 is unimproved, awaiting the reader. Price from $1.25 to $20 per acre. Present population of the State 995,335. It has a school fund of $1,555,360, which is augmenting yearly. There are 4,520 school houses, 6,359 teachers, and 263,576 scholars. Then there are three Normal Schools, for educating teachers; the University at Lawrence, and the Agricultural College at Manhattan. On the line of the Kansas Pacific, there are 76 grain elevators, with storage capacity of 2,515,100 bushels; and 52 flouring mills, with 169 run of stone; capacity, 4,310 barrels per day.

Leaving Topeka, ME-NO-KEN, a small station, is reached in 4.7 miles; SILVER LAKE in 5.9; KINGSVILLE in 2.7; ROSSVILLE in 2.8, and 7.6 miles more to

St. Mary's—an important station, in some respects. The country for the last 25 miles, and surrounding the town, is thickly settled, and the greater portion cultivated. Corn is the principal crop, though much wheat and vegetables are raised.

The Jesuit Fathers visited this country nearly 40 years ago, and established Mission Schools among the Indians. More recently they have erected here large educational institutions; one for ladies, is known as "The Seminary of the Sacred Heart." The building to the north of the railroad, is of brick, with stone trimmings, 100 feet front and four stories in height, completed in 1871. The College for males is adjoining, and can accommodate 1,300 students.

From St. Mary's it is 6.2 miles to BELVUE, a small station 6.9 miles from the end of the Kaw Division, First District, which is at

Wamego.—This is a large and thriving town situated in the midst of a country well watered by numerous small creeks, very fertile and thickly settled. The next station, 6.6 miles, is ST. GEORGE, another growing town of about 700 population, from which it is 7.8 miles to

Manhattan.—Population, about 2,000; County Seat of Riley County, 117 miles west of Kansas City. The town is situated near the junction of the Kansas and Blue rivers, was settled in 1854 by a colony of Ohio "Pilgrims," who purchased a small steamboat at Cincinnati, steamed down the Ohio river, and thence up the Mississippi, Missouri, and Kansas rivers to this place, where they settled, in what was then a wild Indian country, living on their boat until buildings could be erected,

The Kansas State Agricultural College—an experimental farm—is located at Manhattan. Congress, in its benevolent wisdom, endowed this College with a land grant of 81,000 acres, 50,000 of which has been sold, realizing the snug sum of $238,000. The institution has 400 acres fenced and cultivate the greater portion with vineyards and orchards of fruit of every variety. Leaving Manhattan a few miles, the bluffs come close on the right, in places 500 feet in height, covered with trees, rocks and grass alternating, while the river comes in close to the road, on the left, and again receding for miles, along the banks of which ash, oak, hickory, cottonwood and elm trees grow in profusion. Here, too, can be seen some fine farms, surrounded by beautiful osage orange hedge From Manhattan it is 11.1 miles to

Ogden—A town of some historic interest in the annals of the State, as being the place where the first Territorial Legislature, convened by Gov. Reeder, met to "Save the Country" The place was first settled in 1856. Six miles further is the station of

Fort Riley—So called for the Fort of that name, situated upon the high plateau to the right; established in 1852, is in latitude 39° north, 96°30' west. The post was first known as "Camp Center," being situated in the geographical center of the United States.

Junction City.—County Seat of Davis County, is 2.7 miles west of Fort Riley, and is destined to be a place of much importance. It was located in 1859, has grown rapidly and now contains 5,000 population. Here is located the marble quarries before alluded to; here, too, is the northern terminus of the Missouri, Kansas & Texas Railroad, and the Junction City & Fort Kearny Railway, The Republican River unites with the Kansas River at this point, up which is completed the J., C. & F. K. Railway, a branch of the Kansas Pacific, to Concordia, 70 miles northwest. The stations and distances between are: Alder, 7.7 miles; Milford, 5.4 miles; Wakefield, 6 miles; Clay Center, 14 miles; Morganville, 7.6 miles; Clifton, 8.7 miles; C., B., U. P. Crossing, 4.9 miles; Clyde, 0.8 miles; Lawrenceburg, 7.6 miles; Concordia, 7.4 miles.

The valley of the Republican is one of the richest and most productive in the State. It was the Indian's *home*, to retain which he fought the white man long and bitterly, and with the usual result, the Indian *had to* ao! He went! Where once roamed his "pony herd" in thousands, *now* can be counted the dwellings of his successors in equal numbers; where once the Indian's beef (buffalo) ranged in untold millions now range the white man's beef. The buffalo has gone—went with the Indians. Will the time ever come when the "successors" will be succeeded by a *stronger* and more enlightened race? Will they in turn ever be driven out and exterminated?—*Quien sabe*!!

At Junction City the Smoky Hill river comes in from the southward, which, with the Republican, forms the Kansas river. The Smoky will be on our left for the next 47 miles, to Salina.

After leaving Junction City, a peculiar rock formation is noticeable on the right along the summit of the bluffs resembling a long line of fortifications.

Another item, we record for the benefit of the sportsman; feathered game in great abundance are found on the prairies, and along the rivers and small streams in Kansas, such as prairie chicken, quail, ducks, geese, snipe, plover, swans, cranes, pelican, an many other varieties.

Then a run of 5.8 miles to CHAPMAN, 6.2 miles to DETROIT, and 5.2 miles more and we reach

Abilene—county seat of Dickenson county. Population about 2,000. Passenger trains stop 30 minutes—opposite the Henry House—for meals, which are the *best* on the road.

This station was the first great cattle shipping point on the Kansas Pacific Railway. From 1867 to 1870, the number loaded on the cars and sent east, were from 75,000 to 150,000 a year, but as the agriculturalist crowded *in*, the cattlemen were crowded *out*, and we will find them *now*—far to the westward.

We are now in what is called the "Golden Belt"—so named for the wonderful adaptability of the country for raising wheat and other small grains, These "belt" lands, it is claimed, commence near Junction City, and extends beyond Ellis—about 200 miles in length. Wheat is the principal crop, and comprises one half of all the productions. There are several fields of wheat, near Abilene, of 1,000 acres each, one of 3,000, and one of 3,500. Of late years, tree-planting has been quite an industry. Orchards of fruit are numerous, and successfully raised.

From Abilene it is 4.4 miles to SAND SPRINGS, a Signal Station, thence 4.6 miles to

Solomon—situated near the junction of the Smoky Hill and Solomon Rivers, in the midst of a thrifty agricultural section. Population about 500.

The SOLOMON RAILROAD, another branch of the "K. P.," is built up the valley of the Solomon to Beloit, 58 miles northwest from this station. Several Salt Springs are near the town, and the buildings erected for the purpose of manufacturing the salt are quite extensive, and can be seen from the cars after leaving the station.

Leaving Solomon, we cross the river of that name, and 7.8 miles arrive at NEW CAMBRIA, a small station situated on a broad plain, dotted, in all directions with the neat little cottages of the settlers, who are principally engaged raising wheat and corn. Six miles further we reach

Salina—the County Seat of Salina County, settled in 1858. Just before reaching the station we cross the Salina River, which comes down from the north-west.

Salina has a population of about 3,000, some large grain elevators, several good hotels, papers and another railroad, the SALINA & SOUTHWESTERN. This branch comes to McPherson, distant 36 miles to the southwest. Situated on the Smoky Hill River, near the Swedish colony who settled here in 1870.

The principal occupation of the people is agriculture, although there are many herds of cattle and sheep in the county, and some extensive quarries of Gypsum, or Plaster of Paris, also several Salt Springs that are being utelized for the production of salt.

Along all the rivers and streams about this section of country are belts of timber, consisting of cottonwood, oak, mulberry, elm and hackberry.

Bavaria—is the next station 8.4 miles from Salina, where is located a colony from the Western Reserve of Ohio, who settled here in 1869. This colony has been very successful, wheat and corn crops being their reliance. A run of 6.6 miles brings us to the end of the second district of the Kaw Valley Division of the road, at

Brookville.—Here the Railroad Company have the usual division repair shops, good depot buildings, and extensive cattle pens. Population, about 500. The country surrounding the station is a rolling prairie, on which can be seen, besides the usual wheat and corn fields, an occasional herd of cattle and sheep. Leaving Brookville, we pass several small stations in the order, and distances between as follows: 4.2 miles to ROCK SPRINGS; 1.8 miles to TERRA COTTA; 4.4 miles to ELM CREEK; 5.1 miles to SUMMIT SIDING; 2.5 miles to FORT HARKER, and old Government post, on the left, built in 1867-8, abandoned; and 4.7 miles to

Ellsworth—County Seat of Ellsworth County, situated on the north bend of the Smoky Hill River. Settled in 1867; present population 1,100. The town has some good stone buildings, a large grain elevator, several hotels, cattle pens and shutes—the latter not of much use of late, as the farmers are crowding the cattle-men *a little further west.*

The next station is BLACK WOLF, 7.2 miles; then COW CREEK, 2.3 miles.

Wilson's—is 6.5 miles from Cow Creek. This is a thrifty town of 400 population, situated in a rolling prairie country, fast filling up with settlers. From Wilson's it is 6.4 miles to DARRANCE, and 7.3 miles to BUNKER HILL, the County Seat of Russell County, population, 400; first settled in 1871, by a colony from Ohio. Near the station Salt Springs abound, lime stone is plentiful, some coal, and abundance of mineral paint, and pottery clay. Passing on we pass through HOMER in 5.3 miles, and 4.6 miles more to

Russell — population about 800; settled in 1861, by a colony from Ripon, Wisconsin, and is situated about four miles south of the south bend of the Saline River, and surrounded by rich agricultural lands, well cultivated. Leaving Russell it is 9.1 miles to GORHAM; three more to WALKER; and 3.9 miles further to

Victoria—Six miles south of the station is located the "Victoria Colony," established by the late Mr. George Grant, a wealthy scotchman, who bought 30,000 acres of land here, sold a portion to settlers and retained a large estate for himself. The lands have a rich soil, are well cultivated, and dotted in all directions with the homes of the settlers, and their herd of horses, cattle, and sheep. From Victoria it is 10.5 miles to

Hays—County Seat of Ellis County —named for the post established in 1867, about half a mile south of the station on a high plateau. Hays has a population exclusively of soldiers, of about 700, many of whom are engaged in stock-raising—as be it known we have reached the western limits of the agriculturalists, and soon will bid them *good bye*, and grip the hand of the herdsman.

The next station is 13.2 miles distant, and is the end of the third district of the Smoky Hill Division.

Ellis.—We are now on the "Cattle Trail." At this station are immense yards and shutes, for the accommodation of stockmen, many of whom drive up great droves of cattle from Texas, and the country to the southwest, as well as graze them in the surrounding country. In fact, this is the greatest cattle-shipping point on the road. The grasses are mostly "bunch grass" and "buffalo," or "grandma grass," the richest and most nutritious grown. The article on "Western Stock Raising," —in Annex. No. 29—will apply equally as well in this section as the one for which it was written.

Ellis has about 500 enterprising, law-abiding citizens, most of them are engaged in stock-raising, yet, of late, the agriculturalists are crowding in, buying up the lands, and it will not be long before the cattle-men—who do not buy land—will have to *go west.*

From Ellis it is 10.3 miles to OGALLAH, an unimportant station, from which it is 9 miles to

Wa-Keeney,—one of the most enterprising towns on the road. It contains about 500 citizens, many of them are engaged in agricultural pursuits, but the greater portion in the cattle business. Here we find one of the finest depot buildings on the road; it is

SUMMIT OF THE MOUNTAINS, 10,000 FEET HIGH.

100 feet by 30 feet, and 32 feet high, surmounted with a tower 50 feet high, and a platform 27 feet wide. There are many fine stone buildings, principal of which is the Oaks House. Leaving Wa-Keeney, we pass on rapidly through a section of country almost wholly occupied by cattle men, passing the stations, with the distances between as follows: COLYER, 14.1 miles; BUFFALO, 14.3 miles; GRAINFIELD, 5.5 miles; GRINNELL, 9 miles; CARLYLE, 12.1 miles; MON'T SIDING, 9.4 miles; MONUMENT, 2 miles; GOPHER, 9.7 miles; SHERIDAN, 7.6 miles; and 15.1 miles more to

Wallace—A regular eating station, where good meals are served for 75 cts. Population about 250. It is the end of the third district of the Smoky Hill Division, and the commencement of the Denver Division. The station is in the midst of a rolling prairie, two miles north-west of FORT WALLACE, established in 1866. It is situated on the fork of the Smoky Hill River, in latitude 38 deg., 55 min., and longitude 100 deg., 50 min. from Greenwich.

For the last hundred miles the country is almost wholly occupied by the cattle-men, and will continue to be for the next 150 miles, so we shall pass most of the stations, by simply naming them and the distance between: From Wallace it is 8.5 miles to EAGLE TAIL; 8.3 miles to MONNOTONY SIDING; 3.2 miles to MONNOTONY—we are nearing Monnotony on all sides now—12 miles to ARAPAHO, where the State line is crossed, and we enter Colorado; 9.5 miles to CHEYENNE WELLS; 10.5 miles to FIRST VIEW—where, if the day be clear, the first view of Pike's Peak and the Rocky Mountains are to be had—and 14.7 miles more to

Kit Carson—named for the old hunter, trapper and guide of that name—and somewhat famous as being the place where the Grand Duke Alexis of Russia tarried to hunt buffalo, in January, 1872. It was a *big hunt*, and it is said that the Duke killed 40 of the noble animals, and, by the way, we have been in the old buffalo range for the last 250 miles, but, of late years, few, if any, have been seen—went with the Indians. From Kit Carson it is 11.9 miles to WILD HORSE, named for a band of wild horses that once roamed over this country; 10.9 miles to AROYO; 13.4 to MIRAGE, and 11 more to

Hugo—an eating station, from which it is 12.5 to LAKE, where are a few pools of water. 8.8 miles to RIVER BEND, situated on the big bend of the Big Sandy Creek; 6.3 miles to CEDAR POINT; 4.2 miles to GODFRY'S, where there are some coal mines of fair quality; 4.9 miles to AGATE, noted for the moss agates found near the station; 12.2 miles to DEER TAIL, situated on East Bijou Creek; 12 miles to BYERS. From BYERS it is 12.4 miles to BENNET; 9.4 miles to BOX ELDER, situated on a creek of that name; from which it is 12.4 miles to SCHUYLER, and 9.3 miles more to DENVER.

REMEMBER! For full and complete information in regard to Colorado, its wonderful mines of gold, silver, copper and other precious metals, its agricultural, stock-raising and varied resources; its pleasure resorts, lakes, rivers, mountains, parks, sulphur, soda, hot and medicinal springs; its magnificent scenery, railways, etc., buy "*Crofutt's Grip-Sack Guide*" *of Colorado*, a complete encyclopedia of the State, profusely illustrated.

"Tour" No. one gives a complete description of the route and country from Denver to Cheyenne, where connections are made with the Overland trains from Omaha and San Francisco. Sold on all trains.

Leaving Cheyenne, just in the border of the city we cross Crow Creek, and about two miles from the city—by looking to the right, northward—a fine view can be had of Fort Davy Russell, previously described. We are now ascending the eastern slope of the southern range of the Black Hills of Wyoming, which are stretching away in a long rugged line before us.

Colorado Junction—six miles west of Cheyenne, is the first station we reach, and the junction of the Colorado Central Branch. The track turns off at the left of the station and crosses the prairie and hills to the southward. Four miles from the Junction, BORIE, a small side-track, is passed, from which it is 4.2 miles to

Otto—Passenger trains usually meet here, stop a few moments, exchange letters and papers, then pass on

FIRST STEAM RAILROAD TRAIN IN AMERICA.

The above illustration was drawn and engraved from the original painting in the possession of the Connecticut Historical Society, and represents an Excursion Train on the Mohawk and Hudson R. R. from Albany to Schenectady, N. Y., in 1831, the FIRST steam train in America. The engine was the "John Bull," imported from England, as well as the engineer, John Hampton, "expressly for this road, at large expense." Her cylinder was 5½ inches, 16 inch stroke, wheels 4½ feet. The boilers had thirty copper tubes, five feet long, four inches in diameter. Connecting rods are worked on double cranks on front axle. Weight of engine, complete, 4 tons. The tender represents the method of carrying the fuel—wood—in barrels, with a few sticks handy for immediate use. The cars were regular stage bodies set on car wheels. On this grand excursion trial trip were sixteen persons, who were then thought venturesome, many of whom have since filled important positions in the councils of the country. Mr. Sidney Dillon, President of the Union Pacific R. R., it seems, was one of the adventurous few. Here is food for thought and comparison with the improvements of the present day.

—one going East for *light*, the other West for *knowledge*.

We are now 6,724 feet above the sea, and the traveler should note the rapid rise made from this point, in surmounting the Black Hills. Here the heavy grading commences, and snow fences will be numerous till we get over the "Hills." To the north of this place, at the base of the Hills, is a fine valley, where Crow Creek finds its source in many fine springs. The valley contains very superior grazing land, and in conjunction with the adjacent hills, affords ample game for the hunter.

Fifteen miles from this station, to the north, at the eastern entrance of Cheyenne Pass, is the site of old Fort Walbach, now deserted. Near this fort are the head waters of Lodge Pole Creek.

Granite Canyon—is five miles west of Otto, and 574 feet higher. At this point are extensive stone quarries, whence was taken the rock for the company's buildings in Cheyenne, also for the stone warehouses. Limestone abounds in this vicinity, and many kilns have been erected. To the left of the road, and down the canyon a few hundred yards, is a fine spring, from whence the water is elevated to the tank by the roadside. Half a mile to the south are a number of fine springs, which—with others to the westward—are the head-waters of Lone Tree Creek, a tributary of the South Platte River. Along the road now is heavy rock-work, and on the exposed portions of the road may be seen the snow-sheds and snow-fences, built of plank or stone.

Buford—is a small side-track, 6 9-10 miles further. Heavy rock-work, and snow-sheds and fences mark the road. Water for the station is elevated from springs down the ravine, to the southward.

The country here presents a wild, rugged and grand appearance. The level ground or little valleys are covered with a fine coat of buffalo grass, and now and then clumps of stunted pine appear by the roadside. On either hand, near by, high, bold masses of granite rear their gray sides, piled one on the other, in wild confusion. Up, up, still higher, in the background are the rocky, pine-clad peaks of the Black Hills. The scene is peculiarly impressive as we near Sherman, especially if it chances to be one of those days when the clouds float low down the horizon; then the traveler looks over the intervening space between him and the mountain range beyond, and sees naught but floating masses of vapor; no mountains, no valley, no forest, only these fleecy shapes, and a long, dark line rising above them, o'ertopped by the glistening sides of Long's Peak. The altitude gained, we see on the north side of the road, a signboard—"Summit of the Mountains;" and soon after reach

Sherman—*eight thousand two hundred and forty-two feet above level of the sea.* It is named in honor of Gen. Sherman. On a high point just south of the station, a monument is being erected to the memory of Oakes and Oliver Ames. Sherman is 549 miles from Omaha and 1,365 from San Francisco, and is not noted for its size. The trains stop here but a few minutes. The company's buildings consist of a comfortable station, a small repair shop, and a round-house of five stalls. A post-office, telegraph and express offices, one store, two hotels, two saloons, and about twenty houses of all sorts, constitute the town.

Seventy miles to the southwest is Long's Peak, and 165 miles to the south is Pike's Peak, both plainly visible. To the northwest, about 100 miles distant, is Elk Mountain, another noted land-mark. The maximum grade from Cheyenne to Sherman is 88.176 feet per mile. The freight taken on at this station for the East and West is quite extensive, consisting of sawed lumber, telegraph poles, and wood obtained in the hills and ravines but a few miles distant to the northward. On many of these hills, and in the canyons, are found a dense growth of hard spruce pine, which, as to quality and adaptability for being dressed, resembles the hemlock of the Eastern States.

The winters are not as severe at Sherman as many think, neither is the snow-fall as deep as many would suppose from seeing the great number of snow-sheds and fences; snow seldom falls more than a few inches in depth. It is not the depth of snow that causes any inconvenience to the working of the road, but it is the drifting of it into the cuts during the heavy winds. For the purpose of preventing this, the sheds, fences and walls are erected along the road, the latter a few rods away from the banks of the cuts. The fences cause an eddy or current of air, which piles the snow along in huge drifts, keeping it, in a great measure, from the track. Snow-sheds cover the deepest cuts along the road, where obstructions from the snow are most likely to occur. The cold rains and deepest snows come with an east wind; the worst storms come from the southwest.

The thermometer at Sherman ranges from 82 deg. Fahrenheit, in the summer, to 30 deg. below zero in winter. Springs of sparkling water are numerous in the surrounding country, and form many small streams which wind their way among rocks and through gorges until they are lost in the waters of other streams.

At this elevated point, the tourist, if his "wind is good," can spend a long time pleasantly in wandering amid some of the wildest, grandest scenes to be found on the continent. There are places where the rocks rise higher, where the chasms are far deeper, where the surrounding peaks may be loftier, and the torrents mightier in their power, and still they do not possess such power over the mind of man as does the wild, desolate-looking landscape around Sherman. Although the plateau is covered with grass, and occasional shrubs and stunted trees greet the eye, the surrounding bleakness and desolation render this place one of awful grandeur. The hand of Him who rules the universe is nowhere else more marked, and in no place will the tourist feel so utterly alone, so completely isolated from mankind, and left entirely with nature, as at Sherman, on the Black Hills of Wyoming.

At the first the tourist experiences much difficulty in breathing, the extreme lightness of the air trying the lungs to their utmost capacity, but becoming accustomed to the change, and inhaling long draughts of the pure mountain air, will greatly prefer it to a heavier atmosphere.

FISH AND GAME—There is no spot along the line of road which can be compared to the locality around Sherman for trout fish-

PALISADES OF THE HUMBOLDT RIVER, C. P. R.R.

ing. The tiniest rivulets swarm with them, and their speckled sides glisten in every eddy. They weigh from one-fourth to two pounds, and their flesh is as hard and white as that of the mountain trout of Vermont.

Antelope, elk, black-tailed deer, bear, sage hens and grouse abound in the hills and on the plateaus. The angler, hunter, or tourist should never pass Sherman without pausing long enough to fly a hook and try his rifle. Doubtless this point will become a favorite summer resort for travelers, when the hotel accommodations are such as to entice them to remain, as it possesses eminent attractions for hunting and fishing.

From Sherman to Rawlin's, 166 miles, the road runs between the Black Hills and the Rocky Mountain range, presenting varied and impressive scenery at various points.

Leaving Sherman, the road turns to the left, and passes through several long snow-sheds and deep rock cuts to

DALE CREEK BRIDGE—Dale Creek is a noted stream, although a small one, and should have a noted bridge—as it has. When the road was being constructed over these hills, in 1867, the railroad company built a plated wooden frame-work structure 650 feet long, from bluff to bluff and 126 feet high. The bridge stood on trestles, interlaced with each other, and securely corded together and stayed by wire cables, secured to, and sloping from, the bridge on each side to substantial anchorage, down into the valley below, presenting a light and graceful appearance when viewed from the creek below. This old bridge was replaced in 1877 by one of iron, of similar dimensions, built in the most substantial manner—see illustration, page 49.

From the bridge, the beautiful little stream looks like a silver thread below us, the sun glistening its surface with a thousand flashes of silvery light. Anon, the dark walls of the canyon shade it, as though they were envious or jealous of its beauty being rendered common property. A narrow, green valley, half a mile above the bridge, is the site of the former Dale City, where, at one time, were over 600 inhabitants. Now, a few hundred yards above the bridge, can be seen a solitary house—like a lone sentinel in front of a deserted camp. Here, too, as well as around Sherman, and all over the Black Hills, are found countless flowers of every variety and hue, over 300 varieties of which have been classified.

VIRGINIA DALE—is situated fifteen miles southwest of Sherman, in Colorado, at the head of a deep gorge, on Dale Creek, near the Cache-a-la Poudre River. On the east side of the canyon, the wall of overhanging rock rises about 600 feet high, for a mile along the stream, giving a wild and picturesque beauty, a sublimity and grandeur to the scene, rarely surpassed. This point is called the "Lover's Leap," though we never learned that any one ever leaped off; but if the leap was made, we judge that the jar on alighting in the valley, 600 feet below, must have knocked all the love, romance or sentiment out of those making it. In and around this place are numerous dells, grottoes, gorges, canyons, precipices, towering peaks and rugged recesses, enough to employ the tourist for some time in examining their beauties.

Some "yellow-covered novelist" has immortalized Virginia Dale, by calling it the "Robbers' Roost," though failing to inform us what they roosted on. But aside from this questionable honor, Virginia Dale is the most widely known and celebrated of any locality in these mountains. There are a few good buildings around the place, where excursionists, who visit to enjoy the scenery, mountain air, and rare fishing and hunting, are provided for.

See ANNEX, No 10.

We now return to the railroad, cross the bridge, and turn away to the northward, through long snow-sheds and rocky cuts, made through red sandstone, six miles to

Tie Siding—This station is important only in the fact of its being a point where great quantities of ties and wood are brought to the railroad from the hills to the northward. The view to the south is that of a very broken and rugged country. To the west, the southern end of the great Laramie Plains is spread out, almost at our feet, twenty miles in width, with the wondrous Rocky Mountains rising from its western border, range upon range, peak overlapping peak, away up, up into the regions of perpetual snow, over one hundred miles away.

Our train is descending rapidly, and more to the northward; steam is no longer required—only brakes. Onward. 4 1-10 miles, through snow-sheds and deep excavations, brings us to

Harney—an unimportant station Passing on, to the left can be seen the old

CROSSING THE RANGE ON SNOW SKATES—SEE ANNEX NO. 32.

Denver and Salt Lake stage road, the telegraph marking the line for some distance along the railroad. On the right hand, the whole valley has been fenced in for grazing purposes.

The next station is 41.2 miles, denominated

Red Buttes.—This locality derived its name from several ridges and peculiar formations of sandstone lying between the railroad and the Black Hills on the right.

Many of these sandstones rear their peaks from 500 to 1,000 feet above the plain, apparently worn and washed by the elements, into wild, fantastic shapes and grotesque figures. Rocks which, at a distance, might be taken for castles, rise side by side with the wall of an immense fort; churches rear their roofs, almost shading the lowly cottage by their side; columns, monuments and pyramids are mixed up with themselves and each other, as though some malignant power had carried off some mighty city of the olden time, and, wearying of his booty, had thrown it down upon these plains, without much regard to the order in which the buildings were placed.

Some few only of these curiosities, can be seen from the car windows, and those are not the largest. The tourist, by stopping over a day or two at Laramie, would find much of interest in this section of the country.

The Laramie River rises about 50 miles to the southwest from Red Buttes, on the eastern slope of the mountains—its source being composed of almost innumerable springs. Its general course is northeast for 200 miles, when it empties into the North Platte River at Fort Laramie.

During the building of the road, thousands of ties were floated down to Laramie, and thence hauled along the line.

The supply of timber in this region is as near inexhaustible as can well be imagined, where forests do not recover from the cutting. Saw mills will find employment for many generations ere they can lay bare these mountains.

Six and a half miles from Red Buttes is a station for the military post of FORT SANDERS, which is situated on the east side of the railroad, close to the track, and in plain view for miles. The buildings are principally of logs, several of stone and one an ordinary frame—headquarters. This post was established June 23, 1866, by two companies of the Third Battalion U. S. Infantry, under the command of Brevet Lieut.-Col. H. M. Mizner, Captain 18th Infantry. Latitude 41 deg., 13 min., 4 sec. (observation), longitude 105 deg., 40 min. (approximate).

Two and a half miles farther on we arrive at the end of the "Mountain Division" of the road.

Laramie City—the enterprising county seat of Albany county, has wonderfully improved within the last few years. It has a population of 4,000 and is regularly laid out at right angles with the road. A stream of clear, cold water, which rises in a spring a few miles to the eastward, runs through the principal streets; the buildings are small and mostly of wood, with a few substantial structures of stone. The spirit of improvement is manifested on every hand, which has recently completed numerous stores, hotels, banks, churches, schools, dwellings, and other buildings, including a court house and jail. Trees line many of the streets, and present a cheerful and home-like appearance; in fact, most of the Laramie residents are here to stay. For many years after the completion of the road this was a regular eating station for passenger trains from the East and West. The meals were served in the Thornburg House, at present a first-class hotel, in front of which the cars stop. The *Sentinel*, weekly, and *Boomerang*, daily, are published here.

ROLLING MILL—During the year 1874 the railroad company erected a rolling mill at this place, at a cost of $127,500. It is situated to the right of the road, a short distance north of the station, and is in operation night and day, rerolling old rails and other heavy work. The company's division shops are also located here. They are of stone obtained from Rock Creek, 50 miles north. The round-house contains 20 stalls. The machine shop is used for general repairing, and is first-class in outfit.

The railroad was completed to this point June 18, 1868, and for some time Laramie was known as the "end of the track," and during that time it was not only the place from which all freight and supplies for the West were hauled by wagons, but it was the center for all the roughs and abominations which followed the building of the road.

Directly to the east can be seen the Cheyenne Pass wagon road—the old emigrant route—which crosses the plain and river $\frac{1}{2}$ mile above the city, running north along the mouuntains.

Laramie was the first place in America—or in the world even—where a female jury was empaneled. Their first case was that of a Western desperado, and there was no flinching from duty.

Curious passengers will note from this city west, the railroad laborers—section hands—are CHINAMEN. They are said to be very reliable.

LARAMIE PLAINS—comprise a belt of fine grazing lands, about 25 miles wide by 60 long, and the remarks about the grazing lands elsewhere will well apply to this section. Beef can be raised and fattened on these plains as cheaply as in Texas, where, as every one knows, they raise themselves and form the larger half of the population. The peculiar features of these grasses are similar to those already described. The plains are higher, and frost appears earlier in the fall, but the summer sun cures the grass before its arrival, so that the cold weather does not injure it. We need only mention the well-known fact that thousands of buffalo once roamed these plains, furnishing the Indians unlimited beef, to convince any one that the laudations of this as a grazing country are not exaggerated or wild ideas, but substantiated facts, proved by experience.

Stock-raising is now almost the only industry noticeable, and a great many thousand head of cattle, sheep and horses can be seen in almost any direction. It is computed that there are at this time over 90,000 head of cattle, 85,000 head of sheep, and three thousand horses and mules within 40 miles of Laramie, valued at $2,250,000. In 1867, there couldn't be found in the same section 500 head of all kinds—all told. Agriculture is not profitable, yet they have demonstrated that some of the hardy vegetables can be cultivated with success on the bottom lands.

ITEMS OF INTEREST—Crystal Lake is about 40 miles to the westward of Laramie. Sheep Mountain—one of the peaks in the Rocky Mountain range—rears its head for 12,000 feet above the sea. Should the tourist desire to visit the place, he will find the road beyond the plains rough, and the ascent toilsome. Before beginning the ascent of the mountains we enter one of the grandest forests in the country. For ten miles we toil on through the forest, which is so dense that the sunlight hardly penetrates, and the silence is almost oppressive. Bears, mountain lions, and the mountain sheep range here; their haunts, until lately, never having been invaded by the pale face. Emerging from this gloom into the fair sunlight, we find ourselves on the highest point of the mountain, from which we can look over piles of fleecy clouds floating below us to other ranges far beyond. Peak on peak, ridge on ridge, they ascend, until their snow-clad heights are lost in the distance, or in the vast blue dome above.

Looking downward, we behold a vast succession of dark ridges and grey peaks through the rifts in the fog-like vapor floating above them. These dark ridges derive their sombre hue from the forests of pine, which extend for miles and miles in all directions. To the east we see a deep indentation in the mountains, which is Laramie Plains. Across this apparently narrow line, the rugged masses of the Black Hills rise in their grandeur, their black crests closing the scene.

Turn now to the immediate landscape. Here is a green, grassy lawn, dotted with tiny flowers, of varieties such as we never before beheld, or ever read of, and right before us, in the center of this lawn, lies a circular lake nearly a mile wide; its clear, soft, cold water glistening in the rays of the sun, and reflecting, as in a mirror, every object on its banks, transforming them into many fantastic shapes, as the breeze lovingly kisses the silver surface, lifting it into little ripples.

The scene is one of unsurpassed loveliness immediately around you while the view in the distance is grand, aye, sublime—beyond the power of words to depict. Whoever visits this place cannot fail of being impressed with its wonderous beauty, and his mind will take newer and clearer impressions of the power of "Him who hath created all things."

THE SNOWY RANGE—the great backbone of the continent—is covered with snow for a great part of the season; the highest peaks ever wearing their white robes, even when the passes are covered with flowers. This renders them very conspicuous and easily discerned at a great distance. Hence the term "Snowy Range."

CHARACTERISTICS OF THE COUNTRY—In general descriptions we speak of Laramie Plains as including all the country lying between the western base of the Black Hills and the eastern base of the Rocky Mountains—a grand park, similar in formation to the great parks of Colorado, though of much less altitude. These "parks" are immense bodies of table lands, enclosed by the peaks and ridges of the surrounding mountains, sheltered by them from the cold winds, watered by them from the never-failing streams which flow from gorges and canyons among these peaks, from which the snow is never absent. The average elevation of the Laramie Plains or park is about 6,500 feet, though where Laramie City stands it is more. The Black Hill ranges of the Rocky Mountains form the eastern and northern boundary of the "Plains." This range extends nearly due north to Laramie Peak, about 150 miles, thence west, terminating in the Seminole Mountains. On the south, the park or plain is bordered by the Rocky Mountains, which here reach an elevation of from 10,000 to 13,000 feet above the sea—snow-capped always. To the altitude of from 8,000 to 9,000 feet, these slopes are covered with dense pine forests.

In the mountains to the westward, in North Park, Douglass, and other creeks, rich mines of gold, silver, copper, and nearly all the known metals have been discovered, and in several cases, worked to advantage. The Keystone is reported to be a very rich mine, the owners of which are taking out the ore and piling it up, to await the arrival of a stamp mill which will be put up in a few months. Undoubtedly there are vast regions tributary to Laramie equally well-stored with mineral deposits, that have never been prospected or visited by the white men.

On the northwest from out the Elk

Mountains, juts the Rattlesnake Range, extending north to the North Platte, carrying an elevation of near y 8,000 feet.

Through the western range, the North Platte canyons, and. on the east. the Medicine Bow River cuts through the eastern range separating it from the foot-hills of the northerly range of the Black Hills Through the plains flow the Big and Little Laramie Rivers, which, as we before stated, rise in the mountains which border the western rim of the plains. These streams canyon through the Black Hills north of Laramie Peak, and enter the North Platte near Fort Laramie.

Rock Creek rises east of Medicine Bow, and after flowing north to about latitude 42 degs. flows west and empties into the Medicine Bow. This river rises in Medicine Bow Mountains, and flows north to about the same latitude as Rock Creek, thence west, and canyons through the Rattlesnake range of hills, entering the North Platte about 150 miles northwest of Laramie City, in latitude 42 deg. 3 min.

By this showing it will be observed that the immense park, or Laramie Plains, is well watered—sufficiently for grazing and irrigation. We have been more explicit, have dwelt longer on the e points than we should have done, did we not feel a desire to show to the emigrant, or to those who are seeking good locations for grazing lands, that the Laramie Plains possess these advantages in an eminent degree. We have wandered far away from the plains in our descriptions, but the grazing lands end not with the plains. The mountain sides, until the timber belt is reached, the valleys, bluffs, and foot-hills, all present the same feature in point of luxuriant crops of grass. The valleys of the streams mentioned also contain thousands of acres of meadow land, where hay can be cut in abundance, and, if the season will permit, wheat, barley and rye might be grown to advantage, the soil being a black loam, and sufficiently moist to insure good crops without irrigation.

FISH AND GAME—Trout—the finest in the world—can be found in every mountain stream, while every variety of game ranges over the mountains, hills valleys and plains in countless numbers.

With these general remarks, we will return to Laramie, and proceed on our journey. Soon after leaving the city, we cross the Laramie River, and eight miles brings us to

Howell's—an unimportant station, where passenger trains seldom stop. It is then 7 6-10 miles to

Wyoming—on the Little Laramie River. During the building of the road large quantities of ties were received at this point, which were cut at the head of the river and floated down the stream in high water. The country is a broad prairie. At the station we crossed Little Laramie, a small stream which rises in the mountains to the westward and empties into Laramie River. To the next station—

Hutton's—it is 6.9 miles, and 6.6 more to

Cooper Lake—Near the station, to the westward, lies a beautiful sheet of water, about two miles long by half-a-mile wide, for which the station is named.

Lookout—a station with an altitude of 7,169 feet—is 5 4-10 miles from Cooper Lake. We are now entering the rolling prairie country, where, for 25 miles either way along the road, vast herds of elk, deer and antelope are found at different seasons of the year—the elk being mostly found in the winter, when the snow drives them from the mountains. We also begin to find occasional bunches of sage-brush, which tell us that we have entered the country where this more useful than ornamental shrub abounds. Occasionally we pass through cuts and over low fills, by snow-fences, and through snow-sheds, the country growing rougher as we pass along 5.7 miles to

Harper's—from which it is 6.3 miles to

Miser Station—Sage-brush is the rule. Just before reaching the station, we pass through a very deep cut—one of the deepest on the road—where a little spur of the bluffs rises abruptly from the plains, right in the way of the road. Just before reaching the next station, we cross Rock Creek, towards the head of which is good trout fishing. It is 5 1-10 miles to

Rock Creek—a small eating station. on a small creek of the same name. Trains from the east stop for supper, from the west for breakfast, 30 minutes

---

The English language is wonderful for its aptness of expression. When a number of men and women get together and look at each other from the sides of a room, that's called a sociable. When a hungry crowd calls upon a poor station keeper and eats him out of house and home, that's called a donation party.

are allowed for that purpose; and, it is said, the meals served are much improved on those of former years. All travelers appreciate good fare.

Leaving the station, our course now lies to the eastward; the train winds around the spurs of the bluffs, which seem to bar our way by interlocking with each other, on through a rough, rolling country, again turning to the westward, over bridges and fills, through cuts and snow-sheds, for 7 1-10 miles to

Wilcox—an unimportant station, and we continue crossing creeks and ravines for 8 4-10 miles more, of difficult engineering and middling heavy road-work, and arrive at

Aurora—formerly Como, a small place. Soon after passing the station we come to Como Lake, a beautiful little sheet of water, lying to the right of the road. It is about one mile long and half-a-mile wide, and contains a peculiar fish, a "fish with legs." These *fish-animals* possess gills something like a cat-fish; are amphibious, being often found crawling clumsily around on land, miles from the lake. Quite a variety of peculiar fossil shells are found around the lake that are gathered in summer by persons who offer them for sale to the tourists.

MEDICINE BOW RIVER—is crossed a few miles after leaving Como. It rises in the Medicine Bow Mountains, as before stated, and empties its waters into the North Platte River.

This river was long a noted resort for Indians, and several treaties have been made on its banks between the "*noble* red men" and their pale-faced "brothers." The valley of the river, above the railroad, for thirty miles or more, is broad, fine bottom-land, until it reaches the base of the mountain. From thence to its source the course of the river is through immense forests of pine, wh'ch present unrivaled facilities for lumbering. Fish are found in great quantities in the stream, and the various kinds of game which abound in this country are found in the mountains where the river has its source. Soon after crossing the river, and 7 1-10 miles from Como, we come to

Medicine Bow—containing several stores, and saloons, freight house, passenger station, and a five-stall round-house.

Leaving this station, the road is laid over a smooth, level plain, for 7.5 miles, when it enters a rough, hilly, sage-brush country, and stops at

Niles Junction—from which the train, winding around through deep cuts and long snow-sheds, for 3 2-10 miles further, to

Carbon—Here was discovered the first coal on the Union Pacific Ry. Two veins have been opened, averaging about ten feet. This coal is used principally by the Ry. Co, for their locomotives—the quality not being so good for domestic use as that mined further west, at Rock Springs and Evanston.

The coal is raised from the mine and dumped into the flat-cars, while standing on the track—the shaft of the mine being between the main and side track, close to the station; a stationary engine furnishing the hoisting power. Another shaft is to the south of the town, a short distance, reached by a railtrack.

Carbon contains a population of about 800, and is the county seat of Carbon county, which contains a population of about 2,000—most of whom are engaged in stock-raising.

Simpson—a small, unimportant side-track, is reached 6.3 miles from Carbon, after passing through a succession of cuts, many of which are covered with snow-sheds. Passenger cars do not stop. The road now curves around, and runs almost due west for 50 miles. To the next station it is 4 5-10 miles.

Percy—The station was named for Percy T. Brown, an engineer who was killed by the Indians, while employed surveying the line.

During the construction of the road, this was an important station. Ties, telegraph poles, wood and bridge timber, were landed at this point in immense quantities.

They were obtained at Elk Mountain, seven miles to the south. The old stage road winds around the base of the mountain, between that and the railroad. Near the foot of the mountain, old Fort Halleck and one of the most important stations of the Overland Stage Company, were located; both are now abandoned.

ELK MOUNTAIN—is a noted, landmark, and quite a curiosity in its way. It rises to a great height, its top being covered with snow a great portion of the year, and at any time snow can be found in places on the summit. It has the appearance of being an isolated peak, though, really, it is the extreme northern spur of the Medicine Bow Mountains. It is, how-

SEALS AND SEA LIONS AT FARALLONES ISLANDS, BELOW SAN FRANCISCO.

ever, surrounded by rolling prairie land, and seems to rise boldly from it, rough, rugged and alone. On the west side, the summit is easily reached by a good road, made by the lumbermen. The mountain is nearly round, about six miles in diameter at its base. Its sides are covered with dense forests of pine, aspen and hemlock. It is worthy of note, that this is the only point where the latter species of timber is found along the line of the road. It grows in profusion with the spruce in the gorges, near the summit.

To the south is a fine valley, about 15 miles wide and 20 miles long. Pass Creek, which rises in the Medicine Bow Mountains, runs through this valley on its way to the North Platte River. Large quantities of hay are cut in the bottom lands along the creek. This stream, like all others which rise in this range, is full of fine trout and other fish. Antelope abound on the plain, with elk, deer, bears and mountain sheep, while mountain lions find their homes in the dark ravines and gloomy gorges of the mountain.

Dana—is an unimportant station 6 1-10 miles west of Percy. From Percy to the North Platte River, 29 miles, the road is built down the valley of an alkali ravine. Sage-brush and stagnant pools of alkali water are the only objects that greet the eye—perhaps an unpleasant greeting.

Edson—for many years known as St. Mary's—is 7 5-10 miles from Dana. Soon after leaving the station, our train enters the ravine, where the bluffs assume more formidable features; in fact, the ravine becomes a gorge, the rugged spurs shooting out as though they would reach the opposite wall, and bar out farther progress. The first one of these spurs does indeed bar our way, or did until a tunnel was completed. Before this tunnel was finished, the company laid the road around the point of the spur on a temporary track. Emerging from the tunnel, the train rushes down the gorge, the wall now rising close abrupt and high, on either hand, and 7 8-10 miles from St. Mary's we arrive at and pass

Wolcott's—an unimportant station. Down, down we go—the rough spurs point out from either wall of the canyon, an indenture in one bank marking a projection on the other. While looking on this scene, one cannot help fancying that one time this chasm was not; that some fearful convulsion of nature rent the mighty rocks in twain, leaving these ragged walls and fetid pools to attest the fact. Suddenly we whirl out of the mouth of this chasm—out on the level lands of the North Platte River—cross a substantial wooden bridge, and stop at

Fort Fred. Steele—5 9-10 miles west of Wolcot's; elevation, 6,840 feet.

This fort was established June 30th, 1868, by four companies of the 30th Infantry, under command of Brevet Col. R. I. Dodge, Major 30th Infantry. When the posts in the Powder River country were abandoned, the great bulk of the military stores were hauled to this place and stored for future use.

About two miles west of Fort Steele formerly stood

BENTON CITY—now entirely abandoned. The road was completed to this point the last of July, 1868. At that time a large amount of freight for Montana, Idaho, Utah, and the western country was re-shipped in wagons at this point, and during August and September the place presented a lively aspect, which continued until the road was finished to Bryan, the first of October. Benton at that time was composed of canvas tents; about 3,000 people of all kinds made the population; a harder set it would be impossible to find—roughs, thieves, petty gamblers (the same thing), fast women, and the usual accompaniments of the railroad towns, flourished here in profusion. There were high old times in Benton then, but as the road stretched away to the westward, the people "packed up their tents and stole noiselessly away," leaving only a few old chimneys and post-holes to mark the spot of the once flourishing town. Whiskey was the principal drink of the citizens, it being the most convenient, as all the water used had to be hauled from the Platte River, two miles distant, at an expense of one dollar per barrel, or ten cents per bucket-full.

At Benton, the bluffs which mark the entrance to the canyon of the Platte near Fort Steele, are plainly visible and will continue in sight until we near Rawlins. They are of gray sandstone, worn, marked by the waters or by the elements, far up their perpendicular sides. They are on the opposite side of the river, the banks on the west side being comparatively low.

At this point the river makes a bend, and for several miles we seem to be running down the river, parallel with it, though really drawing away from the stream.

To the south is a long, high ridge of grey granite, called the "Hog Back." It is about four miles away from the road, and runs parallel with it for about 15 miles, terminating in the highlands of Rawlins Springs. It is very narrow at the base, not exceeding half-a-mile in width, yet it rises from 1,000 to 3,000 feet high. The ridge is so sharp that cattle cannot be driven across it, and in many places it is all but impracticable for a man to attempt to walk along its summit. Where this ridge reaches the river bank, about two and a half miles above the bridge, the walls are perpendicular and very high, from 1,000 to 1,500 feet. A corresponding bluff on the opposite side shows that the river has cut a channel through this ridge, which at one time barred the progress of the waters.

On the south side of the ridge is a very pretty little valley, through which flows a small creek into the Platte. It furnishes fine grazing and is in marked contrast to the surrounding country.

Many years ago this green and peaceful looking vale was the scene of a fearful battle between the Sioux and their inveterate enemies, the Utes. The Sioux were encamped in the valley, and were surprised by the Utes, who stole on them in the grey light of the morning, and attacked them furiously. Though taken by surprise, the Sioux fought bravely, but were surrounded and overpowered. When trying to escape, they essayed to cross the "Hog Back," but every one who raised his head above the crest was picked off instantly. A portion of the band escaped in another direction, leaving their dead comrades on the field. The Sioux were so badly whipped that from that time forward they have had little use for the Utes.

NORTH PLATTE RIVER—We gave a short description of this river from where we first crossed it near North Platte City, to Fort Steele, so we will now trace it from this station to its source.

From Fort Steele to the head waters of the Platte is about 150 to 200 miles. It rises in the mountains of the North Park, its waters being supplied by many tributaries, which, at present, are mostly nameless. The course of the river, from its source to this point, is nearly due north.

About twenty-five miles above the fort, is the Platte Ferry, on the old overland stage road.

Good bottom lands are found along the stream at intervals. About 100 miles further up, the tributaries of the river begin to empty their waters into the main stream, and the timber land commences.

Douglas Creek and French Creek are tributaries of the Platte, and run through heavy timbered valleys. Gold mines and gulch diggings were discovered here, but not prospected to any great extent. On the west side of the river, Monument and Big creeks empty their waters into the Platte, nearly opposite the creeks first named.

Big Creek rises in a beautiful lake, about three miles long by half a mile wide. A half-mile above this lies another lake, but little smaller. This ground is disputed territory between the Sioux and Utes, rendering it very unsafe for small parties.

Eight miles from Douglas Creek coal is found in abundance, and further on, fine-looking quartz veins crop out on the hillside. Near here are sulphur springs, seven in number, and very hot; while, along side of them rises a clear, sparkling spring of ice-cold water, and we op ne that the time is not far distant when these springs will be taken up, a *narrow gauge* railroad laid down, hotels built, and one of the finest "watering places" in the world opened to the public.

Fish of many kinds, and beavers, are abundant in the streams; the beavers erecting dams often six feet high. The mountains and forests are full of game, and in them and the open valleys can be found elk, deer, antelope, bears, mountain sheep and lions and, occasionally, the bison or mountain buffalo.

The forests are dense and large in extent; from which, during the building of the road, large quantities of ties were cut and floated down the river to Fort Steele. The valleys are fertile and large, and all in all, it is a grand, wild country, where the tourist could enjoy life to his heart's content, in hunting, fishing, and *fighting* the Indians.

Grenuville—is a small side-track station eight miles west of Fort Steele, and it is seven miles further to

Rawlins—(usually called Rawling Springs). This place contains a population of about 800. The Railroad Company

SNOW GALLERIES, SIERRA NEVADA MOUNTAINS.

have built here a fine hotel, a round-house of 20 stalls, and machine-shops for division repairs. The Railroad Company employ 130 men.

The surrounding country is rough and broken, covered with sage-brush and flecked with alkali. Close above the town a fine sulphur spring rises from under the bed of blue limestone, and other springs arise from the surface of a narrow, wet ravine, which extends about a mile above the town. The bed of the ravine, as far as the water extends, is white with alkali, where the pools of stagnant water do not cover it.

From 30 to 40 miles to the northeast of this station, are located the Ferris and Seminole mining districts. The ore is silver, and said by some people to be rich. Several mills were erected some years ago, but the ore proved refractory and little has been done. Stages leave Rawlins tri-weekly for Meeker —150 miles distant.

Rawlins is the county seat of Carbon county, and was named in honor of Gen'l J. A. Rawlins. The principal business in which the citizens are engaged is stock raising and mining.

Two miles north of the station a paint mine has been discovered, which prospects now to be very valuable. It is said to be fire-and-water proof. Two mills have been erected at the station for grinding the paint, with a daily capacity of three and ten tons respectively. The Union Pacific Railroad Company are using it to paint their cars.

Leaving Rawlins, we follow up the narrow ravine spoken of, through a natural pass about 300 feet wide, which leads between two nearly perpendicular bluffs over 200 feet in height, composed of yellowish gray quartzose sandstone, overlaid with carboniferous limestone. This bluff appears to have extended across the ravine sometime in the past. Perhaps a large lake was imprisoned above, which kindly burst these huge walls, and left a natural route for the railroad.

Beyond the pass we follow up this dry lake bed 6 5-10 miles through a sage-brush and alkali country to

Solon—a small station where the passenger trains do not stop, and 6 6-10 miles further arrive at

Separation—This station derives its name from the fact that at this place the various parties of surveyors who had been together or near each other for the last hundred miles, separated to run different lines to the westward; elevation, 6,900 feet. We are rapidly rising, and 15 miles further will be on the summit of the Rocky Mountains.

Artesian wells are quite numerous along the line, most of them having been finished within the past five years. They are from 326 feet to 1,145 feet in depth, flowing from 400 to 1,000 gallons an hour, in one place 26 feet above the surface. By pumping, these wells will supply from 650 to 2,400 gallons of water per hour. The one at this station is 1,103 feet deep, in which the water stands 10 feet from the surface, and by pumping yields 2,000 gallons per hour.

Fillmore—is another station where the cars do not stop. It is 8 3-10 miles west from Separation, and six miles from

Creston—Sage-brush and alkali beds are the rule now, and have been for the last 25 miles, and will be for the next 100 miles. We are now near the summit of the great "back-bone" of the continent— the Rocky Mountains — just 7,030 feet above the level of the sea.

Two and a half miles west of this point a sign-board has been erected on the right of the road, bearing the words:

"CONTINENTAL DIVIDE,"

and marks the summit 7,100 feet above the level of the sea. This point is about 185 miles from Sherman, 737 from Omaha, and from San Francisco, 1,177.

On this wild spot, surrounded by few evidences of vegetation—and those of the most primitive form—this little sign marks the center of the grandest range of mountains on the continent. Amid what seems to have been the wreck of mountains, we stand and gaze away in the vast distance at the receding lines of hill, valley and mountain peaks, which we have passed in our journey. We feel the cool mountain breeze on our cheeks, but it brings no aroma of life and vegetation with its cooling current. We feel and know that the same sky which hangs so warm and blue over the smiling valleys, looks down upon us now—but how changed the aspect; thin, gray and cold it appears, and so clear that we almost expect to see the stars looking down through the glistening sunbeams. We do not seem to be on the mountain height, for the expanse seems but a once level plain, now arched and broken into ugly, repulsive hollows and desolate knobs.

Here, if a spring should rise from this

Foundation of Temple.   Tabernacle.

VIEW OF SALT LAKE CITY, LOOKING SOUTHWEST.

sage-brush knoll, its waters would divide, and the different portions eventually mingle with the two oceans which wash the opposite sides of the continent. We enter the cars and pass on, the track seeming to be lost but a short distance in our front. The view from the rear of the car is the same. The track seems to be warped up and doubled out of sight. The curvature of this backbone gives the track a similar appearance to that witnessed at Sherman. Although much higher at Sherman, still this is the continental divide, but the low, broad pass brings us 1,212 feet below that place. To the north, the Seminole mountains rear their rugged heights, and farther on, and more to the westward, can be seen the long lines and gray peaks of the Sweetwater Range. Still farther to the west and north, the Wind River Mountains close the scene in the dim distance, their summits robed in snow. Away to the south can be seen the hills which form the southern boundary of the pass, near

---

To be "dead broke," or out of money, is "in the cap," "on the bed rock," etc.

"Shooting his mouth off," for one to use defiant or foul language.

by where the Bridger Pass Station is situated on the old overland stage road.

With a last look at this rugged, barren, desolate region, we speed away over the crest, and shall have down grade for the next 108 miles, descending in that distance 1,110 feet.

Latham—is reached 7.3 miles west, but our train does not stop; and 7 6-10 miles more brings to

Wash-a-kie—named after an old chief of the Shoshone Indians, who has always been friendly with the whites. At this place is another artesian well, 638 feet deep, which, at 15 feet above the surface, flows 800 gallons of pure water per hour.

Red Desert—is 9 6-10 miles from Wash-a-kie. The country around here is called the Red Desert, from the color of the barren soil. It is a huge basin, its waters having no outlet. Several alkali lakes are found in it, but nothing lives on its surface. The soil is bad between Table Rock and Creston, the extreme points of the desert, 38 miles apart. It is composed of the decomposition of shale and calcareous clays, and is deep red, showing the presence of an hydrous sesquioxide of iron. The southern margin of the basin is mainly sand, which is lifted up by every passing breeze to fall in drifts and shifting mounds.

Tipton—a side-track, where our train *does not* stop, is 6 1-10 miles west of Red Desert, and 6 4-10 miles further, the train *will stop* at

Table Rock—This station is on the outer edge of the desert, which has an elevation of 6,890 feet. Off to the left can be seen a long line of bluffs, rising from 50 to 500 feet above the surrounding country. They are of red sandstone, which is mainly composed of freshwater shells, worn, cut, and fluted by the action of the elements. One of these bluffs, which gives its name to the station, is level on the top, which rises about 500 feet above the road, and extends for several miles. Heavy cuts and fills are found here, showing that the road is passing through the rim of the desert. After passing through this rim, and by the side-track, called

Monell—we go on, through a rough and broken country for ten miles, when we arrive at a station called

Bitter Creek—At this place the company have a ten-stall round-house, and a machine shop, for repairs.

As we leave this station, we begin the descent of the celebrated Bitter Creek, the valley of which we shall follow to Green River, about 60 miles west. The valley is narrow, the bluffs coming near the creek on either side. The stream is small and so strongly impregnated with alkali as to be almost useless for man or beast. The banks and bottoms are very treacherous in places, miring any cattle which attempt to reach its fetid waters. This section was always a terror to travelers, emigrants and freighters, for nothing in the line of vegetation will grow, excepting grease-wood and sage-brush. The freighter, especially, who had safely navigated this section, would "ring his popper" and claim that he was a "tough cuss on wheels, from Bitter Creek with a perfect education."

From the source to the mouth of this stream, every indication points to the fact that deposits of oil underlie the surface. Coal veins—valuable ones—have been found, and an oil-bearing shale underlies a large portion of the valley. The old overland stage and emigrant road follows this valley from its source to Green River. From the bluffs, spurs reach out as though they would like to meet their jagged friends on the opposite sides; and around the rough points the cars roll merrily on down, down to the Green.

Black Buttes—is 9 1-10 miles down the creek.

Hallville—an unimportant station to the tourist, is 5.1 miles from the Buttes, and 6 2-10 miles to

Point of Rocks—Here an artesian well, 1,015 feet in depth, supplies an abundance of pure water.

Extensive coal mines near this station are being worked by the Wyoming Coal Company, who ship as high as 100 carloads daily. In one bluff, at a depth of 80 feet, five veins of coal have been opened—one upon the other—which are respectively one, three, four, five, and six and a half feet in thickness. On the bluff, just above the coal, is a seam of oyster-shells six inches in thickness, which Hayden says "is an extinct and undescribed species, about the size of our common edible one."

The sandstone bluffs, at points along the road, are worn by the action of the elements into curious, fantastic shapes, some of which have been named "Caves of the Sand," "Hermit's Grotto," "Water-washed Caves of the Fairies," "Sanko's Bower,"

&c. Prof. Hayden, in his geological examination of this section of the creek, reported finding "preserved in the rocks the greatest abundance of deciduous leaves of the poplar, ash, elm and maple." He says further: "Among the plants found is a specimen of fan-palm, which, at the time it grew here, displayed a leaf of enormous dimensions, sometimes having a spread of ten or twelve feet. These gigantic palms seem to have formed a conspicuous feature among the trees of these ancient forests." Several sulphur and iron springs are located near, but little attention has been directed to their special virtues.

Thayer—a small side-track, 5 3-10 miles further west, is passed without stopping, and 6 2-10 miles more we arrive at

Salt Wells—This, until coal was discovered in quantities on the creek, was a wood station. The wood was obtained from five to ten miles south, in the gulches, where also could be found game in abundance—elk, deer, bears, etc.

Baxter—is 6 7-10 miles from the Wells, and 6 8-10 from

Rock Springs—This station was named after a saline spring of water which boils up out of the bluffs, looking very clear and nice, but it is very deceiving—an uncommon thing in this truthful world.

An artesian well has been sunk at this station, 1,145 feet deep. The water flows to the surface at the rate of 960 gallons per hour, and at 26 feet above the surface, flows 571 gallons per hour. The population of this place is 500, mostly engaged in mining and stock raising. Near here are more rich coal mines.

From this point to Green River, the scenery becomes more grand and impressive, the bluffs rising higher and the gorge narrowing, until the hills seem to hang over the narrow valley with their frowning battlements. Through this gorge we rattle on nine miles to

Wilkins—a small station six miles from the end Laramie Division.

STARVATION CAMP, DONNER LAKE—SEE ANNEX NO. 33.

Green River—is the county seat of Sweetwater county, 845 miles west of Omaha, the end of the Laramie and the commencement of the Western Division of the Union Pacific Railway.

The place is a regular eating station, where passenger trains stop 30 minutes—those from the East for breakfast, those from the West for supper. Much taste is displayed at this station in decorating the dining room and office with mountain curiosities, mineral specimens, moss agate and horns of game.

The city has a good court-house—costing $35,000; several dry goods, grocery, clothing and other stores; two hotels, and about 400 population; also, a daily newspaper, the *Evening Press*. The Railroad Company has a round-house of 15 stalls, and machine shops and repair shops, located here, which in the early years of the road, were at Bryan.

It is claimed that the surrounding country is rich in mines, but one thing is certain—it is rich in cattle; it has cattle on *more* than a "thousand hills."

The bluffs near this station present a peculiar formation called, by Prof. Hayden, the "Green River Shales." For a beautiful illustration of the bluffs, the station and the bridge, see ANNEX No. 16.

The walls of these bluffs rise perpendicularly for hundreds of feet, are of a grayish buff color, and are composed of layers, apparently sedimentary deposits of all thicknesses, from that of a knife-blade to two feet. At the base of the bluff the layers are thin and composed of arenaceous clay, with laminated sandstone, mud markings and other indications of shallow water or mud flats; color for 100 feet, ashen brown; next above are lighter colored layers, alternate with greenish layers, and fine

INTERIOR VIEW OF SNOW SHEDS ON THE SIERRA NEVADA MOUNTAINS.

CROSSING TRUCKEE RIVER, C. P. R. R.

white sand. Passing up, clay and lime predominate, then come layers of boulders, pebbles, and small nodules.

There are also seams of very fine black limestone, saturated with petroleum. Near the summit, under the shallow, calcareous sandstone, there are over fifty feet of shales that contain more or less of oily material. The hills all around are capped with a deep, rusty yellow sandstone, which presents the peculiar castellated forms which, with the banded appearance, have given so much celebrity to the scenery about this station.

The point where our photographer stood to take the picture, was about one-half mile below the bridge and immediately opposite the mouth of the noted Bitter Creek, down which, in years past, rolled the wagons of the pioneer-emigrants of the *far* West, on their weary way seeking new El Doradoes towards the setting sun.

OLD TOWN—A short distance from the station to the southward is the site of the old deserted city of Green River, near the old emigrant crossing, and thereby hangs a tale. This city was laid out in July, 1868, and the September following contained 2,000 inhabitants, and many substantial wood and adobe buildings, and presented a permanent appearance. At that time it was thought by the citizens that the Railroad Company would certainly erect their division buildings near the town, and it would become an important station in consequence. But the Railroad Company opposed the Town Company, bridged the river, and as the road stretched away to the westward, the town declined as rapidly as it arose, the people moving on to Bryan, at which place the Railroad Company located *their* city—and sold lots.

Geographical indications *from the first* pointed to the fact that the Railroad Company must eventually select this place in preference to Bryan, which is *now* an accomplished fact.

TWENTY YEARS AGO an important trading post was located near this station just below, on the opposite side of the river. In early days, the Mormons had a ferry here, and as the river was seldom fordable—except late in the fall—they reaped a rich harvest of from $5 to $20 a team for crossing them over the river, according as the owners were found able to pay. Those times were comparatively only yesterday, and we might say with the juggler "Presto!" and we have the "iron horse," and the long trains of magnificent palace cars, crossing the substantial railroad bridge, conveying their hundreds of passengers daily—passengers from every land and clime—and whirling them across the continent from ocean to ocean, on schedule time. Do these passengers, while partaking of a princely meal, lying at ease sipping their wine, (or *possibly* ice water,) and smoking quietly their cigar, ever think of the hardy pioneers who toiled along on foot and alone, many times over *seven months* traveling the same distance that can now be made in *five days?* Thes pioneers suffered *every kind* of hardship, many even unto death, and those that remain are fast passing away. Yet, the fruits of their adventurous and daring intrepidity can be seen on every hand.

GREEN RIVER—This stream rises in the northwest portion of the Wind River Mountains, at the base of Fremont's Peak. The source of the river is found in innumerable little streams, about 200 miles from the railroad crossing. About 150 miles below the station the river empties into the Colorado River. The name "Green River" implies the color of the water, but one would hardly expect to behold a large, rapid river, whose waters possess so deep a hue. The river, for some distance up the stream, commencing about fifty miles above the station, runs through a soil composed of decomposed rock, slate, etc., which is very green, and easily washed and worn away, which accounts for the color of the water. At all seasons of the year the water is very good—the best, by far, of any found in this part of the country. The tributaries abound in trout of fine flavor, and the main river is well stocked with the finny tribe. Game of all kinds abound along the river and in the adjacent mountains.

Fontenelle Creek comes into Green River 40 miles north, and is specially noted for game, trout, etc.

The lower stream presents a very marked feature, aside from the high bluffs of worn sandstone besides sedimentary deposits. These features are strongly marked, above the bridge, for several miles.

From Green River station, the first exploring expedition of Maj. Powell started on the 24th of May, 1869. The party consisted of about a dozen well-armed, intrepid men, mostly Western hunters. They had four well-built boats, with which to explore the mysterious and terrible canyons of Green River and the Colorado. These gorges were comparatively unknown, the abrupt mountain walls having turned the travel far from their sterile shores. Science and commerce demanded a solution of the question: 'Can the upper Colorado be navigated?" and Maj. Powell undertook to solve the problem.

The party encountered hardships, discovered beautiful scenery, and in their report have thrown much light on the mysteries of this heretofore not much traveled country. The result of the expedition afforded the Major the materials for a course of lectures, and demonstrated the important fact that the Colorado canyon *is not* navigable.

We hear that the Major has, since the above, made an expedition to the river, but are not informed as to the results.

A wagon road leads north, up the east side of the river, over which a stage runs regularly to the

SWEETWATER COUNTRY—The principal cities are South Pass, Atlantic and Hamilton. They are situated four miles apart. The principal occupation of the citizens is quartz gold mining. Many of the mines are said to be very rich, but for some reason very unprofitable to work. The principal mines are on Sweetwater River, a tributary of Wind River, which passes through very rich mineral and agricultural country.

Wind River is a tributary of the Big Horn River, which empties into the Yellowstone. The streams abound in fish, including trout of excellent flavor. The valleys and mountains furnish game in abundance, including deer, elk, antelope, mountain sheep, buffalo, cinnamon, brown, black and grizzly bears.

Indian difficulties have retarded mining, agricultural, and business operations very much in the past.

BURNING ROCK CUT, NEAR GREEN RIVER.

to Bryan, September, 1868, and large amounts of freight was delivered here to be re-shipped to the westward. From this station to the northward, it is 80 miles to the Pacific Springs on the old "California trail," and 90 miles to Sweetwater.

At one time stages left this station for the Sweetwater country, but they have been transferred to Green River station. Freight for the Government posts, and country to the northward, Atlantic City, South Pass, &c., is hauled from this station by wagon teams as of old.

Bryan, during its early days, was quite lively, and troubled with the usual number

Leaving the station, we cross Green River on a fine bridge, the cars passing along through heavy cuts, almost over the river in places, affording a fine view of the frowning cliffs on the east side of the river. Twenty miles to the northwest a large barren butte, pilot-knob, stands in isolated loneliness. Soon we turn to the left, leaving the river, and pass

Peru—in 8 miles —and in five miles more, arrive at

Bryan—a deserted old station. The country around is barren, composed of red sand, and uninviting in the extreme. We are again increasing our elevation. The road was completed

DEVIL'S GATE, WEBER CANYON, U. P. R. R.

of roughs, gamblers and desperadoes. When the Vigilance Committee was in session here, in 1868, they wa ted on a noted desperado, and gave him 15 minutes to leave town. He mounted his mule and said: "Gentlemen, if this d—m mule don't buck, I don't want but five." We commend his judgment, and consider that for once "*his head was level.*"

BLACK'S FORK is approached at this station. It rises in the Uintah Mountains, about 100 miles to the southwest, and empties into Green River, below Green River City. The bottom lands of this river, for fifty miles above Bryan, are susceptible of irrigation, and are thought to be capable of raising small grains.

Marston—is an unimportant station, 7.6 miles from Bryan. Soon after passing the station, to the northward, the old Mormon trail from Johnson's Ford on Green River, 12 miles above Green River station, can be seen coming down a ravine. The route is marked for some distance by a line of telegraph poles which leads to Sweetwater.

Soon after leaving this station, a fine view can be had to the left, south, of the Uintah range of mountains. The valley of the Beaver lays at the northern base of the range, and is one of the most productive sections of the territory; corn, potatoes, vegetables, and small grain grow and yield abundantly. Beaver Creek, which flows through the valley, was named for the beavers that inhabit the creek.

As early as 1825, Beaver Creek was known to Bridger and other trappers of the American Fur Co; in after years, it became the headquarters—for years at a time—of Jim Bridger and other trappers. Since trapping beaver has been abandoned the increase in Beaver Creek has been wonderful. Immense dams are here to be seen, from four to six feet high, which flood many thousand acres.

The streams of this section not only abound in beaver, but in fish—the trout here being abundant.

Beyond the Uintah Range is the Great Valley of White Earth River, where is located the Ute Reservation.

Granger—is 9.6 miles west of Marston. The last seven and a half miles of track before reaching this station was laid down by Jack Caseman in one day. The station is named for an old settler, Mr. Granger, who keeps a ranche near by.

HAM'S FORK—which we cross near the station—rises about forty miles to the northwest, in Hodge's Pass. The bottom lands of this stream are very productive of grass; the upper portion of the valleys, near the mountains, produce excellent hay-crops. Up this "Fork" is building the

Oregon Short Line.

—broad gauge—a branch of the Union Pacific. The first survey was made in 1876, but active work of building only commenced in 1881. The road had, at the close of the year, 150 miles of steel rails laid, and its construction is pushed vigorously. The line pierces the Uintah range by a long tunnel, penetrates a region abounding in coal, and in close proximity to the celebrated soda springs of Idaho, and connects with the Utah & Northern branch at Pocotello, 150 miles north from Ogden, Utah. Working parties are now engaged on the line as far west as Boise City, and on a line branching off to the westward of Blackfoot, for Salina, in the Wood River country.

After crossing the bridge we leave Black's Fork and the old stage road, which bears away to the left, to Fort Bridger, while our course is due west, up the BIG MUDDY, which we cross and recross repeatedly before reaching Piedmont, 50 miles distant. The valley of the stream is narrow, producing only sage- and grease-wood.

Church Buttes—is situated on Big Muddy creek, just east of the crossing. The station is 10.5 miles from Granger's, a noted place for moss agates. These beautiful stones are found along the line of the road from Green River to Piedmont; in some places the ground is literally paved with these gems, varying in size from a pea to about five inches in diameter. The outside is a dark gray and a greenish blue in spots. Should the reader conclude to stop over and hunt moss agates, our advise would be: take your time and a hammer with you, crack the rocks and pebbles beneath your feet; and when you find one of the agates, if it looks dull and rusty, do not throw it away in hopes of finding a prettier one, for often the dull-looking stone, when cut and dressed, is very beautiful and valuable; but most of the agates are valueless.

Church Buttes station derives its name from the peculiar formation of the sand-stone bluffs, which extend for many miles on the left-hand side of the road; they are about ten miles distant. At the Old Church Buttes station, on the "old overland stage road,"—about nine miles to the south they rise in lofty domes and pinnacles, which, at a distance, resemble the fluted columns of some cathedral of the olden time, standing in the midst of desolation; its lofty turreted roof and towering spires rising far above the surrounding country; but on nearer approach the scene changes, and we find a huge mass of sandstone, worn and washed by the elements until it has assumed the outline of a church of the grandest dimensions, it being visible for a great distance. Again we go westward 6.9 miles to

JAMES BRIDGER—See following page.

Hampton—a side track, with cattle pens and shute for loading them—large herds of which range in this section, on the hills and in the adjacent valleys.

To the left, after leaving the station, we see high buttes of all fantastic shapes, showing water lines, which indicate that there has been "high water" here some time in the past.

Carter—is 10 miles from the last station. About seven miles north, a large sulphur spring, and near it a calybeta spring has been discovered, and about fourteen miles further a mountain of coal; the total thickness of the veins is 87 feet, traceable for twelve miles. A branch railroad is contemplated to the coal bank, via the springs.

This station is named for Judge Carter, of Bridger. This gentleman has a large warehouse at this point, where freight until recently, was received and shipped to Virginia City, Helena, and Bannock City, Montana Territory. This route was the shortest wagon route from the East, until the **building of the Utah and Northern.**

The series of buttes that has been observed on our left below, continues, but are more of a uniform height—table-topped, with scrub cedars in the gulches and ravines. Some of these buttes look like immense railroad dumps, as they jet out into the valley, round and steep.

On the right, the soil is red-clay, with some rocks of the same color.

FORT BRIDGER—is ten miles east from this station, over the bluffs, out of sight, having been established in 1858, by General A. S. Johnson, latitude 41 deg. 18 min. and 12 sec.; longitude 110 deg. 32 min. and 38 sec.

Black's Fork, which runs through the center of the parade ground, affords excellent water, and with Smith's Fork, a stream five miles southeast, affords as fine trout as there is in the country.

The "good, old-fashioned way" of imparting knowledge to dull pupils—By rule, paddling it in through the pores of the skin.

"Cayotes" are a small species of wolf. "Jack rabbits" are of the hare family.

Infantry soldiers are called, by the Indians, "heap walk men."

This post was named after JAMES BRIDGER, the renouned hunter, trapper and guide, who lived in this country nearly half a century. (See portrait page 77.)

"Jim" Bridger is undoubtedly the most noted of all the old plains men, and early pioneers in our far western country. Through the courtesy of W. A. Carter, of Fort Bridger, we have been furnished with a fine picture of Mr. Bridger, and a short sketch of his eventful life—from which we condense:

"Jim" was born in Richmond, Virginia —sometime about the last of the last century—and while he was very small, his parents emigrated to St. Louis, Mo., where, shortly after their arrival, they both died of an epidemic then prevailing in that city. Having no one to look to or care for him, he engaged to accompany a party of trappers who were then fitting out for a trip to the Rocky Mountains.

Entirely devoid of even the commonest rudiments of education, he crossed the then almost wholly unknown and trackless plains, and plunged into the pathless mountains. Greatly attracted by the novelty of the sport, at that time quite profitable, he entered eagerly upon the business of trading in fur. Being naturally shrewd, and possessing a keen faculty of observation, he carefully studied the habits of the beaver, and profiting by the knowledge obtained from the Indians—with whom he chiefly associated, and with whom he became a great favorite—he soon became one of the most expert trappers and hunters in the mountains.

Eager to satisfy his curiosity, a natural fondness for mountain scenery, and a roving disposition, he traversed the country in every direction, sometimes in company with Indians, but oftener alone; he familiarized himself with every mountain peak, every gorge, every hill, and every landmark in the country. He pursued his trapping expeditions north to the British Possessions, south to Mexico, and west to the Pacific Ocean. In this way he became acquainted with all the tribes of Indians in the country, and by long intercourse with them, learned their language and became familiar with all their signs. He adopted their habits, conformed to their customs, became imbued with all their superstitions, and at length excelled them in strategy. The marvelous stories told by Bridger are numerous, but we have not the space for a "specimen." In after years, when it became necessary to send millitary expeditions through the far western country, the Government employed Bridger as a guide, and his experience was turned to good account as an interpreter of Indian languages.

Mr. Bridger died in 1875, near Kansas City, Mo., having outlived the sphere of his usefulness, there being no longer any portion of the West unexplored, and having reached the period of second childhood.

As this post is one of great historic interest, we publish, in our ANNEX No. 17. Memories of Fort Bridger. To the next station it is 9.5 miles, and is named after that old hunter and trapper,

Bridger—and it is as unpretentious as the original. Scrub cedar in the high rocky bluffs, sage-brush, red sandstone and red clay, with bunch-grass for sandwiching, is the make-up of the surrounding country. It is inhabited by a few wood-choppers, some stock men, with herds of cattle and sheep, a few deer, antelope, coyotes and jack rabbits by the thousands. For agricultural purposes, it is in a high state of desolation.

For the next three stations we shall ascend rapidly. The bluffs are nearer, and we cross and re-cross the "Muddy" very often, the little stream being nearly as crooked as the streets in Boston.

A few miles beyond, on the left, is a towering cliff, which comes to a point, near the road, on the side of which are some notable water-lines.

This cliff is about 500 feet in height, and where it comes to a point is pulpit-shaped, and is known as Pluto's Outlook. A little further south is his Majesty's Stone-Yard, to which the railroad company, years ago, laid a track for the purpose of using the flat stone which lay around scattered all over the "yard," but here a difficulty seems to have arisen. The masons reported that the stone was "bedeviled," and would not lay still; when the stone was laid flat in their work, the next morning they would be found on the edge; when laid on the edge and left alone for a few moments, they were found flatways. This state of things so alarmed the masons that they abandoned their work and the country, and it is not known what has become of them.

Leroy—a side-track, is five miles from Bridger. Near here the old overland road comes down the mountains, crossing the railroad to the west, at Burns' old ranche, the route marked by the line of telegraph poles. Three miles west, on this stage road, are the soda springs.

Piedmont—is ten miles from Leroy; there are a few dozen buildings in sight. The principal business in which the people are engaged is the burning of charcoal for shipment to smelting furnaces in Salt Lake Valley. There are five patent kilns close to the left of the road, the wood being hauled from the Uintah Mountains to the southward, from 15 to 20 miles distant.

Leaving the station, look ahead from the left side, at the track and snow-sheds. The grade is very heavy, the country is rough and broken, and the road is very crooked, almost doubling back on itself in places. The track is laid over many long and high trestle bridges, all of which have been filled in with dirt, within the last six years.

Before reaching the next station, our train will pass through five long snow-sheds. The small houses near the sheds are the habitations of the watchmen who have them in charge. These sheds are built very tight to prevent fine snow from sifting through, which causes them to be quite dark. From Piedmont, it is 9.4 miles to

Aspen—a side-track. Lumber piles and water-tank make up the place. This station is next in height to Sherman, on the line of the Union Pacific. Elevation, 7,835 feet; is 977 miles from San Francisco, and 937 from Omaha, situated on the lowest pass over the Uintah Mountains.

The station derives its name from the high mountain to the north, called "Quaking Asp." The summit of this mountain is covered with snow during most of the year. The "quaking asp," or aspen, a species of poplar, grows in profusion in the gulches and on the sides of the mountain. The old overland stage road winds around the northern base, while the railroad girds its southern borders, nearly encircling it between the old and new; decay and death marking the one, life, energy and growing strength, the other.

Leaving Aspen, the grade is downward to Salt Lake Valley. After rolling through two long snow-sheds and five miles of road, we are at

Hilliard—population 400. At this station *business* can be felt in the air. A "V" flume crosses the railroad track—20 feet above it—in which immense quantities of lumber, ties, telegraph poles, cordwood, etc., are floated down from the pineries of the Uintah Mountains, from 20 to 30 miles distant, south. Just to the right of the station are located rows of the J. C. Cameron bee-hive kilns, for burning charcoal. There are about 30 of them, of two different sizes, some with a capacity for 20 and some 40 cords of wood. These kilns can each be filled and burned three times a month, and from 20 cords of wood 1,000 bushels of charcoal is produced. This coal is mostly shipped to smelting furnaces, to the westward—Salt Lake City, Virginia City, Eureka, San Francisco, etc. One smelting furnace was erected here—at the coal—during the year 1877.

Sulphur springs are located opposite the station, to the north and south, from 10 to 25 miles distant, but *they* are getting *too* common to require a description; and then, owing to late teachings, they possess little interest to *our* readers.

Two miles from Hilliard, to the right of the road, we come to the site of old Bear River City, of early railroad days, but now entirely deserted. It is situated in a little valley at the mouth of a ravine, where the old overland stage road comes down from the north of Quaking Asp Mountain. At one time this place was quite populous, and was supposed likely to become a permanent town. At this point the roughs and gamblers, who had been driven from point to point westward, made a stand, congregating in large numbers. They swore that they would be driven no further; that here they would stay, and fight it out to the bitter end. The town contained about 1,000 law-abiding people, and when the roughs felt that trouble was coming on them, they withdrew to the hills and organized for a raid on the town. Meanwhile some of the roughs remained in the town, and among them were three noted garroters, who had added to their long list of crimes that of murder. The citizens arose, seized and hung them. In this act they were sustained by all law-abiding people, also by the *Index*, a paper which had followed the road, but was then published here. This hastened the conflict, and on the 19th of November, 1868, the roughs attacked the town in force. This attack was repulsed by the citizens, though not until the Bear River riot had cost sixteen lives, including that of one citizen. The mob first attacked and burned the jail, taking thence one of their kind who was confined there. They next sacked the office and destroyed the material of the *Frontier Index*, which was situated in a building close to the railroad, on the south side. Elated with their success, the mob, numbering about 300 well-armed desperadoes, marched over to the north side, up the main street, and made an attack on a store belonging to one of the leading merchants. Here they were met with a volley from Henry rifles, in the hands of brave and determined citizens, who had collected in the store. The mob was thrown into confusion, and fled down the street, pursued by the citizens, about thirty in number. The first volley and the running fight left fifteen of the desperadoes dead on the street. The number of wounded was never ascertained, but several bodies were afterwards found in the gulches and among the rocks, where they had crawled away and died. One citizen was slain in the attack on the jail. From this time the roughs abandoned the city.

The town declined as soon as the road was built past it, and now there is nothing left to mark the place, except a few old chimneys, broken bottles and scattered oyster cans. Passing on, the bluffs are high and broken, coming close to the road, leaving but a narrow valley, until we reach

Millis—a side-track, four miles from Hilliard. Soon after passing Millis, we come to the valley of Bear River, down which we run for two miles and cross that river on a trestle bridge, 600 feet in length.

BEAR RIVER—This stream rises about sixty miles to the south in the Uintah and Wasatch Mountains. It has many tributaries, which abound in very fine trout—and quite a business is carried on in catching and salting them for the trade. The river here runs almost due north, to Port Neuf Gap. Before reaching the Gap, it comes to Bear Lake, from which it takes its name. The lake is about 15 miles long by seven wide, and contains plenty of trout and other fish. There are some pretty Mormon settlements at different points along the river and lake shore.

The Upper Bear Lake Valley is a point of great interest on account of the fertility of the soil, its romantic situation, the beautiful and grand scenery of rock, lake and mountain in that neighborhood. The valley lies in Rich county, the most northern county in Utah Territory, and is about 25 miles long, with a varying width.

At Port Neuf Gap, the river turns, and thence its course is nearly due south, until it empties into Great Salt Lake, near the town of Corinne. The course of the river can best be understood when we say that it resembles the letter U in shape. From where it rises it runs due north to latitude 42 deg. 30 min., then suddenly turning, it runs south to latitude 41 deg. 43 min., before it finds the lake. Within this bend lies the Wasatch Mountains, a spur of the Uintah, a rugged, rough, bold, but narrow range.

The entire region is wild and picturesque, and would well repay the tourist for the time spent in visiting it. About sixty miles distant, to the north, are the far-famed Soda Springs, of Idaho, situated in Oneida county, Idaho Territory.

The old route, by which this northern country was reached, was from Ogden, via Ogden Canyon and Ogden Valley; *now* the best route is via Utah Northern railroad to Franklin, and from thence east; see further on.

We now return to the road, and pass down the valley, cross Yellow Creek, one of the tributaries of Bear River, and 9.5 miles from Millis, arrive at

Evanston—This is a regular eating station, where trains from the East and West stop 30 minutes for dinner; the waiters are Chinese.—The meals, good.

Evanston is the county seat of Uintah county, Wyoming, 957 miles from either Omaha or San Francisco—*just half way* between the Missouri River and the Pacific Ocean. The Railroad Company have erected a 20-stall round-house, repair shops, hotel, freight and passenger buildings, and the place has improved otherwise very much. It now contains about 1,200 white and about 150 Chinese inhabitants. The town boasts of some good buildings—including a fine court-house. The *Age*, a weekly newspaper, is published here.

The citizens of Evanston are mostly engaged in lumbering, coke-burning, coal-mining and stock-raising.

The railroad was completed to this point *late* in the fall of 1868, and a large amount of freight was delivered here for Salt Lake Valley and Montana. Saw-mills supply lumber from the almost inexhaustible pine forests on Bear River to the southward.

About three miles east to the right of the road, and of Bear River Valley, is located the town of

ALMA—Here are located some of the most valuable coal mines on the road, and

STEAMBOAT ROCK, ECHO CANY

UTAH, U P.R.R. (See Annex No. 19.) (6.)

which supply large quantities to the railroad company. The mines are said to be very extensive, easily worked, yielding coal of good quality, and employ about 800 men, most of whom are Chinese. From 150 to 200 car loads are shipped from Alma per day to towns on the line of the Central Pacific railroad, to Virginia City, Gold Hill, and Carson in Nevada and to San Francisco. A branch railroad has been constructed to the mines, leading off about one mile north of Evanston.

Soon after leaving Evanston we leave Bear River to the right, and follow up a beautiful little valley eleven miles to

Wasatch—This station was once a regular eating station, with round-house and machine shops of the company located here, but a change has been made to Evanston, and the place is now deserted.

Four miles west we cross the dividing line between Wyoming and Utah Territories. It is marked by a sign-board beside the road, on which is painted on one side, "WYOMING," the other "UTAH."

Game is found in the hills—deer, elk, and antelope—and in the Uintah and Wasatch ranges, brown, black and cinnamon bear are common, and in all the little streams, fish of different kinds are abundant—*trout particularly*.

On leaving Wasatch, we arrive at the divide and head of Echo Canon, one-half mile distant. Here we find the longest tunnel on the road, 770 feet in length, cut through hard red clay and sandstone. When the tunnel was completed, it was approached from the east by two long pieces of trestle-work, one of which was 230 feet long and 30 feet high; the other 450 feet long and 75 feet high, which have since been filled in with earth. The tunnel opens to the westward, into a beautiful little canyon, with a narrow strip of grassy bottom land on either side of a miniature stream, known as the North Fork of Echo. The hills are abrupt, and near the road, leaving scarcely more than room for a roadway, including the grassy land referred to. Along these bluffs, on the left-hand side of the stream, the road-bed has been made by cutting down the sides of the hills and filling hollows, in some places from 50 to 75 feet deep.

Before the tunnel was completed, the road was laid temporarily from the divide into Echo Canyon by a Z or zigzag track, which let the cars down to the head of the canyon—under the trestles above named. The great difficulty to overcome by the railroad company in locating the road from this point into Salt Lake Valley was the absence of spurs or sloping hills to carry the grade. Every thing seems to give way at once, and pitch headlong away to the level of the lake. The rim, or outer edge, of the table-lands, breaks abruptly over, and the streams which make out from this table-land, instead of keeping their usual grade, seem to cut through

"PRICKEY," THE PET HORNED TOAD OF THE PACIFIC COAST.
SEE PAGE 126.

the rim and drop into the valley below, there being no uplands to carry them.

By the present line of road, the cars enter Echo Canyon proper at the little station of

Castle Rock—8.4 miles from Wasatch. This station derives its name from the long line of sandstone bluffs on the right-hand side of the canyon, which are worn and torn away until, in the distance, they have the appearance of the old feudal castles, so often spoken of, but so seldom seen, by modern tourists. For a long distance these rocks line the right-hand bank of the canyon, their massive red sandstone fronts towering from 500 to 1,500 feet above the little valley, and bearing the general name of "Castle Rocks." The cars descend the canyon amid some of the grandest and wildest scenery imaginable. We do not creep along as though we mistrusted our powers, but with a snort and roar the engine plunges down the defile, which momentarily increases to a gorge, only to become, in a short distance, a grand and awful chasm. About 7.2 miles below Castle Rock—at

Emory—the traveler can see the Natural Bridge, a conglomerate formation, spanning a cleft in the wall on the right-hand side: this "Hanging Rock of Echo" has more than a local reputation. It gave the name to an overland stage station, when the completion of this road was—but in the dreams of its sanguine projectors—an undefined and visionary thing of the future.

The left hand side of the canyon presents but few attractions compared with the bolder and loftier bluffs opposite. The wall breaks away and recedes in sloping, grassy hillsides, while we know not what lies beyond these walls to the right, for they close the view in that direction. Wall, solid wall, broken wall, walls of sandstone, walls of granite, and walls of a conglomerate of both, mixed with clay, rise far above us, and shut from our vision whatever lies beyond.

The beauties of Echo Canyon are so many, so majestic, so awe-inspiring in their sublimity, that there is little use in calling the traveler's attention to them. But as we rush swiftly along, seemingly beneath these towering heights, we can note some of the more prominent features.

The only difficulty will be that one will hardly see them all, as the cars thunder along, waking the echoes among these castellated monuments of red rock, whose towering domes and frowning buttresses gave the name to this remarkable opening in the Wasatch Mountains. Four miles below Hanging Rock the walls rise in massive majesty—the prominent features of the canyon. Rain, wind and time have combined to destroy them, but in vain. Centuries have come and gone since that mighty convulsion shook the earth to its center, when Echo and Weber canyons sprung into existence—twin children—whose birth was heralded by throes such as the earth may never feel again, and still the mighty wall of Echo remains, bidding defiance alike to time and his co-laborers—the elements; still hangs the delicate fret and frost work from the walls; still the pillar, column, dome and spire stand boldly forth in all their grand, wild and weird beauty to entrance the traveler, and fill his mind with wonder and awe.

About six miles below Hanging Rock, up on the topmost heights of the towering cliffs, a thousand feet above the bed of the canyon, can be seen the fortifications erected by the Mormons to defend this pass against the army under Johnson, sent out in 1857 by Uncle Sam. These fortifications consist of massive rocks, placed on the verge of the precipice, which were to be toppled over on the heads of the soldiers below, but the experiment was never made, so the rocks remain to be used on some other foe, or as the evidences of a people's folly.

On goes the engine, whirling us past castle, cathedral, towering column and rugged battlement, past ravines which cut the walls from crest to base in awful chasms, shooting over bridges and flying past and under the overhanging walls (see Steamboat Rock, ANNEX No. 19), when, after crossing Echo Creek, thirty-one times in twenty-six miles, we rush past the Witches' Cave and Pulpit Rock, our engine giving a loud scream of warning to the brakemen, who "throwing on the brakes," bring the train to a stop, and we get out once more to examine the country, Weber River and Echo City station.

Before we take a final leave of Echo Canyon we will relate an incident, thrilling in its nature, but happily ending without serious results, which occurred there during the construction of the road from Echo City to the mouth of Weber, and is known as "Paddy Miles' Ride."—see ANNEX No. 20.

Directly ahead of our train, as it emerges from Echo Canyon, coming in from the south, is

WEBER RIVER—This stream rises in the Wasatch Mountains, 70 miles to the south, its waters being supplied by thousands of springs, many larger tributaries, and the everlasting snows of this rugged mountain range. It empties into the Great Salt Lake, just below Ogden, about 50 miles from Echo City. The valley of the Weber, from Echo City up to its source, is very fertile, and thickly settled by the Mormons. Three miles above this station is Chalk Creek, where a fine coalbank has been discovered. Three miles beyond this point is Coalville, a Mormon settlement of 800 inhabitants—a thriving village. Its name is derived from the carboniferous formations existing there. The coal-beds are extensive, some of the veins being of good quality, others being lignite. Echo & Park City branch is completed from Echo City to Park City—27 miles, with a branch from Coalville, five miles to an extensive coal mine. The track leaves Echo City and passes along close below the Union track at Pulpit Rock.

Seven miles beyond Coalville is the pleasant village of Winship, situated at the junction of Silver Creek and Weber River, containing 1,000 inhabitants. The "old stage road" followed up Weber to this point, thence up Silver Creek via Parley Park, and thence to Salt Lake City, 50 miles distant from Echo.

PARLEY PARK—This is a beautiful valley on the old stage road, about five miles long by three miles wide. It is very fertile, producing fine crops of small grain. Several hundred settlers have located and made themselves homes. There is a fine hotel, once kept as a stage station, now kept by William Kimball, eldest son of Heber C. Fish, in any desired quantity, can be caught in the streams, and game of many varieties, including deer and bears, inhabit the adjoining mountains. It is one of those pleasant places where one loves to linger, regrets to leave, and longs to visit again. We advise tourists to visit it; they will not regret a week or a month among the hills and streams of the Upper Weber. Near this point gold and silver mines have been discovered—which prove very rich, chief of which is the Ontario Mine, the most productive in Utah, and the prospects now are that the "Park" will become a great mining center. Returning, we stop a few moments at

Echo City—The town is situated at the foot of the bluff, which towers far above it, 9.4 miles from Hanging Rock. As the cars enter the city from Echo Canyon, they turn to the right, and close at the base of the cliff, on the right, stands Pulpit Rock (see illustration) and the old stage ranche on the left, just where it appears that we must pitch off into the valley and river below. This city is not very inviting, unless you like to hunt and fish, when a

PULPIT ROCK, MOUTH OF ECHO CANYON, UTAH.

stay of a few days would be passed very pleasantly.

Chalk Creek, Silver Creek, Echo Creek, and Weber River, afford excellent trouting, while antelope are shot near the city. The mountains abound in bears, deer and elk.

Echo contains about 200 inhabitants, including those settlers near by and the railroad employes. Coal beds, extensive ones, are found near by, as well as an indefinite quantity of iron ore, which must possess a market value, sooner or later.

Near Echo City, across the Weber, a ravine leads up the mountain side, winding and turning around among the gray old crags, until it leads into a beautiful little dell, in the center of which reposes a miniature lakelet, shut in on all sides by the hills. It is a charming, beautiful, tiny little gem, nestled amid a gray, grand setting of granite peaks and pine-clad gorges—a speck of delicate etherealized beauty amid the strength and ruggedness of an alpine world.

WEBER CANYON—To give a minute description of this remarkable place we cannot attempt, as it would fill a volume were its beauties fully delineated, and each point of interest noted. But as one of the grand and remarkable features of the road it demands a notice, however meager, at our hands. For about 40 miles the river rushes foaming along, between two massive mountain walls, which close the landscape on either hand. Now, the torrent plunges over some mighty rock which has fallen from the towering cliff 1,000 feet above; anon, it whirls around in frantic struggles to escape from the boiling eddy, thence springing forward over a short, smooth rapid, only to repeat the plunge again and again, until it breaks forth into the plains, whence it glides away toward the lake, as though exhausted with its wild journey through the canyon.

In passing down the canyon, the traveler should closely watch, for fresh objects of wonder and interest will spring suddenly into sight on either hand.

From Echo City, the cars speed along the banks of the Weber for about four miles, when they enter the Narrows of Weber Canyon, through which the road is cut for two miles, most of the way in the side of the steep mountain that drops its base in the river-bed.

Soon after leaving Echo City, on the right, about 100 yards from the road, and 300 feet above it, can be seen the "Wiches' Rocks," a collection of red, yellow and gray conglomerate rocks, standing out from the side of the cliff, varying in height from 20 to 60 feet. Shortly after entering the Narrows, the

ONE THOUSAND MILE TREE is passed— a thrifty, branching pine—bearing on its trunk a sign-board that tells the western-bound traveler that he has passed over 1,000 miles of railway from Omaha. This living milestone of nature's planting has

ONE THOUSAND MILE TREE, U. P. R. R.

INTERIOR VIEW OF MORMON TABERNACLE.

long marked this place; long before the hardy Mormon passed down this wild gorge; long before the great trans-continental railroad was even thought of. It stood a lonely sentinal, when all around was desolation; when the lurking savage and wild beast claimed supremacy, and each in turn reposed in the shade of its waving arms. How changed the scene! The ceaseless bustle of an active, progressive age, the hum of labor, the roar and rush of the passing locomotive, has usurped the old quiet, and henceforward the LONE TREE will be, not a guide to the gloomy past, but an index of the coming greatness of a regenerated country.

Just below this tree, the cars cross a trestle bridge to the left bank of the Weber, thence down but a short distance, before they cross over another trestle to the right-hand side, and then, almost opposite the bridge, on the side of the mountain to the left, can be seen the

DEVIL'S SLIDE, or serrated rocks. This slide is composed of two ridges of granite rock, reaching from the river nearly to the summit of a sloping, grass-clad mountain. They are from 50 to 200 feet high, narrow slabs, standing on edge, as though forced cut of the mountain side. The two ridges run parallel with each other—about 10 feet apart, the space between being covered with grass, wild flowers and climbing vines. (See illustration, page 33.)

Rushing swiftly along past **Croydon**—an unimportant side-track, 8.5 miles from Echo City, we soon lose sight of these rocks and behold others more grand, of different shapes, and massive proportions. The mountains seem to have been dovetailed together, and then torn rudely asunder, leaving the rough promontories and rugged chasms as so many obstacles to bar our progress. But engineering skill has triumphed over all. Where the road could not be built over or around these points, it is tunneled under. Now we shoot across the river, and dart through a tunnel 550 feet long, cut in solid rock, with heavy cuts and fills at either entrance. Just before entering this tunnel, high up to the left, formerly stood "Finger Rock," as seen in the illustration (page 42), but which has been broken away, so as not to be visible now. The frowning cliffs bar our further way, and again we cross the roaring torrent and burrow under the point of another rocky promontory. Here the road stretches across a pretty little valley, known as Round Valley.

Dashing along, with but a moment to spare in which to note its beauties, we enter the narrowing gorge again, where the massive walls close in and crush out the green meadows. Between these lofty walls, with barely room for the track between them and the foaming torrent at our feet; on, around a jotting point—and again we emerge into a lengthened widening of the canyon, and we pause for a moment at

Weber—seven miles from Quarry. This station lies between two Mormon settlements, which, taken in connection, are called Morgan City. The villages are separated by the river which flows through bottom lands, most of which are under cultivation; population about 1,000. There are some good buildings of brick and stone, but the greater number are of logs and adobe—sun-dried bricks. At this station, opposite the depot, the first Z. C. M. I. appears, which, in Mormon rendering, means "Zion's Co-operative Mercantile Institution"—a retail branch of the great co-operative house in Salt Lake City.

This valley shows the effects of irrigation in Utah. Wherever the land is below the irrigating canals, and is cultivated, it yields immense crops. Grass grows all the way to the summit—and on the summit—of nearly all these mountains, affording the best of pasturage all the year round, as the fall of snow is light, and enough of what does fall is blown off by the wind, so that cattle and sheep can find sufficient for their needs at all seasons. The same may be said of the whole slope of the mountains of Utah at the same altitude.

Game of all kinds is numerous throughout the same section, and trout exceedingly plenty, even in the tinyest little streams. The road follows down the right-hand bank through this valley until just below

Peterson—a small, unimportant station, 9.7 miles from Weber, when it crosses to the left-hand side, which it follows for four miles further, between towering mountains, the valley now lost in the narrow, gloomy gorge, when suddenly the whistle shrieks the pass-word as we approach the

Devil's Gate—a mere side-track, soon after leaving which, the brink of the torrent is neared, and the wild scenery of the *Devil's Gate* is before us. Onward toils the long train through a deep cut and across the bridge—50 feet above the seething cauldron of waters, where massive, frowning rocks rear their crests far up toward the black and threatening clouds which hover over this witches' cauldron. With bated breath we gaze on this wild scene, and vainly try to analyze our feelings, in which awe, wonder, and admiration are blended. We have no time for thought, as to how or when this mighty work was accomplished, no time nor inclination to compare the work of nature with the puny work beneath us, but onward, with quickened speed, down the right-hand bank of the stream; on between these massive piles, worn and seamed in their ceaseless struggles against the destroying hand of time; on to where yon opening of light marks the open country; on, past towering mountain and toppling rock, until we catch a view of the broad, sunlit plains, and from the last and blackest of the buttresses which guard the entrance into Weber, we emerge to light and beauty, to catch the first view of the Great Salt Lake, to behold broad plains and well-cultivated fields which stretch their lines of waving green and golden shades beyond

Uintah Station—We have now passed through the Wasatch Mountains, and are fairly in the Great Salt Lake Valley. The elevation at this point is 4,560 feet, 2,319 feet lower than Wasatch, 58 miles to the eastward. Uintah is 4.5 miles from the Devil's Gate.

Near the station, on this broad bottom, in 1862, was the scene of the Morrisite massacre.

Here 500 men of Brigham Young's Mormon Legion, and 500 men who volunteered for the occasion, with five pieces of artillery, commanded by Robert T. Burton, attacked the "Morrisites," and after three days' skirmishing, and after a score or more had been killed, the "Morrisites" surrendered. The *noble* Burton, after the surrender, took possession of everything he could find in the name of the Church; shot down their leader, Joseph Morris—an apostate Mormon—whose only fault was that he claimed to be the true Prophet of God, instead of Brigham Young. This man Burton, at the same time shot and killed *two women* who *dared* to beg him to save the life of their Prophet.

The followers of Morris consisted of about 90 able-bodied men, mostly unarmed, and over 300 old men, women and children. The prisoners were all taken to Salt Lake City, and condemned, and those who were able to work had their legs ornamented with a *ball and chain*, and were put to picking stone to build the Mormon temple. On the 9th of March, 1863, these parties were all pardoned by Hon. S. S. Harding, who had that spring arrived in Utah as Governor of the Territory.

Leaving Uintah, the road winds around to the right and follows the base of the mountains, with the river on the left. The country is fertile and dotted with well-tilled farms. As we run along down the Weber River, and 7.5 miles from Uintah, we reach

Ogden—the junction of Union and Central Pacific railroads. The distance from Omaha is 1,032 miles; from San Francisco 882 miles; from Salt Lake City, 36 miles; elevation, 4,301 feet. Near the station building are the depots of the Utah Central and the Utah & Northern railroads.

All passengers, baggage, mail, and express, "change cars" at this station. Passengers who have through tickets in sleeping cars will occupy the same numbers in the Central as they had in the Union, and those who had their baggage checked through need give it no attention; but those who only checked to this place—to the end of the U. P. road—will need to see that it is re-checked. At this station, trains stop a full hour, and sometimes a little longer—much depending upon the amount of matter to be changed from one train to another.

The station building stands between the tracks, in which passengers will find a dining room, where they can have ample time to eat a good "square meal"—price $1.00. Most of the buildings at the station are of wood, but the necessary grounds have been secured near by for the erection of a "Union Depot." When will it be done? *Quien sabe?*

OGDEN CITY is situated one miles east from the depot, at the mouth of Ogden Canyon, one of the gorges which pierce the Wasatch range, and between the Weber and Ogden rivers. Population, about 6,500. This is the county seat of Weber county, and has amply provided itself with all needful county buildings. The Mormons have a tabernacle, and several other denominations have places of worship here. The citizens are mostly Mormon, and all public improvements are under their supervision. It is a poor place for "carpet-baggers."

The waters of the Ogden River are conducted through the streets, and used in the gardens and fields for irrigating, the result of which is that the city is in the midst of one great flower garden and forest of fruit and shade trees. In the gardens are fruit trees of all kinds, which bear abundantly, and in the fields are raised immense crops of grain and vegetables.

Rich mines of iron, silver and slate are reported near the city, but little has been done towards developing them.

Ogden has several good hotels, chief of which is the Utah House. Two newspapers are published here, the *Junction* and the *Chronicle*.

The Wasatch Mountains rise some thousands of feet above the city, and the tourist would find much of interest in a stroll up the mountain side and along the canyons. Ogden Canyon is about five miles long, and from its mouth to its source, from plain to mountain top, the scenery is grand and imposing. In places the granite walls rise on each side 1,500 feet high, and for a considerable distance not more than 150 feet apart. About six miles from Ogden, up in the mountains behind the town, is a lovely little valley called "The Basin," watered by mountain streams and covered with a luxuriant growth of grass.

Before proceeding further, we will take a hasty glance at

Utah Territory.

This territory extends from the 37th to the 42d parallel of north latitude, and from the 109th to the 114th degree of west longitude, containing a superficial area of about 65,000 square miles, with a population of about 143,907 whites, Indians and Chinese. This area includes large tracts of wild mountainous and barren country. At present, most of the lands under cultivation and the meadow lands are around the lakes and in the neighboring mountain valleys, and are very productive when irrigated; grains, fruits and vegetables maturing readily, and yielding large returns—the aridity of the climate precluding the growing of crops by any other means.

Opposite title page of this book, see illustration—Utah's Best Crop.

Rich veins of gold, silver, coal, iron, copper, zinc, cinnabar, antimony and nearly all the metals found in the "Great West," exist in Utah, and it is the opinion of most men, had it not been for the "Councils" of Brigham Young to his followers, the Mormons, not to prospect for minerals, Utah might to-day be an honored State, in the great family of States, with a *developed mineral wealth*, second only to California, *and possibly the first*. The whole country within her borders would be illuminated with the perpetual fires of her "smelting furnaces," and resound with the thundering echoes and re-echoes of the thousands of descending stamps grinding out the wealth, which, since the completion of the Pacific railroad, and the consequent influx of "Gentiles" has been exported by millions and most effectually demonstrated the fact that Utah, if not the richest, is certainly next to the richest silver-mining country in the world.

Besides the above, brimstone, saltpeter, gypsum, plumbago and soda have been discovered, some of which are being worked, while fire-clay, marble, granite, slate, red and white sandstone, limestone and kindred formations exist to an almost unlimited extent. Salt can be shoveled up in its crude state on the shores of Salt Lake, and in the southern part of the Territory, is found by the mountain, in a remarkably transparent and pure state.

Iron ore exists in large quantities in Iron, Summit and Weber counties. Coal abounds in various parts, but the principal mines now worked are at Coalville, in Summit county and in San Pete. The latter yields a good quality of blacksmith coal, in large quantities.

At this time there are about 30 organized mining districts in the Territory. We have not the space to devote to a description of the mines, were we able; they appear to be inexhaustible and very rich. Many are producing large quantities of ore.

Fish culture has, since 1874, been receiving some attention, and a fish farm with a superintendent thereof, is located a few miles from Salt Lake City.

There are quite a number of smelting furnaces in operation in various parts of the Territory, and in Salt Lake City.

Utah was first settled in 1847. On the 24th of July, the advance guard of the Mormon emigration, numbering 143 men, entered Salt Lake Valley; five days later 150 more men arrived under Captain Brown, and on July 31st, Great Salt Lake City was laid out. At that time the country belonged to the Republic of Mexico, but by the treaty of Gaudaloup Hidalgo, in 1848 it was ceded to the United States.

The summers are very warm and dry; the winters mild and open. The fall of snow is light in the valley and heavy in the mountain, the melting of which affords ample water for irrigating the foot-hills and valleys. Vegetables of all kinds grow astonishingly large, and of superior quality.

Timber is not very plenty, and *then*, is only found in the mountains of difficult access. Returning to business; at Ogden, we will step into the cars of the

Utah Central Railroad,

The principal offices of which are at Salt Lake City.

SIDNEY DILLON......................*President*
JOHN SHARP.....................*Vice-President*
FRANCIS COPE.........*Freight and Ticket Agent*

The Utah Central is 36.5 miles in length and the pioneer road of Utah, excepting the through line. May 17, 1869, just one week after the "love feast" of the Union and the Central at Promentory, ground was broken at Ogden, and the enterprise was inaugurated with due ceremonies: President Brigham Young and the chief dignitaries of the Mormon church being in attendance.

In about half an hour after the overland trains arrive at Ogden Junction, the cars of this road roll up to the depot for passengers. When leaving, the train crosses the Weber River, on a fine bridge; just to the north of the depot passes through a deep cut and comes out on a bench of land that gradually slopes from the mountains on the left, to the waters of the lake on the right, six and four miles distant, respectively.

From the car window, on the right, a good view can be had of a portion of Great Salt Lake, but the *best* view is to be had from the top of Promontory Mountain. See ANNEX No. 21. The first station from Ogden is 16 miles distant, along the sloping land named, which is covered above the line of irrigation, with sage, but below with the thrifty Mormon farmers. A wide strip of land near the Lake is valueless, owing to the salt in the soil.

KAYSVILLE—is an incorporated town in Davis county, and is surrounded with well-cultivated farms, finely kept gardens, with water running through the streets, and has fruit and shrubbery in profusion.

The county is comprised of five towns, all, with one exception, traversed by our road, within the next 15 miles. The county has about 7,000 population, seven flouring mills and three saw mills.

FARMINGTON—is the next station, five miles distant, being the county seat of Davis county, and contains good county buildings, several flouring mills, and the usual beautiful surroundings of fruit trees and orchards, for which *all* Mormon settlements are noted.

BRIGHAM YOUNG.—For sketch of life see Annex No. 25.

CENTERVILLE—is the next station, four miles from Farmington. The description of *one* Mormon village will do for nearly all; good farms and crops are the rule, where the land is irrigated, and none where it is not.

WOOD'S CROSSING—comes next, two miles further, being the station for the little village of Bountiful, on the left, and is in the midst of the best cultivated and best producing land in the Territory.

The course of our road from Ogden to Salt Lake is almost due south, while the Wasatch Mountains, for 30 miles, describe a huge circle in the middle to the eastward. The lower point of this circle we are fast approaching, and will reach in about two miles, just at the point of the mountain ahead, where steam is rising. *There,* under the point of that huge rock, boils up a hot spring, in a large volume, forming a creek several feet in width, with a depth of six inches, and it is *very hot.* There is no nonsense about this spring; it sends forth a never-failing stream.

The highest peak in the mountain, close to the eastward of these springs alluded to, is 1,200 feet above the valley, and is called Ensign Peak—the "Mount of Prophecy,"—where the late Prophet, Brigham Young, was wont to wrestle with the Lord.

Just beyond, on the right, is Hot Spring Lake, which is formed from the waters of this and others of lesser volume, near by. This lake freezes over in the winter, except near the shore on the northeastern end, and is a great resort for skating parties from Salt Lake City. Great Salt Lake, *never* freezes over—it's too salt.

Passing the lake, our road keeps straight across a broad bottom, while the mountains on the left again curve away to the eastward.

The Warm Spring buildings, where are located the city baths, can be seen beside the mountain on the left, marked by a continuous column of steam, rising near the buildings.

These are the disputed springs, to obtain possession of which, it is supposed by many, Dr. Robinson was murdered. The baths are well patronized by invalids, who visit them for health, relying on their medicinal qualities to remove their ailments. The following is an analysis of the water, as made by Dr. Charles T. Jackson, of Boston:

Three fluid ounces of the water on evaporation to entire dryness in a platina capsule gave 8.25 grains of solid, dry, saline matter.

Carbonate of lime and magnesia	0.240	1.280
Per oxide of iron	0.040	0.208
Lime	0.545	2.937
Chlorine	3.454	18.421
Soda	2.877	15.348
Magnesia	0.370	2.073
Sulphuric acid	0.703	3.748
	8.229	43.981

It is slightly charged with hydro-sulphuric acid gas and with carbonic acid gas, and is a pleasant saline mineral water, having valuable properties belonging to saline sulphur springs. The usual temperature is 102 degrees F.

They are one mile north of Salt Lake City, and are reached by street cars.

Rolling on through the northern suburbs of the city, a little over eight miles from Wood's Crossing, we stop at the depot in Salt Lake, the City of Zion. Passengers arriving at the depot will find a "Bus" at the eastern gate that will take a passenger and his baggage to any hotel or point in the city for 50 cents; or, at the same gate, street cars, that pass the door of every prominent hotel in the city; fare, 10 cents, or ten tickets for 50 cents.

Salt Lake City—or "Zion," as the city is often called by the Mormon faithful, is one of the most beautiful and pleasantly located of cities. It is situated at the foot of a spur of the Wasatch Mountains, the northern limits extending on to the "bench" or upland, which unites the plain with the mountain. From the east two wagon roads enter the city, via Emigrant and Parley Canyons.

The streets are wide, bordered with shade-trees, and laid out at right angles. Along each side of the streets is a clear, cold stream of water from the mountain canyons, which, with the numerous shade-trees and gardens, give the city an indescribable air of coolness, comfort and repose. The city contains a population of full 25,000, is the capital of the Territory and county seat of Salt Lake county. It has 21 wards within its limits, and is the terminus of four railroads. It contains some as fine business blocks, hotels, and private residences—many lit by electricity—as can be found in any city west of the Missouri River.

The Mormon church, besides its Tabernacle, has a bishop located in every ward of the city, who holds ward meetings regularly. The other churches hold services in four or more places in the city. The Masons have five lodges in the city; the Odd Fellows four, and some of the *other fellows* several. There are 38 mining and smelting offices, five sampling and smelting works, five iron foundries, boiler and brass works, two flouring mills, one woolen mill, nine hotels, six breweries, two extensive marble works, and a score or more of small manufacturing establishments.

There are four daily newspapers. The *Deseret News* is the church organ, the *Herald* claims to be independent, the *Tribune* strong *opposition* Mormon, and the *Times*. Each of these issue weeklies. Newspaper business is *very* precarious in Utah. It's as fine an opening for a young man to get his "teeth cut," as we know of in the world—he can soon get a double and single set all around.

In the mercantile line, Salt Lake City

TERMS HEARD ON THE PLAINS.—"Lariat" is the Spanish name for rope. "Bronco," California or Spanish pony. "Bueano," (wa-no) good. "Esta Bueano," (star wa-no) very good, no better. "No sabe," (sarvey) don't understand. "Quien sabe," (kin sarvey) who knows, or do you understand.

has several establishments that would do credit to any city in the Union, one of which is

The above cut represents the Mormon "Co-operative Sign"—called by the Gentiles the "Bull's Eye." At the Mormon Conference, in the fall of 1868, all good Mormon merchants, manufacturers and dealers who desired the patronage of the Mormon people, were directed to place this sign upon their buildings in a conspicuous place, that it might indicate to the people that they were sound in the faith.

The Mormon people were also directed and *warned* not to purchase goods or in any manner deal with those who refused or did not have the sign. The object seemed to be only to deal with their own people, to the exclusion of all others.

The result of these measures on the part of the church was to force many who were Gentiles or apostate Mormons to sacrifice their goods, and leave the Territory for want of patronage. However, the order was not very strictly enforced—or complied with yet many of these signs are to be seen in Salt Lake City and other parts of the Territory on buildings occupied by the faithful.

To more effectually carry out the plan of co-operation, one great company was to be formed to purchase goods in large quantities and establish branches throughout all the Mormon settlements. Such a company was organized, and incorporated with many high Mormon dignitaries as either stockholders or officers, and it is now known as "Zion's Co-operative Mercantile Institution"—with headquarters in Salt Lake City.

The "Z. C. M. I," undoubtedly have the finest and largest building in the city. It is of brick, 318 feet long, 53 feet in width, three stories and cellar, and finished throughout in the best manner. It also has an addition 25 by 195 feet, and used for a warehouse; cost, $175,000, built of iron, stone and glass, and is now being enlarged. The Walker Bros. have the largest Mercantile business in Utah, requiring *five* different departments, each occupying a large building.

Think of it, "O ye people!" 35 years ago this whole country 1,000 miles in any direction, was uninhabited and almost unknown to the white race. Now annual sales of *these two* establishments exceed $5,000,000, and with their goods, gathered here from all parts of the world, stand forth as monuments of American enterprise, IN AN AMERICAN DESERT.

The late President Brigham Young's residence (see illustration, page 89, also of "Eagle Gate," page 109), tithing house, printing office and business offices connected with the church occupy an entire block, on the bench of land overlooking the city, which is one of the first objects of interest visited by the traveler on arriving in Salt Lake City.

The traveler who visited this city some years ago—before the discovery of the rich silver mines—would be surprised by a visit now, at the remarkable changes noticeable on every hand; all is life and energy; everybody seems to have a pocketful of certificates of mining property, and you hear of extensive preparations making on every side with a view to a vigorous prosecution of various mining enterprises.

The public buildings are not very numerous. They consist of a court-house, city hall, city prison, theatre, and

THE TABERNACLE—an immense building—the first object one beholds on entering the city. The building is oblong in shape, having a length of 250 feet from east to west, by 150 feet in width. The roof is supported by 46 columns of cut sandstone, which, with the spaces between, used for doors, windows, etc., constitute the wall. From these pillars or walls, the roof springs in one unbroken arch, forming the largest self-sustaining roof on the continent, with one notable exception, the Grand Union Depot, New York. The ceiling of the roof is 65 ft. above the floor. In one end of this egg-shaped building is the organ, the second in size in America. The Tabernacle is used for church purposes, as well as other large gatherings of the people. With the gallery it will seat 8,000 people. See illustration, p. 69, also interior view, 85.

The Temple.—This building is not yet completed, but work is progressing steadily, and it is up about 30 ft, The dimensions of the foundation are 99x186½ feet. The site of the Temple is on the eastern half of the same block with the Tabernacle.

Since the advent of railroads into Utah and the discovery of rich mines, church property has not accumulated very rapidly. Within the past few years nearly all the religious denominations have secured a foothold in this city.

Fort Douglas—a military post, established Oct. 26, 1862, by Gen. E. P. Conner, Third Regiment of California Volunteer Infantry, is situated on the east side of the Jordan, 4 miles from that stream, 3 miles east of the City of Salt Lake, and 15 miles southeast of Salt Lake. Latitude 40 deg. 46 min. 2 sec.: longitude, 111 deg. 53 min. 34 sec. Its location is on a sloping upland or bench at the base of the mountains and overlooking the city, and affords a fine view of the country to the west and south.

Jordan River.—This stream, which borders Salt Lake City on the west, is the outlet of Utah Lake, which lies about 40 miles south. It empties into the Great Salt Lake, about 12 miles northwest of the city.

There are a great many hotels in Salt Lake City, but the principal ones are the Walker, Continental, White,

MORMON TEMPLE, SALT LAKE CITY.

Cliff, Valley and Overland. The two former are under the management of G. S. Erb, Esq., and we *know* them to be *first-class*.

The picture of the late President Brigham Young on page 89, was the last one ever taken. It was made by Mr. Savage of Salt Lake City, an eminent artist and is said to be a very accurate picture.

For sketch of the life of Brigham Young, see Annex 25.

We will now take a run over all the railroads in Utah, commencing with the

Utah Southern Railroad.

This road was consolidated in 1881 with the Utah Central. It was commenced May 1st, 1871, and built thirteen miles during the year, to Sandy, and then extended from time to time until at this time, January, 1882, it is completed 226 miles south, to Frisco.

The cars start from the same depot as the Central.

We will step on board and roll southward through the city—passing fine

gardens, thrifty or-
ited fields, with the
right, the Oquirrh
ar in the distance,
atch Mountains on
a Valley. This val-
Lake City, south, to
iles distant, with a
two to twelve miles.
uth, on the left, is
melting Works, on
lown from Big Cot-
e Parley Canyon,
he east, just below
located the State
Utah Woolen Mill.
id passing on seven
come to
o—the first station.
Wasatch Smelting
yond the American
; after crossing the
k, on the right, is
lting and Refining
of Germania, con-

; right, can be seen
c—"matte" as it is

mountain view is
rd, the canyons of
g and Little Cotton-
are all in view.
shows what irriga-
tere can be no finer
re here found. In
1 is covered with a
d ditches, conduct-
o all vegetation be-
lly causing the land
. honey."
o come to
station, where all
Mountain, Bingham
e vicinity, "change

on Railroad.

ame a branch of the
stern, and is operated
hrough line.]

'oot narrow gauge,
h of Bingham Can-
d had cars running
Let us take a roll
, our course is due
cultivated section of

JORDAN STATION—is one mile distant, where there is a postoffice, and a small collection of cottages. Near by, a track leads off to the Old Telegraph Smelter, the dressing works of which are a short distance below the road, on the opposite side of the Jordan River, which we cross soon after leaving the station.

Just as we raise on to the west bank of the Jordan, we come to the residence of Bishop Gardner, who is the "better half" of *eleven wives*. The Bishop appears to stand it pretty well, although they *do* say that he is occasionally found singing, "On Jordan's stormy banks I stand," with a tear accompaniment.

WEST JORDAN—is on the west bank of the river and contains a few hundred people. From this station, the grade increases, and soon we reach a high tableland, too high for irrigation by ditches, without great expense. Bunchgrass, white sage, sheep, some cattle, and Jack-rabbits abound—the latter are very numerous.

The road, about five miles from the river, enters the long, broad ravine that leads to Bingham, up which we roll—the ravine gradually becoming narrower as we ascend. Occasionally we pass a little farm-house, and a few acres of farm and garden land. Nearing the mountains, the ravine narrows, to a few hundred feet, and finally to only sufficient room for the railroad and a little creek, between the bluffs on each side.

These bluffs are from 250 to 1,000 feet in height, covered with small stone, sage, and a few small pine trees.

As we ascend, the bluffs are more precipitous, higher and pierced in numerous places with "prospect holes." In places the grade is 120 feet, and *then*, MORE, and finally it becomes too much for our iron horse, and we stop at the end of the steam road, one mile below

BINGHAM CITY—population about 2,000. Just below the city is located on the left, the Winnemucca mill and mine. We *know* it is there, as we "prospected" it once—about 30 feet.

Bingham City is built along the canyon for two miles, and contains a number of mills and works connected with mining. From the station, a tramway up which small ore cars are hauled with mules, extends up the canyon for three miles, with a branch running back from about half the distance up to a mine on the top of the mountain, about one and a half miles further, making of tramway 4½ miles; whole length of road 20½ miles.

The tramway is built on the south side of the canyon, away up on the side of the mountain. From the cars can be had a fine view of the canyon, Bingham, the mines and mills in the neighboring ravines and on the opposite mountain side, and the miners at the bottom of the canyon, working over the old "placer diggings."

At the end of this tramway is located the old Telegraph mine, one of the richest in the Territory, from which over 200 tons of ore a day is shipped, down over the tram and railroad to the smelters in the valley. The cars are hauled up by mules, and lowered down to the "iron horse" below Bingham by the car brakes.

The mines are numerous in and around Bingham, but we have not the space for a description of them, but will return to the junction on the Utah Southern, and one mile further arrive at

SANDY—This station is 13 miles south of Salt Lake City, and one of considerable importance.

At Sandy is to be seen immense quantities of ore—ore in sacks, ore loose by the car load, ore in warehouses 500 feet long, with a train unloading on one side and another loading on the other; in fact, *this* is the greatest shipping, smelting and sampling point in all Utah.

At Sandy are three sampling works, and two smelting works, and a lively town of 700 inhabitants, the greater portion employed in the handling and manipulation of ores. Here we find another railroad branching off; this time it is the

Wasatch & Jordan Valley.

[In 1882 this road passed to the control of the Denver & Rio Grande Western, and is now operated in connection with their through line.]

This road is a three-foot narrow-gauge, 16 miles in length, running to Alta, at the head of Little Cottonwood Canyon. It is operated by narrow gauge steam engines for 8.5 miles, and the other 7.5 by *broad gauge* mules. The road was commenced in 1870, finished to Wasatch in 1872, and to Alta in 1876. Let us take a trip over it and note a few of the sights.

From Sandy the train runs north a short distance, and then turns to the east, directly for the Wasatch Mountains, leaving the old Flagstaff smelter on the left-hand side of the track, just above the station. The grade is heavy, the soil is stony, and covered more or less with sage-brush, and traversed by irrigating ditches conveying the water to a more productive and less stony soil below.

Nearing the mountains, about six miles from Sandy, we come to a deep gorge on the left, through which Little Cottonwood Creek has worn its way to the valley. From this point we bear away to the southward around a low butte, then turn again to the east and northward and run along on an elevated plateau where a most beautiful view can be had. On the west, the Jordan Valley, in all its magnificent shades of green and gold, is at our feet, with the brown old mountains bordering the horizon in the distance. To the north, fifteen miles away, over as beautiful a succession of little streams, well-cultivated fields, white cottages, orchards and gardens, as are to be found within the same number of miles in this country—sleeps "Zion" in full view, embowered in green, with the dome of the monster Tabernacle glistening like some half-obscured "silver moon," sinking at the mountain base; while *far* beyond, and more to the westward, lays the Great Salt Lake—a mysterious problem. Away to the south, is Utah Lake, looking like one large sheet of burnished silver, surrounded by a net-work of green and gold, while to the east looms up towering granite walls, cleft from summit to base, forming a narrow gorge only sufficiently wide to allow our little road to be built beside a little rippling creek of crystal water.

Rolling along, our train rounds the head of a ravine, through a deep cut, passes the old Davenport Smelting Works on the left, enters the mouth of the canyon between great walls of granite, crosses and recrosses the little creek, and soon stops at

WASATCH—the end of the steam road, 8.5 miles from Sandy Junction. This is a small station with postoffice, store, and a few dwellings containing a population of about 100, more than half of whom are engaged in the stone quarries on the north side of the station.

At Wasatch all the granite is got out and shaped for the Temple in Salt Lake City. The stone is the best yet discovered in the Territory, being of close, fine grain, of light gray color, and of beautiful birds-eye appearance. The granite on the south side appears much darker than that on the north side of the canyon.

From almost every nook and crevice of these mountain cliffs—from the station away up the canyon—grow small pines, cedars, ferns, and mosses, which, in connection with the gray walls, snow-capped mountains, glistening waterfalls, pure air and golden sun, presents a picture of rare beauty.

Just above, on the left of the station, away up on a projecting cliff, 1,000 feet above the road, stands a granite column which measures 66¾ feet in height, from the pedestal-like cliff on which it stands. On each side of this column, and receding from its base, is a little grotto-park, filled with nature's evergreens, and surrounded on three sides and on the top with rocks of every size and shape.

Finding that this granite column has had no name, we name it "Humphry's Peak," in honor of the very gentlemanly late superintendent of the road.

At Wasatch we "change cars," taking those of about the size of an ordinary hand-car, fitted up with seats that will comfortably accommodate about nine persons, besides the knight of the whip—who chirrups the "broad gauge mules."

About a half-mile above the station we enter the snow-sheds, which will continue for *seven* miles, to the end of the track at

ALTA—a small mining town, at the head of little Cottonwood Canyon. The end of the track is on the side of the mountain about 200 feet *above* the town of Alta, and about 500 feet *below* the mouth of the celebrated Emma Mine, which is a little further to the east, and opposite the Flagstaff Mine, which is about the same height above the road.

The town of Alta is at the bottom of the canyon 200 feet lower than the end of the railroad surrounded with mountain peaks, which are covered with snow eight months of the year, and at all times surrounded with an eternal mantle of evergreen. It contains about 500 population, all of whom are engaged in mining and kindred pursuits. There are several stores, express, telegraph, and postoffice, besides several small hotels, chief of which is the Adolph.

To the north, over the mountain two miles is the Big Cottonwood Canyon; to the south, three miles, is the Miller Mine, and American Fork Canyon; Forest City is four miles. Three miles east by trail is Crystal Lake, a beautiful sheet of water —the angler's paradise.

The principal mines near Alta are, the Emma, Flagstaff, Grizzly, Nabob, Kate Hays, Consolidated Alta, Laramie, Prince of Wales, and 1,800 others, located within five miles. The business of the railroad is the transportation of ores and supplies to and from the mines. Hundreds of cars are loaded *daily* with ore that is taken to the valley to be smelted or are sent to San Francisco, the East, or to Swansea, Wales. For novel methods of hauling ore to the depot, see ANNEX No. 24.

The sheds over the railroad are seven miles in length, and are made in various styles of architecture, more for *business* than beauty, the style being adopted according to circumstances. They are, however, in all places constructed of heavy material, rocks, round or sawed timber, and built in the most substantial manner. In one place they are in the shape of a letter A, sharp peaked; in other places, nearly upright on each side, one side higher than the other, with a sloping roof. Again the lower hillside is built with a little slope toward the up hill side, and long heavy timbers from the top of these uprights slope up onto the mountain side, resting on a solid granite foundation leveled to a uniform grade, for that purpose.

Where the latter plan has been adopted, there is danger of snow-slides which are more likely to occur, in fact, have occurred a number of times since the sheds were constructed, and each time, the snow and rocks passed over the shed into the canyon below, without causing one cent's worth of damage to the road or shed.

The length of this road, where it is operated with mules, is seven miles long. As before stated, the grade is 600 feet to the mile; the curves are in places 30 degrees, and not, as once stated in the "*Railway Age*," 30 curves and 600 foot *gauge*. But we suppose that Col. Bridges, when he wrote that, was thinking about those "broad gauge mules."

Returning, the mule power that took us up is no longer in demand; the knight of the whip now mans the brakes, and away we go around the *Age's* 30 curves, to the valley below, "change cars" at Sandy, and are once more headed for the south, on the Utah Southern railroad. A short distance south, we pass the McIntosh Sampling mill, on the west and another on the east.

Sampling is testing such ores as are presented in quantities sufficient to enable the sampling company to give certificates of their value, and then the ore is sold at the certificate rates. One mile further is the Mingo Smelter of the Penn. Lead Co.

The land is more rolling, as we approach DRAPER—This is not a very important station to the tourist, but to the few villagers of Herramon, at the mouth of a little canyon beside the mountains on the left, it is a *big* institution. Draper is four miles from Sandy and seventeen from Salt Lake City.

Leaving Draper, our course is east, and after crossing South Willow Creek, turns more to the south, and finally to the west, having kept around the foot of the mountains, which here make a full half-circle. In the distance around, there are many cuts and some hard work, and we queried, *why* the road was built around, when the work was so heavy and the distance much further than across where there was very little work to be done? In answer, we were told that President Brigham Young laid out the road around the side of the mountain, by "*revelation.*" If that is so, we conclude that the revelation came from the same "deity" that took our Savior up on the Mount, but as it is not "our funeral," we will not criticise.

The lower point of the great curve is called the "Point of the Mountain." At the point where the railroad is built around, the track is about 300 feet above a little round valley to the west, in which is located a hot spring, marked by a brown burned patch of land and rising steam.

As our train curves around this point, a most charming view can be had; one of the *finest* on the road. The valley is here nearer, to the northward the view in unobstructed for 50 miles; to the south, Utah Lake, a gem in rich setting, and the great Lower Basins.

Passing through numerous cuts and around the point, the train curves again to the eastward, and starts again on another grand curve around the rim of the basin, in which is located Utah Lake, in plain view. Nearly opposite the "point of the mountain" is a low divide in the Oquirrh Range on the west, over which the road leads to Camp Floyd.

Continuing along through sand cuts, sage and an occasional farm, 14 miles from Draper we arrive at

LEHI—This town is situated in the midst of a perfect forest of fruit trees, orchards and gardens, with the waters of Dry Canyon Creek running through all the streets, and contains a population of about 1,500, including those living in the immediate vicinity. The good results of irrigating sage-brush land, are here demonstrated by the large crops of wheat, oats, barley and vegetables produced, where, before the land was irrigated, nothing but sage-brush and greasewood were to be seen.

Three miles further is

AMERICAN FORK—a station 34 miles from Salt Lake City—the "banner" town for free schools; the *first* in the Territory, having been established here in 1860. The streets are wide, with the waters of Deer Creek, which comes down the American Fork Canyon, running through them, and the orchards, gardens and farms in the neighborhood making an attractive and beautiful town. The population numbers about 1,600, the greater portion of whom are engaged in agricultural pursuits. The American Fork House, opposite the station, is the principal hotel, and Robert Keppeneck is one of the jolliest of German hosts.

To the southwest of the station, a company is engaged in building a dam across the Jordan River for irrigating purposes. The canal is to be 22 feet wide on the bottom and 30 inches deep, and when completed will extend north 20 miles, winding around the base of the Wasatch, near our road, keeping as far up on the side of the mountain as possible. From the dam, a canal will be taken out for the west side of the Jordan, with a view of taking the water all over the lands as far north as Salt Lake City, and if possible, reclaim the vast tract of sage land between the Jordan River at Salt Lake City, and the Oquirrh Range, at the foot of Salt Lake.

From this station a road branches off to the eastward, up American Fork Canyon, called the

American Fork Railroad.

[*Since the following description was written, the iron track of this road has been removed, and the road abandoned.*]

This was a three-foot, narrow gauge railroad, 15 miles in length; commenced May, 1872, and completed 12 miles during the year. The grade for the whole distance is heavy, in places 312 feet to the mile.

Leaving the station at American Fork, the road turns directly to the east, and follows up Deer Creek, through a general assortment of sage brush, sand and boulders, for six miles to the mouth of the Canyon. On the way up, to the right, a fine view can be had of Mt. Aspinwall, rising from the lower range of the Wasatch to an alti-

THE WINDSOR.
DENVER, COLO.

Centrally located, with a commanding view of the Rocky Mountains. Turkish, Sulphur, Vapor, and Mercurial Bath Rooms in connection with the Hotel.

Parties visiting Colorado, either for Business, Pleasure, or Health, will find at the Windsor accommodations unsurpassed.

THE LARGEST AND MOST ELEGANTLY APPOINTED HOTEL IN THE WEST.

BUSH, TABOR & CO.,
PROPRIETORS.

HERALD PUBLISHING HOUSE.

GIBSON, MILLER & RICHARDSON,

PRINTERS,

Stationers, Lithographers,

WOOD ENGRAVERS AND ELECTROTYPERS.

Herald Building, cor. 15th and Harney Sts.,

OMAHA, NEBRASKA.

HENRY GIBSON.
MILLER & RICHARDSON.

THE LATE BRIGHAM YOUNG'S RESIDENCE.

tude of 11,011 feet above the sea.

From the mouth of the canyon, about two miles north, is the little village of Alpine, containing about 250 agriculturalists.

Entering the canyon, the passage is quite narrow between the towering cliffs, which rise up in sharp peaks 600 feet in height, leaving only about 100 feet between, through which the road is built, and a sparkling little stream comes rippling down; the road, on its way up, crossing and re-crossing the stream many times.

Our train is rapidly climbing, but the canyon walls seem to be much more rapidly rising, and at a distance of one, two and three miles, gain an additional 500 feet, until, in places, they are full 2,500 feet above the road bed. In places these cliffs are pillared and castelated granite, in others, of slate, shale and conglomerate, seamed in places as though built up from the bed of the canyon by successive layers, some as thin as a knife blade, others much thicker; then again, the rocks have the appearance of iron slag, or dark colored lava suddenly cooled, presenting to the eye every conceivable angle and fantastic shape—a continuous, ever-changing panorama.

Imagine, then, this canyon with its grottoes, amphitheatres, and its towering crags, peaks, and needle-pointed rocks, towering *far* above the road, overhanging it in places, with patches of eternal snow in the gloomy gorges near the summit, and clothed at all times in a mantle of green, the pine, spruce and cedar trees growing in all the nooks and gulches and away up on the summit; then countless mosses and ferns clinging to each crevice and seam where a foothold can be secured, together with the millions of flowers of every hue; where the sun's rays are sifted through countless objects on their way to the silvery, sparkling stream below, with its miniature cascades and eddies. We say imagine all these things, and then you will only have a faint outline of the wild and romantic, picturesque and glorious American Fork Canyon.

Proceeding on up, up, around sharp crags, under the very overhanging mountains, we pass "Lion Rock" on the right, and "Telescope Peak" on the left. In the top of the latter is a round aperture, through which the sky beyond can be plainly seen; this hole is called the "Devil's Eye."

About three miles from the mouth of the canyon, on the left, we come to Hanging Rock. (See illustration page 15.) Close above, on the same side, is a very large spring, and almost immediately opposite "Sled-runner Curve;"—an inverted vein of rock in the side of the perpendicular cliff, resembling a sled-runner—possibly this is the Devil's sled-runner; who knows? Along here the rock seams are badly mixed, and run at all angles—horizontal, longitudinal and "through other." Half-a-mile farther we come to "Rainbow Cliff," on the right; opposite, a narrow peak rises sharp, like a knife-blade, 300 ft.; a little farther on to the right, comes in the South Fork, on which are several saw mills. Keeping to the left, and soon after passing the South Fork a look back down this wonderful cañon affords one of the grandest of views; we cannot describe it, but will have it engraved for future volumes. One mile farther, and the train stops at the end of the track, at

DEER CREEK.—Near this station the hills are bare of trees, but covered with shrubs of different kinds, sage and moss predominating; the gulches and ravines bear stunted pine and aspen trees. The chief business of this road was in connection with the mines above, among which are the Smelter's Sultana, Wild Dutchman, Treasure, and Pittsburgh. The Miller Smelting Works are four miles farther up and the mines seven.

Opposite the station, in a cosy little nook, is located the Mountain Glen House, where the tourist will find his wants anticipated, and plans can be matured for a ramble over the mountain peaks; and there *are* a number of little tours that can be made from this point each day that will well pay for a week's time devoted to this locality.

When returning to the valley, then it is that the view is most grand, and the ride one beyond the powers of man with his best goose-quill to describe. Make the tour of the American Fork, and our word for it, it will live in pleasant memory while the sun of life descends upon a ripe old age.

At American Fork station we again enter the cars on the Utah Southern, and start once more for the south. Rolling along three miles brings our train to

DONNER LAKE BOATING PARTY.

FOREST VIEW—THE FOOT HILLS OF CALIFORNIA.

PLEASANT GROVE — properly named. In early days it was known as "Battle Creek"—so-called from a fight the early settlers once had here with the Utes. It is a thriving place of 1,000 inhabitants, and like all other Mormon towns, is surrounded with orchards and gardens of fruit, with water flowing through every street. Herds of cattle are now to be seen grazing on the surrounding hills.

Eleven miles around, on the rim of the basin, across some sage and some well-cultivated land, our train stops at

PROVO—This is a regularly incorporated city, with all the requisite municipal officers; is also the county seat of Utah county, which was first settled in 1849.

Provo is 48 miles south of Salt Lake City, at the mouth of Provo Canyon, and on the east bank of Utah Lake, and contains an increasing population of 4,000. This place has several fine hotels, chief of which is the Excelsior House.

The court-house and public buildings of the city are very good, and all kinds of business is represented here. The principal manufactories are the Provo Woolen Mills, three flour and three saw mills.

Provo River, which is formed by numerous small streams, to the eastward, affords the best water power of any stream in Utah.

The woolen mill is a noted feature of the city; the buildings number four, are built of stone, four stories high, and cost, complete, ready for business, $210,000. There are in the mill four "mules" with 3,240 spindles, machinery for carding, dyeing and preparing 2,000 pounds of wool per day, and 215 looms, which turn out superior fabrics, in amount exceeding $200,000 per annum.

The Mormons have a very capacious tabernacle, and the Methodists a fine church, and schools are ample. The Brigham Young Academy is located here.

which was amply endowed by president Brigham Young some years before his death. A regular stage leaves for Provo Valley, 20 miles to eastward on the arrival of trains.

UTAH LAKE—is a body of fresh water, 30 miles long and 6 miles wide; is fed by Provo river, American Fork, Spanish Fork, Hobble, Salt and Peteetweet creeks, having its outlet through Jordan river, which runs north and empties into Great Salt Lake. Utah Lake abounds in trout, mullet and chubs.

Passing along through a well cultivated section of country, for five miles, we arrive at

SPRINGVILLE—This place was named from a warm spring which flows from Hobble Cañon, above the town. Pop. 1,500. The water from this spring is utilized to run a flouring mill, whereby the mill is enabled to run all seasons of the year. So much for a hot spring. In Pleasant Valley, 50 miles east, are located vast beds of coal, said to be of the best coking quality, large quantities of which are used at the various smelting works in Utah, and for domestic purposes in Salt Lake City and adjoining towns. In 1878 the Utah & Pleasant Valley R'y Co. was organized for the purpose of handling this coal, and the road was soon built to the mines. In 1881 this road was bought by the Denver & Rio Grande Western Ry., who are extending the road to the Valleys of Kanab, the Gunnison and ultimately connects with the Colorado system of narrow gauge roads, forming another through line from Salt Lake Valley eastward. Grading is being done northward from Springville, parallel with the Utah Southern, to Salt Lake City, and everything that money and muscle can do is being done to complete the road in 1883.

Hobble Creek Cañon, just east of Springville, was so named by the first Mormons in 1847, who found there a set of old Spanish hobbles.

Rolling along for five miles further through a well-cultivated land, we arrive at

SPANISH FORK—a village of 1,800 population, most of whom are engaged in agricultural and pastoral pursuits. The town is to the left of the road on the banks of the Spanish Fork River.

Butter and cheese are quite a specialty with many of the citizens; on the tablelands vineyards are numerous, and wine is made to some extent; wheat is also a good crop. Duck shooting is said to be exceedingly fine, and trout are found in great numbers in all the mountain streams, as well as in the lake.

Continuing on through rich farm land, eight miles brings our train to

PAYSON—This is an incorporated city of about 2,200 population, situated to the left of the road, and near the southern end of Utah Lake. The people appear to be well-to-do, and do not trouble themselves much about the "war in Europe," or the "Chinese question." Large quantities of ore are hauled here for shipment to the smelting furnaces at Sandy and other places.

Three miles further, and two and a half miles eastward, is a beautiful little place called Spring Lake Villa, nestling cosily in beside the mountain and a little lakelet of similar name. This villa is noted for its abundant and superior fruit of various kinds, where is located a large canning establishment.

Five miles further, through less valuable lands than those to the northward, and we arrive at

SANTAQUIN—which is a very important point. It contains a population of about 2,000, and is a point from which all passengers, mails, express and freight, leave for the Tintic mining regions, to the westward. Here, too, will be found stage lines for the different mining towns and camps. To Goshen the distance is six miles; Diamond City, 13 miles; Silver City, 16 miles; and Eureka, 21 miles.

The Tintic district furnished at this station, in 1879, 20,000,000 pounds of hematite iron ore for shipment to the different smelting furnaces to the northward, for a flux in the manipulation of ores.

YORK—is 75 miles from Salt Lake City, and is a station of very little importance, four miles from Santaqin. A few miles further, to the right are the Hot Springs in which were found the bodies of the Aiken party who were murdered in 1857.

To the south, rises Mount Nebo, with his cap of snow, to an altitude of 12,000 feet.

MONO—To the left, is a small hamlet.

The Juab Valley commences at York, averages about three miles in

YO-SEMITE FALLS, 2,634 FEET FALL, YO-SEMITE VALLEY.

width, and is 36 miles in length, generally good land and well cultivated.

NEPHI—Is a city of 2,000 population, from which stages run regularly to San Pete, 80 miles, and Kanab, 195 miles, passing through many small villages and mining camps.

JUAB—is an eating station. 30 miles from York. Here a large amount of freight is shipped on wagons for the villages to the eastward, and stages leave regularly for Sipio, 22 miles; Filmore, 47 miles, and Corn Creek, 60 miles

Juab is the end of the Utah Southern, and the commencement of its Extension.

Soon after leaving Juab—named for the county of Juab—we cross Chicken Creek, and in about three miles come to the Sevier River, where the hills come close together, forming a canyon

The Sevier is a crooked, muddy, sluggish stream, down which the road is built through a worthless country crossing it often for 52 miles, to Deseret, a station situated a few miles east of the Sevier River, which is here dammed for irrigating purposes.

Leaving Deseret—where breakfast is served going north—we pass over a broad, level stretch of desert country, traversed by great numbers of irrigating ditches, from the dam aforesaid, but the waters are so strong and the soil so impregnated with alkali, that the aforesaid wilderness fails to blossom, except with sage and greasewood. The road crosses the edge of Sevier Lake, on a raised track, the salt deposits of which are *very strong*. The scenery along this road, below the Sevier canyon, is not very striking—unless one is anxious to be struck.

MILFORD—is reached, 69 miles from Deseret and 226 miles south from Salt Lake City. It is on Beaver River. population about 200, the end of the Utah Southern Railroad, from which large quantities of freight are shipped for Southern Utah. Stages run to Minersville, 16 miles; Marysville. 16 miles; Beaver, 37 miles; Silver Reef, 96 miles; St. George, 114 miles, and Pioche, 120 miles. At Milford is located one quartz mill and one smelting furnace. To the westward, 16 miles, by a branch railroad is

FRISCO—a mining town of about 1000 population, near the celebrated Horn Silver Mine. The "Frisco mines" are said to be exceedingly rich in silver and lead. The ores are galena, yielding from $15 to $1,500 per ton of silver, and from 20 to 40 per centum of lead. Heavy investments of eastern capital have been made in these mines, and vigorous efforts are making to soon work them by the latest and most improved methods. The Horn Silver, Carbonate and Mountain Queen are the best known and developed mines.

Returning to Zion we will take a run over the

Utah Western Railroad.

This road is a late acquisition of the Union Pacific Railway Co. It is a three foot narrow gauge, commenced in 1874, and was completed 12 miles during the year. In 1875 about 13 miles more were finished, and in 1877 it was extended to within two miles of Stockton, 37.5 miles from Salt Lake City.

The depot in Salt Lake City is located one-half mile west of the Utah Central, on the same street.

The route is due west, crossing the Jordan River the first mile, about half a mile south of the wagon road bridge, thence 12 miles to the Hot Springs, at the northeast point of the Oquirrh Mountains. This 12 miles is built across the level bottom land, the major portion of which is covered with sage-brush and greasewood, with an occasional patch of "bunch" and alkali grasses. The soil in most parts is a black vegetable mold with a mixture of fine sand. Some sand beds are noticeable, and near the Hot Springs a deposit of alkali with yellow clay.

The length of this land belt is about 50 miles, of which the first 15 will average ten miles wide, the balance averaging five miles wide, and extending south to Utah Lake, and when properly irrigated—as we have heretofore noted, a plan now being carried out for so doing—it will be as productive as the same number of acres in the valley of the famous River Nile, in Egypt. Herds of cattle and sheep now roam over these bottom lands, as well as jack rabbits by the legion.

Near the hot springs, on the left, noticeable from the amount of steam rising and the brown burned appearance of the ground, are some comfortable little farm-houses, and a few well appearing farms. The hot springs spoken of are fresh and produce a large creek of water. Near, are several store houses, and a place called Millstone, from the fact that at this point the first millstones were quarried in the Territory. There are no accommodations, at present, near, for tourists to stop over. Proceeding along, around the side of the mountain, our train gradually approaches the lake, and five miles from Millstone we are at

BLACK ROCK—This station is just after passing a high rocky cliff on the right, and derives its name from a black-looking rock sitting out in the lake 300 feet distant, and 50 feet high.

Near the station is Lion's Head Rock, the highest cliff is known as "Observation Point," so named from the unobstructed view which can be had from its summit. Antelope, or Church Island, to the northeast, is 14 miles distant, Kimball's, 22; Goose Creek Mountains, northwest 100; West Mountain, west 15; Oquirrh, close to the south, while the view to the southwest extends to the great rim of the basin, 17 miles distant. On Church Island large herds of cattle range, and some mines of gold, slate, and copper have been discovered. On Carrington Island, opposite Black Rock, a slate mine of good quality has been discovered, which has been traced 4,500 ft. Opposite the station, away up in the side of the mountain, is the

"GIANT'S CAVE—an opening extending several hundred ft. into the mountain side, with a ceiling ranging in height from 10 to 75 feet, from which hang stalactites of great beauty and brilliancy. Remains of some of the ancient tribes of Indians, it is said, are still to be found scattered around the floor of the cave. The presence of these remains is explained by a tradition among the Indians to the effect that "many hundred years ago, two tribes of Indians were at war with each other, and that the weaker party was forced to take refuge in the cave, but were followed by the enemy, who closed the entrance with huge boulders, forming an impenetrable barrier to their escape"—and thus their place of refuge became their grave.

Leaving Black Rock, our train skirts the lake for a distance of one mile and stops at

GARFIELD.—Of all the bathing places in and about Salt Lake, this is the *best*. The veteran Cap. Douri's—who by the way has become quite a "land-lubber" —is located here, having dismantled his steamer, Gen. Garfield, and converted it into a first-class floating hotel. To take a run out from Salt Lake City, secure a state-room on the Garfield, sleep on the bosom of the "Dead Sea," and with the "Captain's gig" explore its mysteries, bathe in its wonderful waters, is one of the luxuries that the traveler visiting Utah should never miss. In fact it is worth a long journey to enjoy. Baron von Humboldt, in speaking of the marvelous grandeur with which this inland sea abounds, said: "Here is the beauty and grandeur of Como and Killarney combined."

LAKE POINT—is two miles from Garfield, is another bathing place, where the traveler will find fair accommodations at the "Short Branch Hotel."

Black Rock, Garfield, and Lake Point, are in summer great resorts for pic-nic parties from Zion, who come out, take a trip over the lake, have a swim and a ramble up the mountains, "make a day of it," and return to the city in the evening.

Game in the mountains and on the plains, such as deer, antelope, bears and smaller game, are to be had for the necessary effort; ducks are abundant six miles to the eastward, and fish,—*nary one.*

The mountains are about 1,000 feet above the road, have rounded peaks, covered with small trees, in places, sage and grass in others, and large timber in the inaccessible gulches and ravines, near the summit.

Leaving the Point, our course is more to the southward, along the side of the lake, by a few well-cultivated farms, irrigated by water from the mountain on the left.

Turning more to the left, and drawing away from the lake, the road follows along a few miles from the base of the mountains, beside which is located the small Mormon village of "E. T. City"—named after E. T. Bensen, one of the early settlers. Four and a half miles from the "Point" comes the

HALF-WAYHOUSE—near, is a flouring, and woolen mill. On the opposite side of the valley, west, is the town of Grantsville, eight miles distant. It lies in one of the richest agricultural sections of the state; population, 2,000. In the background is the West Mountain Range, which rears its peaks full 2,000 ft. above the town, and in which are located some very rich mines of silver. Beyond these mountains is Skull Valley—so named for an Indian fight which once occurred there, after which the ground was left covered with bones. Passing on, to the left, note the waterlines on the side of the mountain.

TOOELE— is six miles from the last, and is the nearest station to the thriving town of Tooele, which is situated to the left about two miles, beside the mountain. The principal business of the citizens is agriculture and fruit raising. It is considered the best fruit and vegetable district in the Territory.

Tooele is the county seat of Tooele county; population about 2,500. Along the base of the mountain the land is irrigated from little springs and creeks in the mountain gorges, the waters of which seldom find their way to the lake below. About 10 miles over the mountain, to the southeast, is located Bingham City. Leaving Tooele, sage small cedars, bunch-grass and herds of stock abound.

The road is on a high plateau, curving with the mountain more to the westward, and some miles below the the lower end of the lake. As we near the lower portion to the great valley, which lays on our right, the land rises, rim-like, and a few hundred yards below the end of the track, rises 500 feet, completely locking in the valley by a mountain range or semi-circle extending in a great arch from Oquirrh Range on the east, to meet the range on the west, one great bend, full five miles in curvature. Here, at **the base** of this rim, terminates the **railroad.**

"THE GRAND" YELLOWSTONE PARK.

tricts to the south and west. Distances from Stockton to Ophir, southeast, 10 miles; to Dry Canyon, southeast, 12 miles; to Salt Lake City, 39.5 miles.

Stockton is in Tooele county, in the northeast corner of Rush Valley, and about one-half mile east of Rush Lake—a sheet of fresh water two miles long and half mile wide. The town contains three smelting furnaces, several stores, hotels, and about 80 dwellings, with a population—by taking in the surroundings—of 600. The Waterman Furnace is close in the eastern edge of the town; the Jacob's Smelter about half a mile west, at the head of the lake, and the Chicago Smelter about one mile southwest, on the eastern bank of the lake. The ores come from the several mining districts in the vicinity.

Rush Valley is one of the class of valleys so often found in the Salt Lake and Nevada Basins—only varying in size. This is 10 miles in length and about three in width—land-locked, surrounded by mountain ranges, with a lake in the center and no visible outlet.

NEW RAILWAYS. — We should judge from present appearances that all Utah will soon be "riding on a rail," as the "boom" for railroad building struck the territory in 1881. From the various documents on file with the Auditor of the Territory, it would seem that every canyon and watercourse would be paralleled with a rail track, and there would hardly be a "sheep ranch" without a railway station. It is said the Union Pacific Railway Company will construct 1000 miles of track at an early day, and that they are now "throwing dirt" in several places. July 21, 1881, the Denver and Rio Grande Western Railway Co. filed on routes aggregating 2,370 miles. This new company is virtually the same as the Denver & Rio Grande of Colorado, and it is designed to connect the two lines at an early day. At this time work is being pushed vigorously, both from the east and west.

On the south side of this rim, which, on the top, is less than one-half a mile in width, is located the city of

STOCKTON—two miles distant from the end of the railroad. To reach Stockton by rail a 1000-foot tunnel must be drove through this rim, exclusive of approaches.

Stockton is now reached by stage, which also extends its route to Dry Canyon, and the Ophir mining dis-

Returning to Salt Lake, "change cars" for Ogden, and again we take a look at the Great Overland trains. But we cannot think of neglecting to take a trip over the

Utah & Northern Branch
Union Pacific Railway.

(IDAHO DIVISION.)
W. B. DODDRIDGE.................*Ogden, Supt.*
R. BLICKENSDERFER, *Pocatello, Idaho, Div. Supt.*

This road is a three-foot narrow gauge, commenced March 29th, 1872, and extended at different times to Franklin, 78 miles, in 1874. In the spring of 1878 work was again commenced and the road completed 181 miles to Blackfoot, on Snake River, ten miles above old Fort Hall, and during the year 1879 to Beaver Canyon, 93 miles, 274 miles from Ogden. Work has continued since, and the trains are now (Jan., 1883,) fully equipped with palace cars and all modern improvements, running to Deer Lodge, 442 miles from Ogden.

The "Oregon Short Line," noted on page 76, when completed will connect with this road at Pocatello, 158 miles north from Ogden. Another branch has been surveyed and will leave this road at Blackfoot and pass through Idaho via the Wood River Mines to Oregon.

Trains leave Ogden opposite the Union Depot to the eastward and skirt the western edge of the city, across rich, broad, and well-cultivated fields, orchards and gardens, with the Wasatch Mountains towering to the right.

From Ogden depot it is five miles to HARRISVILLE, an unimportant station, from which it is four miles to HOT SPRINGS, where will be found a large hotel and extensive bathing accommodations. Here is one of the many hot springs which abound in the Great Salt Lake and Nevada basins. In cold weather it sends up a dense cloud of vapor, which is visible a long distance. It is strongly impregnated with sulphur and other mineral substances, and the odor arising is very strong, and by no means pleasant for some people to inhale. This spring is close on the right of the road, and besides the steam continually arising from it, is marked by the red-burnt soil, much resembling a yard, where hides are tanned.

From the cars an occasional glimps of Salt Lake can be obtained, with its numerous islands, lifting their peaks far far above the briny waters. The views will be very imperfect; but as we near Promontory Point, and after leaving that place, excellent views can be obtained. On the left, only a few hundred yards away, can be seen the track of the Central Pacific—and near, the unimportant station of Bonneville on that road. Near are some fine farming lands, which yield large crops of wheat, barley and corn.

With the rugged mountains on our right and the waters of the lake seen at times on our left, we find objects of interest continually rising around us. Far up the sides of the mountain, stretching along in one unbroken line, save where it is sundered by canyons, gulches, and ravines, is the old water-mark of the ancient lake, showing that at one time this lake was a mighty sea, washing the mountain sides several hundred feet above us. The old water-line is no creation of the imagination, but a broad bench, whereupon the well-worn rocks, the rounded pebbles, and marine shells still attest the fact that once the waters of the lake washed this broad upland. Beneath the highest and largest bench, at various places, may be seen two others, at about equal distances apart, showing that the waters of the lake have had three different altitudes before they reached their present level.

We are gradually rising up on to a high bench and will continue along near the base of the mountains for the next thirty miles. In places the view will be grand. The Great Lake at the southwest with its numerous islands in the distance, the well-cultivated fields in the foreground, together with the orchards and rippling rills from the mountain springs, which we cross every few minutes, make a beautiful picture; then back of all, on the east, rises the Wasatch, peak upon peak, towering to the skies.

From the last station it is 15 miles to

WILLARD—This is a quiet Mormon town of 700 inhabitants, and contains some fine buildings, but the greater portion are built of logs and adobe, yet neat and cosy. Most of the fences are of small willows interwoven through large willow stakes stuck in the ground. The mountains near this town present indications which would as-

ENTERING THE PALISADES OF THE HUMBOLDT.

sure the "prospector" that they were rich in various minerals. Strong evidences also exist of the great volcanic upheaval which once lit up this country with its lurid fires, most effectualy demolishing many philosophical theories, leaving their originators to study nature more and books less.

Near the city, in the first range of hills, is the crater of an extinct volcano, which covers several acres. The masses of lava laying around, its bleak, barren, and desolate appearance would seem to indicate that, comparatively speaking, not many years had elapsed since it was in active operation.

Leaving Willard, our course is more to the left, with broad fields and some fine dwellings; then a strip of sage and alkali; and seven miles north we reach

BRIGHAM—This is the county seat of Box Elder county, situated near the mouth of Box Elder and Wellsville Canyon. Like Willard, it nestles close under the shadow of the Wasatch, and is embowered in fruit trees. Population, 1,800. The buildings are mostly of adobe. A thriving trade and rapidly increasing population attest the importance of the place. The public buildings include a court-house and tabernacle, two hotels, and no saloons.

From Brigham our course is more to the left, following around the great arc of the mountains, as well as the old Montana stage road.

CALL'S FORK — is 7 miles from Brigham, and is a little collection of houses, close in beside the mountain on the right. All around this mountain base are, at intervals, springs—some are cold and some are very hot-water—well-cultivated fields and alkali beds, little lakes, and sage-brush knolls, rich soil and large crops; then occur barren waste and *nary* shrub.

Two and a half miles further is

HONEYVILLE—Ah! here we have it! a dozen stone and adobe houses on a *sage-brush honey*. Bear River and valley is now on the left, as is also the city of

Corinne, about six miles distant to the southwest.

When this road was first built, a track extended to Corinne, which has in later years been taken up and abandoned, the *why!* I will *never tell you.*

DEWEYVILLE is five miles further, around which, are some good farms and a grist mill. Curving around the point of the mountain and heading for the north, up Bear Valley, the grade increases; sage is the rule, pines and cedars appear in the mountain gorges, and up we climb. To the west on the opposite side of Bear River, about five miles above the station, is located a village of Shoshone Indians, about 100 in number. Their tepees—lodges—can be plainly seen. These Indians took up this land in 1874, under the pre-emption laws of the United States, and abandoned their tribal relations. They own some large herds of cattle and bands of horses, and are very quiet and peaceably disposed.

Passing on up a heavy grade through deep cuts for six miles and we are at

COLLINSTON—formerly Hamptons, a side track station of no importance to **the tourist.**

Just before reaching this station, the road cuts through a spur of the mountain that juts out to the westward into the valley, leaving a high, isolated peak. Let us climb this peak and take a look. To the north, six miles the Bear River canyons through a low spur of the Wasatch which reaches away to the northwest. To the west of this spur lies the Malad Valley, and Malad River; the latter and the Bear come close together into the valley, immediately to the west of where we stand; then flow close together down the valley to the south parallel for ten miles before they unite, in some places not more than 20 feet apart. To the west of this valley rise the long range of the Malad Mountains, which, commencing near Corinne, runs nearly north to opposite this point, and then bears away to the northwest.

Only a small portion of the lands in the Bear or Malad valleys are cultivated; cattle and sheep are plenty. Leaving Collinston, our road is up a 100 foot grade, curving around to raise the spur of the Wasatch above alluded to, through which Bear River canyons a few miles to the northward. Finally the

SUMMIT—is reached and passed four miles from Collinston and we curve to the east and then to the south, around the narrow spur alluded to, which separated **Bear** Valley from Cache Valley.

From the Summit we have been rapidly descending into Cache Valley, which is on our left, and is one of the most productive in Utah Territory. The valley heads in the Wasatch Mountains, northeast of Ogden, and is 40 miles long with an average width of six miles, to where it intersects Marsh Valley on the north, five miles distant. The Logan River runs through the lower portion of this valley, and is composed of the Little Bear, Blacksmith Fork, and Logan creeks, making a stream of ample volume to irrigate all the land in the valley, much of which is yet open for pre-emption.

In an ordinary season the shipments from this valley average 500 car-loads of wheat, 200 car-loads of oats, and 100 car-loads of potatoes, most of which go to California. Wheat often yields 50 bushels to the acre.

MENDON—is the first station from the Summit, 5.5 miles distant, on the west side of the valley, and contains about 700 population.

From Mendon our course is due east to Logan, across the valley, which runs north and south, but before we start, let us note the towns situated on the arc, around the upper portion of the valley. The first is Wellsville, six miles south, on the west side, population 1,300. Paradise comes next, with a population of 500. Continuing around to the east and then north, is Hyrum, population 1,400. Next comes Millville, population 600; and then Providence, population 550. This latter village is the first south of Logan.

Looking north from Mendon, northeast of the point where we crossed the ridge at Summit, and eight miles from Mendon, is located the village of Newton, population 300; three miles further is Clarkston, population 500; next six miles is Weston, population 500; next is Clifton, ten miles, population 300; then Oxford, seven miles, population 250. These are all Mormon villages, are all surrounded with well-cultivated lands, orchards, vines and gardens, with the sparkling waters from the adjoining mountains rippling through all the streets, fields, gardens and lands, and with crops and fruits of all kinds abundant; and, taking them all in all, they are prosperous and thriving communities, in which each one of the community seems to strives to advance the good of all. They are an industrious, hard-working, self-reliant and apparently contented people, always living within their means. The population of the valley is upwards of 15,000.

Leaving Mendon to cross the valley, we pass through a farm of 9,643 acres, upon which were 30 miles of fencing, houses and out buildings, which were deeded by President Brigham Young, just before his death, to trustees, in trust to endow a college at Logan City, to be called "Brigham Young College." The trustees are leasing the lands—of which there are no better in the Territory—for the purpose of creating a fund to carry out the bequest. These lands are the most valuable in the Territory. Crossing Logan River, our train stops at the city of

LOGAN—This city is the county seat of Cache county, situated on the east side of Cache Valley, just below the mouth of Logan Canyon. It is the largest place in the valley—containing a population of about 3,000, most of whom are engaged in agricultural and pastoral pursuits. Water runs through the streets from the mountains and orchards; gardens, fruits and flowers abound.

The city contains two flouring mills, a woolen mill, the railroad machine and repair shops, one hotel—the Logan House—and a branch of the Z. C. M. I., besides various small mechanical establishments. The new Tabernacle is of cut stone, and seats 2,500 people.

On the east side of the city, a round plateau rises 300 feet above the streets, projecting out from the average front of the mountain range 2,000 feet, into the valley. This plateau is about 500 feet in width, and shaped like the end of a monster canal boat, bottom upwards. Standing on the point, and looking west, the city is close at our feet, the broad valley beyond, and in the distance the spur of the Wasatch, over which we came from Bear Valley. To the right and left, the valley is spread out in all its beauty, and no less than 14 towns and villages are in sight, surrounded with mountain ranges, which rise, range upon range, and peak overtopping peak, the highest of which are robed in a perpetual mantle of snow. The view is one of the most beautiful that one could conceive.

Upon this plateau, the Mormon people who reside in Cache and the four adjoining counties, have elected to build a magnificent temple, in which to conduct the rites and ordinar church. The ma slate stone, 171 f and 86 feet high, feet high from bas cost, when comple $450,000 to $500,0(

Around the out double row of tree the water from th ducted in little dit(the entire grounds.

Leaving Logan, along the base of —having made a from the summit.

From Logan it i HYDE PARK S Hyde Park is to th mountain, one mi! a population of ab(Two mile furthe Summit Creek, wh wood trees, comes

SMITHFIELD—T! population, a short the road. Six m

RICHMOND — an people, on the righ

These towns are (roundings, and the description of all running through 1 gardens, and are (streets, by the sid(trees and good' wal

LEWISTOWN—is 400 people, situatec the valley, four mi.

Nine miles fu the station for

FRANKLIN—Thi of the line, betwee consequently, is in lation about 400. Valley, Oneida co Creek about one the northeast, at th Mountains.

The county sea Malad City 40 m lages of Weston, (ford, and Clifton— are to the westwar(to twenty miles.] east, over the mou Paris, Montpelier gating a populatio:

From Franklin

cross Chubb and Worm Creeks, along which are some fine farming lands; pass through a number of deep cuts and find Bear River on our left, far below our road, with narrow bottom lands on each side. The road turns north and runs up on the east bank of the river a few miles where it crosses to the west and stops at a small side track called

BATTLE CREEK—twelve miles from Franklin. Soon after leaving the station the road turns west up CONNOR'S CANYON, where, in the winter of 1863-4, Gen. Connor had his celebrated fight with the Shoshone Indians. At the time of this fight there was two foot of snow on the ground, and the weather very cold. The Indians—some hundreds—were hid in the Canyon among the willows along the Creek, and in the cedars to the right along the bluffs. By a vigorous charge of the troops, the Indians were completely overcome, and with few exceptions, none were left alive to tell the tale. The bones of the dead are still to be seen near the station.

In ascending the Canyon the grade is heavy, deep cuts are numerous, sage brush abounds, and the country is very broken, only adapted to stock raising. About

OXFORD—in Marsh Valley, eleven miles from Battle Creek, are a few well cultivated farms, and herds of cattle and sheep range around the bluffs.

SWAN LAKE—is the next station, just below a small sheet of water of that name, in which sport, at certain seasons of the year, numbers of swans. The Malad Mountains border the valley on the west, beyond which is Malad Valley and river of same name, also Malad City, 20 miles distant. Pass on down the valley, north 21 miles, we come to

ARIMO—a small town of perhaps fifty people. The famous Soda Springs of Idaho, are 80 miles east of this station, where are ample hotel accommodations for tourists, but the facilities for reaching them are limited, as there is no regular stage line; livery team must be procured at Arimo

Leaving Arimo a low cut in the mountains about five miles distant to the northeast, marks the passage of Port Neuf River through Port Neuf Gap. The old stage road is on our right, along the base of the mountain. After crossing a number of small creeks, and 9 miles from Arimo come to BELLE MARSH, on Port Neuf River, down which we go 36 miles.

Along this river are many peculiar rock formations. In places the rocks rise like a solid wall, from 20 to 100 feet from the ground in a line of uniform height for miles in extent, resembling huge fortifications. In several places along the road there are two and sometimes three of these walls running parallel with each other. Proceeding down the river we come to "Robbers' Roost" on the right, about four miles before reaching the next station. It is the point where the Montana stage robbery was committed in 1864.

EAGLE GATE.

PORT NEUF—once known as BLACK ROCK, so named for the ridge of slate rock to be seen just east of the station. It is 12 miles to Pocatello and ten to

ROSS FORK.—This is a small station on the river of the same name. The lands are mostly covered with sage brush, very rich, and with irrigation, water for which is abundant—could be made very productive. Stock raising is about the only occupation the few settlers are engaged in.

Game of all kinds abounds in the valleys and in the mountains, while along the water courses, wild geese and ducks are legion. The streams, little and big, are full of fish, notably the trout, which are very abundant and bite with a snap that makes an old sportsman feel happy.

Fruit, apples, peaches, pears, cherries, plums, currants, and, in fact, all kinds of fruit are raised by the Mormons, in this and adjacent valleys in great abundance. Although we are now in Idaho Territory, we shall speak of the chief towns and the routes to them in another place.

The direction of our road from this point is north; about three miles brings us to

EAGLE ROCK BRIDGE, SNAKE RIVER, IDAHO.—THOS. MORAN.
(Known by old-timers as Taylor's Bridge).

BLACKFOOT—named for the Blackfoot Indians. It is situated on a broad, sage-covered plain, one mile north of Blackfoot River, and two miles southeast from Snake River, which is here marked by a dense growth of trees and willows. The place has about 200 population and some good stores and other buildings. Trains stop thirty minutes for meals—breakfast and supper. Stages leave Blackfoot for Challis daily—distance 70 miles northwest, for old Ft. Hall, 10 miles west; new Ft. Hall, 8 miles east.

RIVERSIDE—is a side-track station on the bank of the Snake, 12 miles north from Blackfoot and 13 miles south of

EAGLE ROCK—known by old-time pilgrims as Taylor's Bridge, at the crossing of Snake River. See illustration opposite: both railroad and wagon bridge are shown, the old and the new.

At Eagle Rock is located a railroad round-house and repair shops, several stores, hotels, and a few comfortable private dwellings of stockmen who make this place their headquarters.

Crossing the river, just below the old bridge, 18 miles, brings us to

MARKET LANE station—unimportant except as a shipping point for stock—cattle and sheep.

The whole country, now, has a volcanic appearance—valueless for agricultural purposes—but, in and along the base of the mountains, on each side from five to ten miles distant, the grasses are very good, and all kinds of stock do well. The "Three Tetons" are to be seen to the eastward. They overlook the Yellowstone National Park.

LAVA SIDING—a small station, comes next in 10 miles, from which it is 11 miles to

CAMAS.—Freight in large quantities is shipped on wagons from Camas to Challis—60 miles west—and to the Salmon River mines to the northwest—130 miles. Stages also leave daily for Salmon City. Camas is the nearest point on the railroad to the Yellowstone National Park. A wagon road has been completed and stages put on the route. The distances are estimated by Col. Norris, Superintendent of the National Park, to be:—Camas to Henry's Lake, 60 miles; Henry's Lake to Junction, 25 miles; Junction to Mammoth Hot Springs, 45 miles—making 130 miles—which includes quite a tour of the Park, en route. (See ANNEX No. 26.)

Rolling on through sage brush and barren wastes of volcanic deposites, we pass DRY CREEK in 12 miles, and 17 miles more reach

BEAVER CAÑON station—274 miles north of Ogden. Passenger trains from the north and south meet here for dinner.

The road for 12 miles up Beaver Cañon to

MONIDA—is built through some beautiful scenery—to the summit of the Rocky Mountain Range, altitude 6,869 feet—and is the first railroad to cross the "Rockies" from the westward

The station of Monida is named for the two territories on the line between which it is located—Mon-ida, Montana, Idaho. The "Continental Divide," marked by a sign-board on the west side of the track, is just south of the station buildings.

From Monida the descent is gradual down a little valley, a kind of natural road-way, with magnificent snow-capped mountain scenery in the distance, and on all sides, herds of cattle, sheep and varieties of game.

WILLIAMS—a small station, is 11 miles, and six more to

SPRING HILL—A small unimportant station, situated in the southern portion of Red Rock Valley. This valley is nearly fifty miles in length, followed by our railroad the entire distance, and also by the river of the same name. The valley is dotted at intervals with comfortable farm houses, many herds of cattle and sheep, varieties of game, and some well fenced and cultivated lands. Bordering the valley on the east are high rolling, grass covered bluffs, with some timber in the higher ravines, while on the west, extend as far as the eye can see the Continental Divide, rising from the valley, the lower portions timber-clad, peak upon peak, to the region of perpetual snow, where their white heads stand forth as veterans of their kind, indicating age at least, if not respectability.

RED ROCK STATION—Is reached after passing several small side-tracks twenty-three miles north from Spring Hill. About midway between these two stations is the somewhat noted RED ROCK, from which the valley, river and station derives its name. This rock is a bold cliff, probably five hundred feet in height and half a mile long, projecting out into the valley from the eastward—of a bright red color, and can be seen for a distance, up and down the valley, for over twenty miles each way. The old wagon road follows this valley for the entire length, and this Rock was a well-known landmark for the "Pilgrims" in early days.

GRAYLING—Is a small station eleven miles from Red Rock Station, near the mouth of Beaver Head River, which comes in from the westward. Rolling down Ryan's Cañon we come to a sign, "Soda Springs," on the right, near a small house. By looking up we discover a large stream of water pouring over the cañon walls, which is here 200 feet in height. As our train stopped at a tank near by for water we commissioned our Pullman porter, a very accomodating boy by the way, although his name was Vinegar, to fill a flask. It was about blood-heat, but not very strong with mineral.

About eight miles from Greyling we come to Beaver Head Rock, at the gateway or mouth of the Cañon, which here opens out into the Beaver Head Valley. This valley is nearly round, about twenty miles in diameter, in the centre of which, eight miles from Beaver Head Rock, is situated the new town of

DILLON—Named for the President of the Union Pacific Railway. This is a busy place. Passenger trains going north stop for supper, and those for the south breakfast. Present population, about 500, but increasing rapidly. Large amounts of freight are shipped from this Station on wagons for the cities, towns, and mining camps to the eastward. The "Corinne" and "Valley" are the two principal hotels. Stages leave here daily on arrival of trains for the following places: Salisbury, 35 miles, fare, $8; Virginia City, 60 miles, fare, $12; Helena, 120 miles, fare $24; Bozeman, 140 miles, fare, $24.

The valley of the Beaver Head is nearly round, and not far from twenty miles in diameter, about one-fourth of which is under cultivation, producing good crops of small grains and vegetables; the balance is occupied by stock raisers, some of whom have large herds of cattle and sheep. The mountains on the west are high, many of the most elevated peaks covered with snow. This range is a continuation of the Continental Divide—heretofore noticed.

Ten miles from Dillon the Railroad bears away more to the northwest. We pass several small stations while rolling along down a beautiful little valley for 31 miles to

MELROSE—This is a small place of several hundred population, situated on the Big Hole or Windom River, which comes down from the west. The town was named for Miss Melrose, daughter of Mrs. Blow, who keeps at this place one of the best hotels in Montana, and, by the way, the Madam has an extraordinary history in connection with the hostile Indians who infested this country many years ago, one worthy to rank with the most heroic deeds of bravery recorded of mothers in the annals of frontier life in this country, but we have not the space to record it.

Butte.—The first of the year, 1882, this was the "end of the track," but it is very hard for a guide-book to tell just where the terminus of any western road will be next month, next week, or even to-morrow. Yes, and it is difficult for one to keep up with the older portions of our trans-Missouri country as everybody appears to be running a foot race to settle upon and gather up the numberless good things that are laying about all over this fair land.

The new town of Butte is surrounded by rich mineral prospects, with a bright out-look for the future. Stages leave here daily for Boulder, 37 miles, fare $4.50; Jefferson, 50 miles, fare $6.00; Helena, 72 miles, fare $8.00; New Chicago, 76 miles, fare $9.50; Missoula, 132 miles, fare $15.

Montana Territory—For many years was considered solely as a mining country, but there was never a greater mistake. That it does con-

FALLS OF THE YELLOWSTONE.   (See Annex No. 36.)

tain mines, of all kinds, in great numbers, rich, and inexhaustible is well known, but the agricultural and stock raising resources are immense. At one time it possessed excellent "placer" mines and "gold diggings," but they have been to a great extent, worked out. Yet there are still some camps where *good pay* is being taken out, and many of the "old diggings" are being worked over by the "heathen Chinee," and with good results. The mining is now mostly confined to quartz, some of which are of extraordinary richness.

No section of our country at the present time offers greater inducements to the immigrant and capitalist than Montana. Its population by the late census was 39,157, but the completion of the railway to the heart of the territory, the low rates of transportation, quick transit, both passengers and freight, as compared with wagons and stage, will promote and *assure* rapid development. Aside from the mining advantages the valleys of the Missouri, Madison, Gallatin, Yellowstone, and many other rivers, possess the very best of farming and grazing lands in quantities sufficient to support millions of industrious people.

Labor of all kinds is in demand, and the wages paid are double the amount, for the same services, current in the east.

Game, of all kinds is abundant all over the territory, and for scenery, the equal of Montana is yet to be discovered. For articles on the Yellowstone National Park, see Annex Nos. 26, 35, 36, and illustrations on pages 24, 104, 113, 146, and the large double page plate of Yellowstone Falls, No. 8.

Guides and all equipage necessary to a thorough enjoyment of the trip to the Park can be procured at Virginia City, or in Lower Geyser Basin. Fare from Virginia City to Lower Geyser Basin, $20. Parties of ten or more will be carried from Virginia City to Lower Geyser Basin and return for $30 each.

Again returning to Ogden, we take up the Overland Route.

GIANTESS GEYSER IN ERUPTION.

HON. LELAND STANFORD.

Ex-Governor Leland Stanford, President of the Central Pacific Railroad of California, was born in the town of Watervliet, Albany county, N. Y., March 9, 1824. His ancestors were English, who settled in the Valley of the Mohawk about the beginning of the last century. Josiah Stanford, father of Leland, was a farmer and prominent citizen of the county, whose family consisted of seven sons—Leland being the fourth—and one daughter. Until the age of twenty, Leland's time was passed at study and on the farm. He then commenced the study of law, and in 1845 entered the law office of Wheaten, Doolittle & Hudley, in Albany, N. Y. In 1849 he moved West, and commenced the practice of law at Port Washington, Wisconsin. Here, in June, 1850, he was married to Miss Jane Lathrop. In 1852, we find him following many of his friends to the new El Dorado. He landed in California July 12, 1852, proceeded directly to the mines, and settled at Michigan Bluffs, on the American River, Placer county, and in a few years he had not only realized a fortune, but so far won the confidence of the people as to secure the nomination for State Treasurer, in 1859, on the Republican ticket. At this time the Democratic party had never been beaten, and the canvass was made on principle. He was defeated; but in 1861—a split-up in the ranks of the dominant party having taken place—he was nominated for Governor, and elected by a plurality of 23,000 votes. How he performed the trust, is well known. Suffice it to say, he received the thanks of the Legislature and won the approval of all classes. Governor Stanford early moved in the interest of the Pacific Railroad; and on the 22d of February, 1863, while Sacramento was still staggering under the devastating flood, and all was gloomy in the future, with the whole country rent by civil war, he—all hope, all life and energy—

shoveled the first earth, and May 10, 1869, drove the last spike at Promontory, Utah, which completed the first Great Pacific Railroad across the American continent.

Central Pacific Railroad,

Official headquarters, corner Fourth and Townsend Streets, San Francisco, Cal,

LELAND STANFORD	...President...	San Francisco.
C. P. HUNTINGTON	....Vice-Prest......	New York.
CHAS. CROCKER	.......Vice-Prest..	San Francisco.
E. W. HOPKINS	........Treasurer..	"
E. H. MILLER, JR.	.....Secretary ..	"
A. N. TOWNE	..........Gen'l Mg'r.	"
J. A. FILLMORE	.......Gen'l Supt.	"
R. H. PRATT	.........Ass't Supt..	"
T. H. GOODMAN	.......G.P.& T.A.	"
R. A. DONALDSON	.....Assistant ..	"
J. C. STUBBS	..........F. T. Mg'r.	"
RICHARD GREY	.......G. F. A.....	"
S. S. MONTAGUE	......Chief Engr.	"
	....................Land Com'r	"
W. H. PORTER	........Auditor....	"
O. C. WHEELER	......Gen'l B. Agt	"
F. KNOWLAND, Gen'l Eastern Ag't, 287 B'way N. Y.		
M. T. DENNIS, Gen'l Eastern Ag't for New Eng., Boston, Mass		

As most of the people who read this book, we conclude, are familiar with the history of the building of the Pacific R. R., and as we have, for *13 years* past, published a condensed account of it—the trials, struggles and final triumph of the enterprise—it must suffice for this time to give a few facts and figures, and then pass on to our review of the cities, towns and objects of interest along the road and in the country adjacent. The first survey was for the Central, over the Sierra Nevada Mountains, by Theo. D. Judah, in the Summer of 1860, followed in 1861 by a more thorough one, when a passage was discovered and declared feasible.

In 1862 Congress granted the Pacific railroad charter, and the first ground was broken for it by the Central, at Sacramento, Cal., Feb. 22, 1863, two years and eight months before ground was broken for the Union, at Omaha,

THE WAY WE ONCE WENT TO VIRGINIA CITY.

Neb. The following will show the number of miles completed by the Central during each year: In 1863-4-5, 20 miles each year; in 1866, 30 miles; in 1867, 46 miles; in 1868, 364 miles; in 1869, 190½ miles, making 690½ miles from Sacramento to Promontory, where the roads met, May 10, 1869.

The whole length of the Pacific railroad proper, from Omaha to Sacramento, is 1,776¼ miles, of which the Union built 1,085 and the Central 690½ miles. By a subsequent arrangement, the Union relinquished 53 miles to the Central, and in '69 the latter purchased the whole of the Western Pacific, from San Francisco to Sacramento, 137½ miles in length, which gave the Central Pacific 882 miles of road, from Ogden to San Francisco, and made the entire line from Omaha to San Francisco 1,914 miles.

"All aboard," is now the order, and our train glides northward through the western suburbs of Ogden, crossing broad bottom lands, largely under cultivation. The Weber River is on the left, the long high range of the Wasatch Mountains on the right. Within a few miles the Ogden River is crossed, and also many irrigating canals. The track of the Utah & Northern is on the right, and will be for the next 24 miles, near the foot of the mountains; and as the towns and objects of note were described on the trip over that road they will be passed in this place.

Bonneville—is the first that we *pass* on the Central. It is 9.9 miles from Ogden, near Willard, in the midst of good farming land, which yields large crops of wheat, barley and corn.

Brigham—comes next, 7:14 miles further. The town is to the eastward, near the base of the mountains, heretofore described under the head of the Utah & Northern. Passing Brigham, the road inclines to the left, west, and crosses Bear River on a trestle bridge 1,200 feet long, the piles of which were driven in water 18 feet deep; and half a mile further, and 7.14 miles from Brigham, we stop at

Colorado was first visited by white men—Spaniards—in 1540. Explored by Z. M. Pike, who gave his name to Pike's Peak, in 1806; by Col. S. H. Long in 1820, who named Long's Peak; by Gen. Fremont in 1843; by Gov. Wm. Gilpin in 1840, who has traversed the country more or less until the present time.

Corinne—This city is not as prosperous in its mercantile and forwarding business as it was several years ago, owing principally to the fact that the Utah Northern has been extended north too far; and then the taking up of the branch track from the city has entirely cut off the freighting business to Montana and the northern settlements, that formerly went from this place. However, the citizens are by no means blue, but have built a canal from a point 11 miles to the northward, and now conduct the waters of the Malad River down to the city, and not only use it for irrigating thousands of acres of land, but for city and manufacturing purposes, chief of which is a flouring mill which produces about four tons of flour a day. Corinne has three churches, a good school, several hotels, and a weekly newspaper, the *Record*.

Many of the citizens have embarked in the stock-raising business, and are doing well; the range to the northward is very good. Around the town are many thousand acres of land, which only require irrigation and culture to render them productive in the highest degree.

Again *Westward!* The farming lands gradually give way to alkali beds—white, barren, and glittering in the sun. Now the road curves along the bank of the lake, crossing the low flats on a bed raised several feet above the salt deposits. The channel along the road, caused by excavation, is filled with a reddish, cold-looking water. Taste it at the first opportunity, and you will wish that the first opportunity had never offered.

Quarry—is 7.64 miles further west, being a side-track where trains seldom stop, but skirt along the base of the mountains with the lake and broad alkali bottoms on the left. The cars pass over several long and high embankments, and reach the high broken land again at

Blue Creek—which is 11.96 miles from Quarry. During the construction of the road, this was one of the hardest "Camps" along the whole line.

Leaving the station, we cross Blue Creek on a trestle bridge 300 feet long and 80 feet high. Thence by tortuous curves we wind around the heads of several little valleys, crossing them well against the hillside by heavy fills. The track along here has been changed, avoiding several long trestle bridges, and running on a solid embankment.

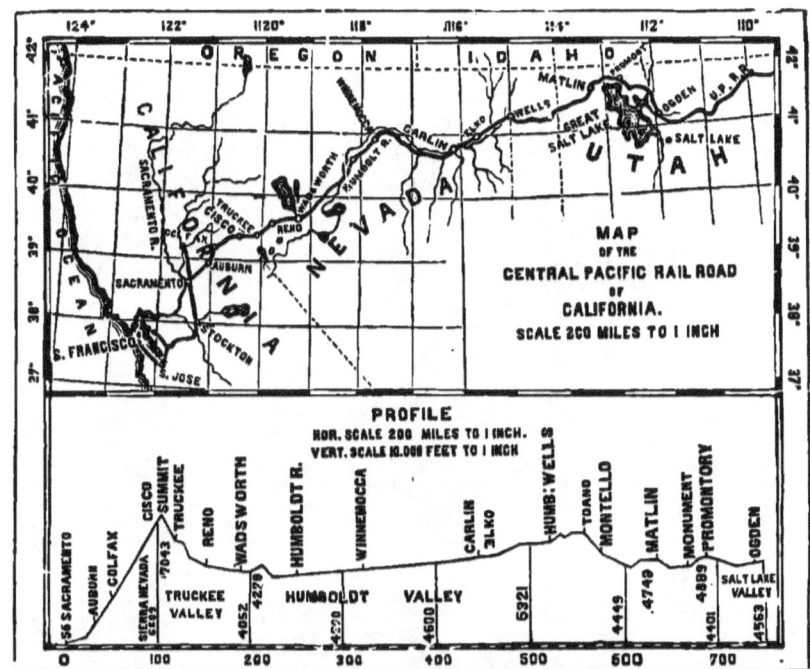

Through more deep rock cuts we wind around Promontory Mountain until the lake is lost to view. Up, up we go, the engine puffing and snorting with its arduous labors, until the summit is gained, and we arrive at the former terminus of the two Pacific railroads—8.93 miles from Blue Creek.

Promontory—elevation, 4,905 feet; distance from Omaha 1,084 miles; from San Francisco 830 miles—is celebrated for being the point where the connection between the two roads was made on the 10th of May, 1869.

This town, formerly very lively, is now almost entirely deserted. The supply of water is obtained from a spring about four miles south of the road, in one of the gulches of Promontory Mountain.

The bench on which the station stands would doubtless produce vegetables or grain, if it could be irrigated, for the sandy soil is largely mixed with loam, and the bunch grass and sage-brush grow luxuriantly.

The Last Spike—On Monday, the 10th day of May, 1869, a large party was congregated on Promontory Point, Utah Territory, gathered from the four quarters of the Union, and, we might say, from the four quarters of the earth. There were men from the pine-clad hills of Maine, the rock-bound coast of Massachusetts, the everglades of Florida, the golden shores of the Pacific slope, from China, Europe, and the wilds of the American continent. There were the lines of blue-clad boys, with their burnished muskets and glistening bayonets, and over all, in the bright May sun, floated the glorious old stars and stripes, an emblem of unity, power and prosperity. They are grave, earnest men, most of them, who are gathered here; men who would not leave their homes and business and traverse half or two-thirds of the continent only on the most urgent necessity, or on an occasion of great national importance, such as they might never hope to behold again. It was to witness such an event, to be present at

THE EAST AND THE WEST.
THE ORIENT AND THE OCCIDENT MEETING AFTER DRIVING THE LAST SPIKE.

the consummation of one of the grandest of modern enterprises, that they had gathered here. They were here to do honor to the occasion when 1,774 miles of railroad should be united, binding in one unbroken chain the East and the West. (Sacramento at that time was the western terminus.)

To witness this grand event—to be partakers in the glorious act—this assemblage had convened. All around was excitement and bustle that morning; men hurrying to and fro, grasping their neighbors' hands in hearty greeting, as they paused to ask or answer hurried questions. This is the day of final triumph of the friends of the road over their croaking opponents, for long ere the sun shall kiss the western summits of the gray old monarchs of the desert, the work will be accomplished, the assemblage dispersed, and quiet reign once more, broken only by the hoarse scream of the locomotive; and when the lengthening mountain shadows shall sweep across the plain, flecked and mottled with the departing sunbeams, they will fall on the iron rails which will stretch away in one unbroken line from the Sacramento to the Missouri River.

The hours passed slowly on until the sun rode high in the zenith, his glittering rays falling directly down upon the vacant place between the two roads, which was waiting to receive the last tie and rails which would unite them forever. On either road stood long lines of cars, the impatient locomotives occasionally snorting out their cheering notes, as though they understood what was going on, and rejoiced in common with the excited assemblage.

To give effect to the proceedings, arrangements had been made by which the large cities of the Union should be notified of the exact minute and second when the road should be finished. Telegraphic communications were organized with the principal cities of the East and West, and at the designated hour the lines were put in connection, and all other business suspended. In San Francisco the wires were connected with the fire-alarm in the tower, where the ponderous bell could spread the news over the city the instant the event occured. Baltimore, Philadelphia, Boston, New York, Cincinnati, and Chicago were waiting for the moment to arrive when the chained lightning should be loosed, carrying the news of a great civil victory over the length and breadth of the land.

The hour and minute designated arrived, and Leland Stanford, President, assisted by other officers of the Central Pacific, came forward; T. C. Durant, Vice-President of the Union Pacific, assisted by General Dodge and others of the same company, met them at the end of the rail, where they reverently paused, while Rev. Dr. Todd, of Mass., invoked the Divine blessing. Then the last tie, a beautiful piece of workmanship, of California laurel, with silver plates on which were suitable inscriptions, was put in place, and the last connecting rails were laid by parties from each company. The last spikes were then presented, one of gold from California, one of silver from Nevada, and one of gold, silver and iron from Arizona. President Stanford then took the hammer, made of solid silver—and to the handle of which were attached the telegraph wires—and with the first tap on the head of the gold spike at 12, m., the news of the event was flashed over the continent. Speeches were made as each spike was driven, and when all was completed, cheer after cheer rent the air from the enthusiastic assemblage.

Then the Jupiter, a locomotive of the C. P. R. R. Co., and locomotive No. 116, of the U. P. R. R. Co., approached from each way, meeting on the dividing line, where they rubbed their brown noses together, while shaking hands, as illustrated.

To say that wine flowed freely would convey but a faint idea of the good feeling manifested and the provision made by each company for the entertainment of their guests, and the celebration of the event.

Immediately on the completion of the work, a charge was made on the last tie (not the silver-plated, gold-spiked laurel, for that had been removed and a pine tie substituted) by relic hunters, and soon it was cut and hacked to pieces, and the fragments carried away as trophies or mementoes of the great event. Even one of the last rails laid in place was cut and battered so badly that it was removed and another substituted. Weeks after the event we passed the place again, and found an enthusiastic person cutting a piece out of the *last* tie laid. He was proud of his treasure — that little chip of pine—for it was a piece of the last tie. We did not tell him that three or four ties had been placed there since the first was cut in pieces.

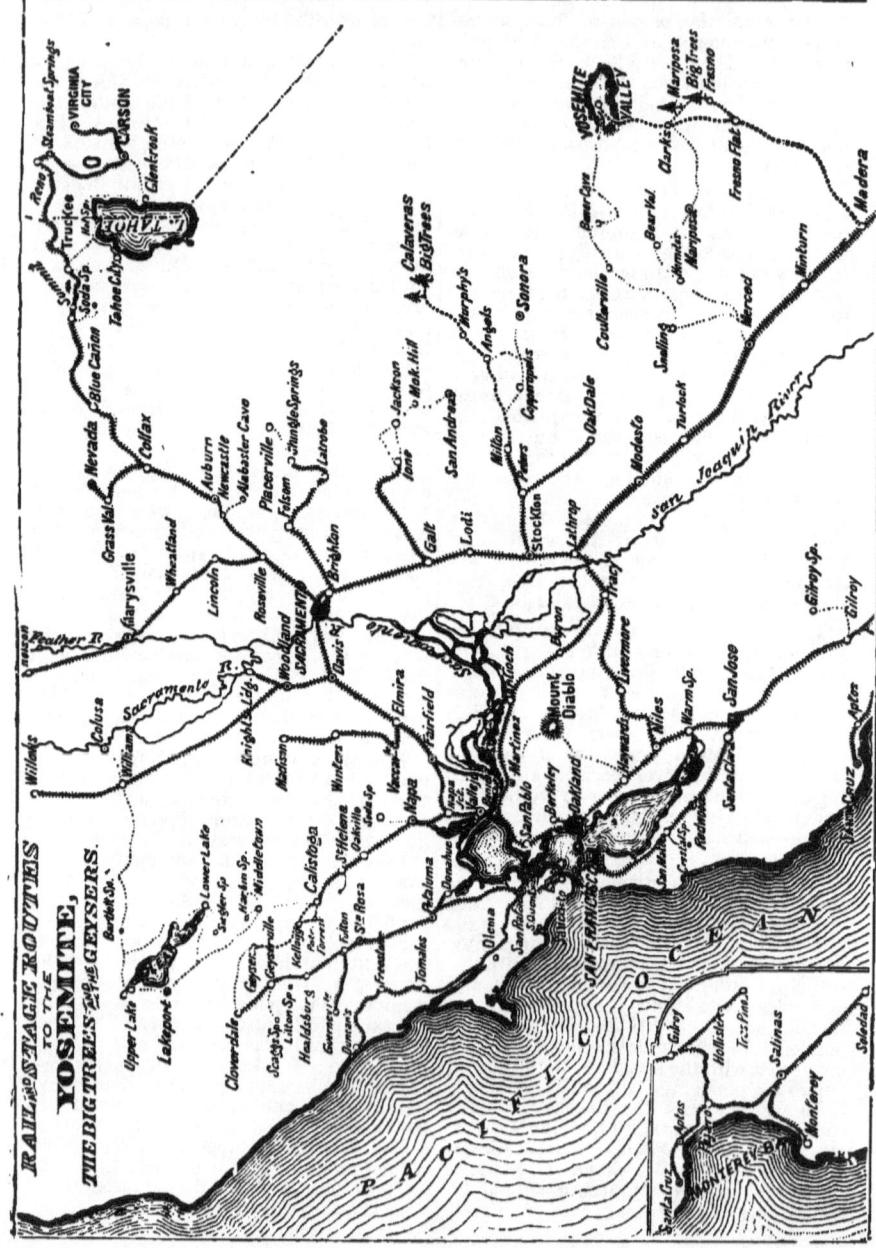

In the cars belonging to each line, a sumptuous repast was served up to the invited guests. Then, as the sun sank low to the western summit of Promontory Point, the trains moved away with parting salutes from locomotives, and the celebration was ended, the participants speeding away to their far distant homes, and so closed the eventful day on Promontory Point.

For Sketch of Great Salt Lake, see ANNEX No. 21.

For Hudnut's Survey of route to Oregon, see ANNEX No. 28.

We now resume our westward journey from Promontory. Four miles west (near a gravel track on the north side) can be seen close to the road, on the *south side*, a sign-board, which reads,

"TEN MILES OF TRACK IN ONE DAY."
Again, on the same side, ten miles further west, another with the same inscription will appear. These boards mark the track which was laid by the track layers of the Central Pacific company in *one day*, under the immediate charge of J. H. Strowbridge, Superintendent of Construction, H. H. Minkler, track layer, and James Campbell, Superintendent of Division. This undoubtedly is the most extraordinary feat of the kind ever accomplished in this or any other country.

WHY IT WAS DONE—During the building of the road, a great rivalry existed between the two companies as to which could lay the most track in one day. This rivalry commenced early in the year 1868. The "Union" laying six miles; soon after the "Central" laid seven miles, and then again the "Union" seven and a half miles. The "Central" men, not to be outdone, announced that they could lay ten miles in a day. Mr. Durant, Vice-President of the "Union" offered to bet $10,000 that it could not be done, and the "Central" resolved it *should* be done. Consequently, on the 29th of April, 1869, when only fourteen miles of track remained to be laid to meet the "Union" at Promontory Point, and in the presence of Gov. Stanford and many prominent men from the East and West, and a committee from the "Union" to note the progress, the work commenced.

HOW IT WAS DONE—When the car loaded with rails came to the end of the track, the two outer rails on either side were seized with iron nippers, hauled forward off the car, and laid on the ties by four men who attended exclusively to this. Over these rails the car was pushed forward, and the process repeated. Behind these men came a gang of men who half-drove the spikes and screwed on the fish-plates. At a short interval behind these came a gang of Chinamen, who drove home the spikes already inserted and added the rest. Behind these came a second squad of Chinamen, two deep on each side of the track. The inner men had shovels, the outer ones picks. Together, they ballasted the track. The average rate of speed at which all these processes were carried on was one minute and 47½ seconds to every 240 feet of track laid down.

MATERIAL REQUIRED—Those unacquainted with the enormous amount of material required to build ten miles of railroad can learn something from the following figures: It requires 25,800 cross ties, 3,520 iron rails, 55,000 spikes, 7,040 fish-plates, and 14,080 bolts, the whole weighing 4,362,000 lbs. This material is required for a *single* track, exclusive of "turnouts."

To bring this material forward and place it in position, over 4,000 men, and hundreds of cars and wagons were employed. The discipline acquired in the four years since the commencement of the road enabled the force to begin at the usual time in the morning, calm and unexcited, and march steadily on to "VICTORY," as the place where they rested at 1:30 P. M. was called, having laid *eight miles of track in six hours*. Here this great "Central" army must be fed, but Campbell was equal to the requirements. The camp and water train was brought up at the proper moment, and the whole force took dinner, including many distinguished guests. After the "*hour nooning*," the army was again on the march, and at precisely 7 P. M. 10 *miles and 200 feet had been complet d*.

When this was done, the "Union" Committee expressed their satisfaction and returned to their camp, and Campbell sprang upon the engine and ran it over the ten miles of track in *forty minutes*, thus demonstrating that the work was *well done*.

Soon after leaving Promontory, the grade of the road descends, and 7.93 miles we reach

Rosel—situated almost on the edge of Salt Lake. It is an unimportant station, where passenger trains never stop, unless signaled. A few miles further, and we pass the sign-board where commenced the work of laying the "ten miles of track in one day." Continuing along on the lake shore, with large bluff on the right, for 9.49 miles further, we pass

Lake—another side-track, and 6.98 miles more arrive at

Monument—Here, many times, the lake breeze sweeps by, bearing the heavy alkaline and saline odors peculiar to this locality, and peculiarly offensive to invalids. Monument Point, a slim, tapering promontory, stretches far out into the lake, covered with excellent grass. We shall not see much more of the article for some time to come, for we are now on what might well be called the American Desert. Leaving Monument, it is 7.34 miles to

Seco—another side-track of no account, as all is sage-brush. Descending a heavy grade, we sweep around the head of the western arm of the lake, nearing and leaving its waters for the last time. Another run of 7.1 miles brings us to

Kelton—or Indian Creek, as it is sometimes called. This is a station of more importance than any yet passed since leaving Promontory. There are large water-tanks by the road-side, supplied from a spring in the foot-hills some miles to the northward. Here the Railroad Co. fill their water-cars—a train of which run daily to supply many of the stations on this division of the road. The Red Dome Mountains show their scattered spurs to the north, and to the southeast Pilot Knob or Peak can be seen lifting its rocky front far above the desert.

From this station a daily line of coaches leaves for Idaho and Oregon, on arrival of the cars. The route passes through Idaho and the eastern part of Oregon, connecting with the steamers of the Oregon Steam Navigation Company at Umatilla, on the Columbia River—through to Boise in two days; Walla Walla in four days; Portland in five and a half days.

The BOISE COUNTRY, to which the line of stages spoken of conveys the adventurous passengers, lies in the southeastern portion o. Idaho Territory, bordering on Oregon. Extensive mines of gold have been worked ther for years, and still continue to attract m.i .ttention, as rich mines of gold-bearing q   have been discovered and worked since th placer mines have been partially exhausted. The principal mining country i i that portion generally designated as the "oise Basin, which comprises a scope of country about 150 miles north and south, by a length of about 200 miles. The Boise mines lie north of the Snake or Shoshone River. The principal streams in the mining section are Boise River, Fayette River, Wind Creek, Moore's Creek and Salmon River. The Owyhee mines lie south of the Snake River and War Eagle Mountains. This portion of the mining belt of Idaho is not as extensive as the one just mentioned. The ores are mostly silver.

BOISE CITY—is the capital of the Territory and county seat of Ada county. Population 6,000. The town site was laid out in 1863, and now contains about 700 buildings, mostly brick and stone. The town is situated in a fine agricultural valley, about two miles wide by 50 long. It is the center of several stage routes, and also of trade for a large section of country. The *Statesman*, a tri-weekly paper, is published here.

IDAHO—is the second city in size in the Territory, population about 2,500. It lies 36 miles northeast of Boise City, with which it is connected with stage, and also with Umatilla, Oregon. The *World*, newspaper, is published here—semi-weekly.

SILVER CITY—contains about 2,000 inhabitants. The buildings are mostly granite. The *Avalanche*, a weekly paper, represents the interests of the town.

We now return to the railroad, and 11.43 miles further, arrive at

Ombey—Passenger trains seldom stop here, but roll on 9.87 miles further, to

Matlin—This station is on the highland, which sweeps out from the Red Dome Mountains. Here these mountains—low sandstone ridges—are nearer the track, breaking the monotony of the scene. The road lies on the northern border of a vast waste whereon we see few signs of verdure. The station is midway from east to west of the

AMERICAN DESERT — which extends over an area of 60 square miles. Over this vast extent the eye wanders in vain for some green object—some evidence that in times gone by this waste supported animal life, or will eventually in years to come. All is desolate in the extreme; the bare beds of alkali, or wastes of gray sand only meet the vision, if we except now and then a rocky hill more barren than the plains, if such things were possible. Evidently this desert was once the bed of a saline lake, perhaps a portion of the Great Salt Lake itself. The sloping plain sweeps off towards that body of water, and in places bends down until its thirsty sands are laved by the briny flood. There are many evidences in support of the theory that it was once covered by those waters, although much higher than the present level of the lake. The saline matter is plainly discernible in many places, and along the red sandstone buttes which mark its northern border. The long line of water-wash, so distinctly seen at Ogden, and other points along the lake shore, can be distinctly traced, and apparently on the same level as the bench at those places. The difference in the altitude of the road is plainly indicated by this line, for as we journey westward, and the elevation of the plateau increases, we find that the water-wash line blends with the rising ground and is seen no more.

Matlin is an unimportant station, 10.78 miles from

Terrace—Here the railroad company have erected work-shops and a 16-stall round-house. To the northward the hills which mark the entrance to the Thousand Spring Valley are plainly seen; they are brown, bare and uninviting as the country we are passing through. Some mines are reported near, but have not yet been developed. From Terrace it is 10.54 miles to

Bovine—Here there is little of interest to note, the face of the country remaining about the same, though gradually improving. Spots of bunch-grass appear at intervals, and the sage-brush seems to have taken a new lease of life, indicating a more congenial soil.

Continuing on 10.85 miles further we reach

Lucin—At this point we find water tanks supplied by springs in the hills at the outlet of Thousand Spring Valley, which lies to the north, just behind that first bare ridge, one of the spurs of the Humboldt Ridge, but a few miles distant. The valley is about four miles wide, and not far from 60 miles long, taking in its windings from this point to where it breaks over the Divide into Humboldt Valley. It is little better than one continual bog in the center—the water from the numerous brackish springs found there standing in pools over the surface. There is good range of pasturage for the cattle in the valley and hills beyond. The old emigrant road branches off at or near the station, one road passing through the valley, the other following nearly the line of railroad until it reaches the Humboldt *via* Humboldt Wells.

Goose or Hot Spring Creek, a small stream which courses through the valley its entire length, sinks near by the station, rising and sinking at intervals, until it is lost in the desert.

Before reaching the next station we leave Utah and enter the State of Nevada. Passing over 11.75 miles of up-grade, our train arrives at

Tecoma—In 1874 quite an excitement was created among the mining operators by the discovery of rich silver and lead mines, situated about five miles south of this station in the Toano range of mountains. A new town was laid out at the mines—called Buel. A smelting furnace was erected at the mines and a run of 200 tons of bullion produced, valued at $360,000, which was shipped to San Francisco on one train, creating no small excitement on California street. Indications of coal mines have been found in the vicinity, but no systematic effort has yet been made to develop them.

Stock-raising is now the principal business of this country. To the northward of this station, and in fact for the last two stations, large herds of cattle can be seen, and at the stations, pens and shutes for shipping.

PILOT PEAK, a noted landmark which has been visible for the past fifty miles, lies almost due south of this station—distance 36 miles. It is a lofty pile of rocks —the eastern terminus of Pilot Mountains —rising about 2,500 feet above the barren sands. For about half-way from the base to the summit the sides are shelving piles of shattered rock—huge masses crushed to atoms. Above that it rises perpendicularly the summit looking like some old castle when seen at a distance. From Promontory Point looking westward, this vast pile can be seen on a clear day—a dark mass amid the blue haze which bounds the western horizon. To the emigrant, in early days, before the railroad, it was a welcome landmark, pointing his course to Humboldt Wells or Thousand Spring Valley, where he was sure to find water and feed for his weary teams, after crossing the barren waste.

From Tecoma it is 9.56 miles up-grade to

Montello—elevation 4,999 feet. The general aspect of the country is changing with the increasing elevation. We approach nearer the long, rough ridge of the Goose Creek Range, the sides and gulches of which afford pasturage and water at intervals. We are leaving the barren sands behind us, and the country looks more capable of supporting animal life.

Continuing the up-grade—over 550 feet within the next 9.6 miles—we arrive at

Loray—a station of little importance to the traveler.

From Loray, up we go for 7.1 miles further to

Toano—until recently the end of the division.

The company have here erected workshops and a 14-stall round-house. Toano is centrally located as regards many mining districts in Eastern Nevada, among which are Egan Canyon, Kinsley, Kern, Patterson, Ely, Pahranagat and Deep Creek—all of which are under rapid development. A stage line is in operation from this place to Egan Canyon and the Cherry Creek mines, a distance of 90 miles south. Soon after leaving Toano we beg n the ascent of Cedar Pass, which divides the Desert from Humboldt Valley. The country is more broken, but possessing more vegetation. We have passed the western line of the desert, where, in early days, the travel-worn emigrant wearily toiled through the burning sand, his journey unenlivened by the sight of water or vegetation. One word more, regarding this desert: The term sand is generally applied, when speaking of the soil of the barren wastes which occur at intervals along the road. With one or two exceptions it is a misnomer, though it well applies to the desert we have crossed. Most of the surface of this waste is sand, fine, hard and grey, mixed with marine shells and fossilized fragments of another age. There is no evidence on which to found a hope that this portion of the country could be rendered subservient to the use of man, consisting, as it does, of beds of sand and alkali, overlaying a heavy gravel deposit. Ages must pass away before nature's wondrous changes shall render this desert fit for the habitation of man. Continuing on up the ridge, 9.91 miles, we pass

Pequop—and 5.83 miles further

Otego—both side-tracks of little importance. Then we commence to descend, and 5.6 miles further arrive at

Independence—Independence Springs, from which this station derives its name, are near by, and supply an abundance of very good cold water.

Independence, Clover and Ruby valleys, lie to the southward. The two first named are small and valueless except for grazing purposes. From Cedar Pass a spur, or rather a low range of hills, extends far to the southward. About 70 or 80 miles south of the pass, is the South Fork of the Humboldt which canyons through this range, running to the east and north of another range until it reaches the main Humboldt. Although the range first mentioned, after having united with the western range south of the South Fork, extends much farther south, we will follow it only to Fort Ruby, which is situated in the south end of Ruby Valley, near to the South Fork. From this fort to the pass is about 65 miles, which may be taken as the length of the valley. The average width is ten miles from the western range mentioned to the foot-hills of Ruby Range, which hems in the valley to the east. A large portion of this valley is very productive and is occupied by settlers—mostly discharged soldiers from Fort Ruby. In the southeastern portion of the valley is Ruby and Franklin lakes, which are spoken of under the general term of Ruby Lake, for in high water they are united, forming a brackish sheet of water about 15 miles long by seven in width, which has no outlet. It is—like Humboldt, Carson and Pyramid lakes in the Truckee Desert—merely a reservoir, where the floods accumulate to evaporate in the dry summer. The old stage road, from Salt Lake to Austin, crosses the foot of the valley at Ruby station. About 20 miles east of the Ruby Range lies Goshoot Lake, another brackish pond, with two small tributaries and no outlet, rather wider and about the same length as Ruby Lake. About half-way between Goshoot and the railroad lies Snow Lake, about five miles in diameter, possessing the same general characteristics as the others. With the exception of the valleys around these lakes and along the watercourses, the country is valueless except for stock-raising. In the Ruby Range rich silver lodes have been discovered, some rock of which has been found to assay as high as $6000 per ton.

Returning to Independence, we again proceed westward—the country is rolling and broken—and the up-grade continues 6.1 miles to

Moore's—on the summit of Cedar Pass. We now have down-grade for 311 miles to the Nevada Desert.

In general outline this pass resembles a

rather rough, broken plateau, bent upward in the middle, forming a natural roadbed from the desert to the Humboldt Valley. It was once covered with scrub cedar, which was cut off for use by the railroad company and others. Some is still obtained in the mountains to the north. About 15 miles to the north a high, craggy peak marks the point where Thousand Spring Valley bends to the south, and from its divide slopes down to the valley of the Humboldt. Descending 2.65 miles is

Cedar— a small side-track, and six miles further brings our train to

Wells—Here are located the usual round-house and machine shops of a division. The station is 1,250 miles from Omaha and 664 from San Francisco; elevation, 5,628 feet. Owing to the location of railroad shops at this place much improvement is noticeable in the last few years. The chief points of interest around the station are the celebrated.

HUMBOLDT WELLS — around which the emigrants, in early times, camped to recruit their teams, after a long, hard journey across the desert. The wells are in the midst of a beautiful meadow or valley, which slopes away until it joins the Humboldt or main valley. The springs or wells—about 20 in number—are scattered over this little valley; one from which the company obtain their supply of water being within 200 yards of the road, and about that distance west of the station. A house has been built over it, and the water is raised into the tanks by an engine. These wells would hardly be noticed by the traveler unless his attention was called to them. Nothing marks their presence except the circle of rank grass around them. When standing on the bank of one of these curious springs, you look on a still surface of water, perhaps 6 or 7 feet across and nearly round; no current disturbs it; it resembles a well more than a natural spring, and you look to see the dirt taken from it when dug. The water, which is slightly brackish, rises to the surface, seeping off through the loose, sandy loam soil of the valley. No bottom has been found to these wells, and they have been sounded to a great depth. Undoubtedly they are the craters of volcanoes long since extinct, but which at one time threw up this vast body of lava of which the soil of Cedar Pass is largely composed. The whole face of the country bears evidence of the mighty change which has been taking place for centuries. Lava in hard, rough blocks; lava decomposed and powdered; huge blocks of granite and sandstone in the foothills, broken, shattered and thrown around in wild confusion, are some of the signs indicative of an age when desolation reigned supreme. The valley in which the wells are situated is about five miles long by three wide, covered with a luxuriant growth of grass. The low hills afford an excellent stock "range." The transition from the parched desert and barren, desolate upland to these green and well-watered valleys, redolent with the aroma of the countless flowers which deck its bosom, seems like the work of magic.

Rich mineral discoveries have been made about 35 or 40 miles southeast of Wells—east of Clover Valley—in the Johnson & Latham district. The veins are reported large, well-defined, and rich in silver, copper and lead; large deposits of iron ore have also been found. The district is well supplied with wood and water, and easy of access from the railroad. A stage runs through the district, extending 100 miles south to Shelburn, near the old overland stage road, in the Shellcreek mining district. A stage line is also in operation to the Bull Run district.

Stock-raising occupies the attention of most of the settlers about this section and to the northward.

Leaving the Wells we proceed down the valley for a few miles; when we enter the main valley of the Humboldt, which is very rich, but the seasons are too short for agricultural purposes. The soil is a deep black loam, moist enough for all purposes without irrigation, from one to two feet deep. This portion of the Humboldt is about 80 miles in length, averaging 10 miles in width, nearly every acre being of the quality described. From Osino Cañon to the headwaters of the valley is occupied by settlers who have taken up hay ranches and stock ranges. The river abounds in fish and the foothills with deer and other game.

THE HUMBOLDT RIVER rises in the Humboldt Mountains, northwest of Cedar Pass, and courses westerly for about 250 miles, when it bends to the south, emptying into Humboldt Lake, about 50 miles from the Big Bend. It is a rapid stream for most of the distance, possessing few fords or convenient places for crossing. The railroad follows down its northern bank until it reaches Twelve-Mile Cañon, about 16 miles west of Carlin, where it crosses to the south side of the river and continues about 170 miles, when it crosses again and leaves the river, skirting the foot-hills in full view of the river and lake. The main stream has many varieties of fish, and at certain seasons its waters are a great resort for wild ducks and geese. Where it enters the lake the volume of water is much less than it is 100 miles above, owing to the aridity of the soil through which it passes. Of the valleys bordering it we shall speak separately, as each division is totally distinct in its general features. The "old emigrant road" can be distinctly traced along the river from its head to its source. From Wells, continuing down grade, it is 7.5 miles to

Tulasco—a small side-track, five miles from

Bishop's—This is another unimportant side-track, where Bishop's valley unites with the Humboldt. This valley is 70 miles long, average width about five miles. It is very fertile, being watered by Bishop's Creek, which rises in the Humboldt Mountains, near Humboldt Cañon, about 70 miles to the northeast, winding through the valley.

Deeth—is passed 7.7 miles from Bishop's, and 12.9 miles farther to

Halleck—At this station Government stores are left for Fort Halleck, a military station on the opposite side of the river. At the foot of the mountain—about 12 miles from the station—can be seen some settlers' buildings, situated on the road to the post. The military post is hid from view by the intervening hills. It is situated on an elevated plateau, which lies partially behind the first range, debouching thence in a long upland, which extends some distance down the river. The valleys along the hills and much of the upland, are settled, and for vegetables and cerals not affected by the early frost, prove very productive. A ready market is found along the railroad.

Peko—is an unimportant station, 3.3 miles west of Halleck. Just after leaving the station we cross the north fork of the Humboldt on a truss bridge This river, where it unites with the main stream, is of equal size. It rises about 100 miles north, and receives as tributaries many small creeks and rivulets. The valley of the North Fork is from five to seven miles wide and covered with a heavy growth of grass, and, like the main valley, is not susceptible of cultivation to any great extent. Some kinds of vegetables yield handsome returns. The seasons are long enough, and the absence of early and late frosts insures a crop. Around the head of this valley are many smaller ones, each tributary stream having its own separate body of valley land. Some are perfect gems, nestled among the hills and almost surrounded by timber. Here game in abundance is found—quail, grouse, hare, deer and bear, and sometimes a "mountain lion," and the tourist, angler and hunter will find enough to occupy them pleasantly should they visit this region. In these valleys are many thousand acres of Government land unclaimed, excepting that portion owned by the Railroad Company.

The Humboldt and its tributary valleys, as a range for stock, have no superior west of the Rocky Mountains. The winters are mild—snow rarely sufficiently deep to render it necessary to

No. 52 ANNEX. "Prickey," the Horned Toad.—This singular little member of the lizard species is certainly a native Californian. It is found upon nearly every dry hill or gravelly plain; and although it is rare in some districts, in others it is still common. There are several varieties and sizes of it, and all perfectly harmless. It lives chiefly on flies and small insects. A California friend of ours had a pair of these picketed in front of his cabin for over three months; and, one morning, the male toad wound itself around the picket pin and strangled to death, and the same day the female followed his example. Upon a *post mortem* examination of the female 15 eggs were found, about the size and shape of a small wren's egg. (See illustration, page 81.)

The Mammoth Snow Plow—owned by the Central Pacific Railroad, rests upon two four-wheel trucks, is 28 feet long, 10 feet 6 inches wide, 13 feet 3 inches high, and weighs 41,860 pounds. It was once propelled by ten locomotives, at the rate of 60 miles an hour, into a snow drift on the Sierra Nevada Mountains, resulting—in a big hole in the snow.

feed the stock. Wild cattle are found in the valleys and among the hills, which have never received any attention or care. Stock-raisers are turning their attention of late to this country and find it very remunerative. The range is not confined to the valley alone, the foot-hills and even the mountain sides produce the bunch grass in profusion. Wherever sage-brush grows rank on the hillsides, bunchgrass thrives equally well.

Osino—Is 11.8 miles down the valley from Peko—a signal station at the head of Osino Canyon, where the valley suddenly ends.

At this point the northern range of mountains sweeps to the river bank, which now assumes a tortuous course—seeming to double back on itself in places—completely bewildering the traveler. Across the river the high peaks of the opposite chain rise clear and bold from the valley, contrasting strongly with the black, broken masses of shattered mountains among which we are winding in and out, seemingly in an endless labyrinth. Now we wind round a high point, the rail lying close to the river's bank, and next we cross a little valley with the water washing against the opposite bluffs, half a mile away. A dense mass of willow covers the bottom lands through which the river wanders. On around another rocky point and we are in a wider portion of the canyon, with an occasional strip of meadow land in view, when suddenly we emerge into a beautiful valley, across which we speed, the road curving around to the right, and 8.8 miles from the last station we arrive at

Elko—The county seat of Elko county; population about 1,200. Elko is a regular eating station for all trains from East and West. The town consisted of wood and canvas houses—which latter class is rapidly being replaced by something more substantial. In the last few years the town has improved materially. The State University, which cost $30,000, is located here, just to the northward of the town. At this station—and almost every one to the westward—can be seen representatives of the Shoshone or Piute Indians, who come around the cars to beg. Any person who wishes to tell a big "whopper" would say, they are clean, neatly dressed, "child-like and bland," and perfumed with the choicest attar of roses, but an old plainsman would reverse the saying in terms more expressive than elegant.

Near the town some WARM SPRINGS are attracting attention. The medicinal qualities of the water are highly spoken of. A hack lies between the hotel and the springs, making regular trips for the accommodation of visitors.

The rich silver mining district of Cope is about 80 miles due north of Elko, near the head waters of the North Fork of the Humboldt, bordering on the Owyhee country. Some very rich mines have been discovered and several quartz mills erected, in that district, but the more recent discoveries are in Tuscarora district about 50 miles north, and are said to be very rich.

Stages leave Elko daily for Mountain City—north, in Cope district—80 miles distant, and all intermediate towns and camps. Stages also run to Railroad district—south 25 miles, and to Eureka district, 100 miles; also a weekly line to the South Fork of the Humboldt and Huntington valleys. Large quantities of freight arrive at, and are re-shipped from this station on wagons, for the various mining districts to the north and south.

Another important business that has sprung up at Elko, within the last few years, is cattle-raising. Elko county contains more cattle than any other two counties in the State, and Elko ships more cattle than any four stations on the road, being amply provided with all the facilities —roomy yards, shutes, etc., for a business that is rapidly increasing, and is destined, before many years, to *far* exceed all others in the State.

This section is well watered by rapid mountain streams, and the country abounds in game of all kinds—a hunter's paradise. The valley of the Humboldt, for twenty miles above and below Elko, cannot be ranked as among the best of its bottom-lands, though it is susceptible of cultivation to a considerable degree. But a narrow strip is meadow, the remainder being higher, gravelly land, covered with sage-brush and bunch-grass. Without irrigation it is useless for agricultural purposes.

Passing down from Elko—the valley dotted with the hamlets of the rancher for about nine miles—we come opposite the South Fork of the Humboldt. This stream rises about 100 miles to the southeast. It canyons through Ruby Mountains, and then follows down the eastern side of one of the numerous ranges, which, under the general name of the Humboldt Mountains, intersect the country.

For portions of the distance there is fine valley land along the stream, ranging from one to seven miles wide, adapted to early crops, but, as a body, it is inferior to either the Main or North Fork valleys. However, the land is *all* admirably adapted for grazing purposes.

Moleen—is a signal station, 11.8 m les west of Elko. After leaving this station the valley presents a changed appearance. The meadow lands are broad and green, extending over most of the valley; on the right the bluffs are high and covered with luxuriant bunch-grass. Soon the meadows are almost entirely closed out, and we enter Five Mile Canyon. Through this the river runs quite rapidly; its clear waters sparkling in the sunlight as they speed along, while occasional narrow strips of meadow land are to be seen at times.

The scenery along this canyon is hardly surpassed by the bold and varied panorama presented to our view along the base of the snow-capped mountains through which the river and railroad have forced their way. Soon after entering the canyon we pass several isolated towers of conglomerate rock, towering to the height of nearly 200 feet. Leaving this canyon, we find Susan Valley, another strip of good bottom land, about twenty miles long, by four wide, bordering the East Fork of Maggie's Creek. Among the foot-hills of Owyhee Range, to the northward, are many beautiful, little valleys, well watered by mountain streams, waiting only the advent of the settler to transform them into pleasant homes. Timber is pleanty in the ravines and on the hill-sides—sufficient for the wants of a large population. Passing on to near the next station, we cross Maggie's Creek, which empties into the Humboldt from the north. This stream is named for a beautiful Scotch girl, who, with her parents, stayed here for a time "recruiting their stock" in the old times when the early emigrants toiled along the river. It rises in the Owyhee Mountains, about 80 miles to the northward.

The valley through which the stream flows is from three to five miles wide and very rich. It extends to the base of the mountains, about 70 miles, and is now mostly occupied by stockmen. The stream affords excellent trout fishing, and game of various kinds abounds on the hills bordering the valley. Some time since, a wagon road was surveyed and located up this valley to Idaho Territory.

From Moleen, it is 11.6 miles to

Carlin—This is quite a busy station, of about 600 population. Here are located the offices of Humboldt Division, and the division workshops. The latter are of wood and consist of a round-house of 16 stalls, a machine, car and blacksmith shop. The railroad was completed to this place Dec. 20, 1868.

To the south of Carlin, from 15 to 60 miles, are located mines rich in gold, silver, copper and iron. To the northward, rich discoveries have been made, extending to the Owyhee country. In both these sections new mining districts have been located, and the attention of experienced capitalists is being attracted thereto.

MARY'S CREEK—rises three miles north and enters the Humboldt at Carlin. It rises in a beautiful lakelet nestled among the hills and bordered by a narrow slip of fine valley land. The valley of the stream, and that portion surrounding its head waters, is occupied by settlers.

Proceeding down the river from Carlin, for some distance the green meadows continue fair and wide; then the sloping hills give place to lofty mountains, which close in on either hand, shutting out the valley.

From the appearance of this mountain range one would suppose that it had extended across the valley at one time, forming a vast lake of the waters of the river, then some mighty convulsion of nature rent the solid wall asunder, forming a passage for the waters which wash the base of the cliffs, which are from 500 to 1,000 feet high. This place is generally known as

THE PALISADES—Humboldt or Twelve Mile Canyon, although it does not possess points of interest with Echo or Weber canyons, in many particulars the scenery is equally grand. The absence of varied colors may urge against its claims to equal with those places, but, on the other hand, its bleak, bare, brown walls have a majestic, gloomy grandeur. which coloring could not improve. In passing down this canyon, we seem to pass between two walls which threaten to close together ere we shall gain the outlet. The river rolls at our feet a rapid, boiling current, tossed from side to side of the gorge by the rocks, wasting its fury in vain attempts to break away its prison walls. The walls in places have crumbled, and large masses of crushed rocks

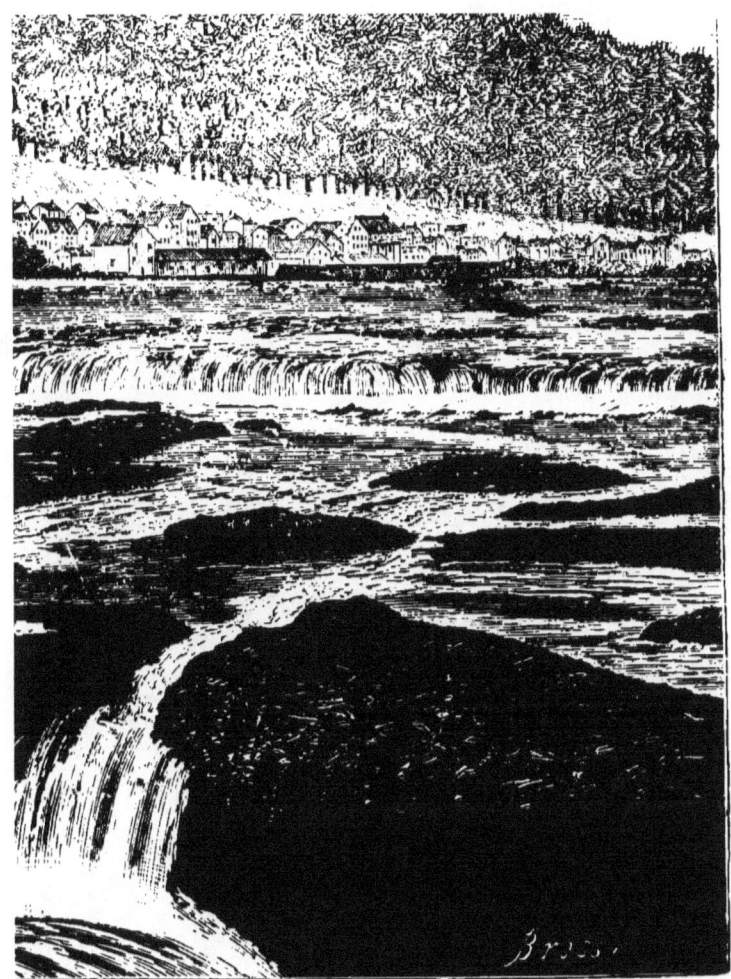

R, OREGON. (See Annex No. 37.)

slope down to the river brink. Seams of iron ore and copper-bearing rock break the monotony of color, showing the existence of large deposits of these materials among these brown old mountains. Now we pass "Red Cliff," which rears its battered frontlet 800 feet above the water. A colony of swallows have taken possession of the rock, and built their curious nests upon its face. From out their mud palaces they look down upon us, no doubt wondering about the great monster rushing past, and after he has disappeared, gossiping among themselves of the good old times when his presence was unknown in the canyon. Now we pass "Maggie's Bower," a brown arch on the face of the cliff, about 500 feet from its base. We could not see much bower – unless it was the left bower, for we *left* it behind us.

Twisting in and around these projecting cliffs, 9.1 miles from Carlin, we reach

Palisade—a station, in the midst of the Palisades, and apparently locked in on all sides. This is a busy place for a small one of only 200 population, as it is the junction of the Eureka & Palisade railroad, where are located their machine and workshops. Most of the box and flat cars of this company are made here in their own shops. The amount of freight handled at this station is enormous. Passengers can, almost always, see large piles of base bullion pigs piled up at the freight house, awaiting shipment. This bullion is mostly freighted here from the smelting furnaces at Eureka, by the Eureka and Palisade railroad, which alone handled over 35,000,000 pounds annually.

One great item of freight taken down over this road—the E. & P.—is timber from the Sierra Nevada Mountains, for use in timbering up the mines at Eureka.

Palisade, beside the machine shops above named, has several large buildings, used by the Railroad Company, for freights and storage and one a fine, commodious passenger station; these with several stores, hotels, restaurants and saloons make up the town.

The station is supplied with water from a huge tank, situated upon the mountain side, to the north, 800 feet above the station. This tank in turn is supplied from springs situated further up the mountain, that never fail in their supply.

Now, as we started out to see what was worth seeing, let us take a run down over the road that comes in here and note what can be seen.

Eureka & Palisade Railroad.
Principal Office, Eureka, Nevada.

E. MILLS,........*President*,..............*Eureka*.
P. EVERTS ....*Gen'l Sup't*..............*Eureka*.
J. L. FAST.....*Gen'l F. P. & T. Agt*...*Palisade*.

This road is a three-foot narrow gauge, commenced in December, 1873, and completed to Eureka in October, 1874, and is 90 miles in length. Passenger trains leave for the South on arrival of trains on the Central, and arrive in time to connect for either the East or West. The trains make full 20 miles an hour, and the cars are as commodious and nice as on any road in this country.

Leaving Palisade we cross the Humboldt River and start for the West, the C. P. on the north side of the river and our little train on the south side. But a few hundred yards from the station we curve around to the left, while the C. P. makes a similar one to the right and is soon lost to view. The general course of our train is south, following up Pine Valley, which is, for the first ten miles, covered with sage-brush—as is also the surrounding hills.

BULLION—is the first station on the bills, eight miles from Palisade, but we pass it, and the valley widens and 4.25 miles further is over one mile in width. Now our train is at

EVANS—a simple side-track, important only to a few settlers near, who are cultivating small fields and watching herds of cattle and sheep, which find good ranges on the hills, ravines, and neighboring valleys. Some fields are fenced, for the protection of the grass, which is cut for hay. The Cortez Mountains are on the west—the right side—and the Diamond range on the left.

WILLARD'S—is 15 miles from Palisade, and four miles from

HAY RANCH—This station is the first from Palisade where anything like business is to be seen. Here the Railroad Company have 2,500 acres of bottom land fenced, on which they cut annually about 1,000 tons of hay, which they bale and store away in those long warehouses to be seen on the right of the road. The company run freight teams from the end of their road at Eureka, and—in connection with it—to Pioche and all intermediate places. These teams are composed of 18 mules each, with three and sometimes four

AMERICAN RIVER CANYON.
Near Shady Run, Sierra Nevada Mountains.

wagons coupled together—as illustrated on page 28—employing from 300 to 400 mules, each team hauling from 30,000 to 40,000 lbs. In winter, when their mules are not in use, they are kept at this station, and the hay is harvested by the company and used for their own stock.

Continuing on up the valley 9.31 miles, we come to

BOX SPRINGS—but trains do not stop only on signal, and nine miles further stop at

MINERAL—This is a regular eating station, and in fact is the only one on the road. The meals cost $1.00, and are of the substantial order, that makes bone and sinew.

To the east is located the Mineral Hill Mine, once the most productive in the State, but it proved to be a "pocket" and the original owners, after taking out several millions of dollars, sold their mine to an English company—who, until within the last year, have allowed it to remain idle. It is now being worked with good prospects.

From Mineral, it is 5.5 miles to

DEEP WELLS—so named for a deep well that was dug near by, to procure water for the freight teams that were in the habit of traveling this road. The enterprising "Pilgrim" who dug the well was wont to charge $1.00 for sufficient water to water an 18 mule team, or "four skins full for a quarter."

From Deep Wells the route is over a sage-brush country, rough and bluffy, for seven miles, to

ALPHA—a small station, near where kilns of charcoal are burned for use at the smelting furnaces. To the west of this station is a broad valley, over which range large herds of cattle. Willow Creek, a small stream, is crossed, and ten miles from Alpha is

PINE STATION—another place where charcoal burning is the principal business. Cedar trees are noticeable along on the bluffs as we pass by; sage is abundant, and jack-rabbits are numerous and very large—they call them "narrow-gauge mules" in this country.

CEDAR—is the next station, three and-a-half miles from Pine. The country is rough and broken, and sage predominates; the grade is heavy and the road crooked—twisting and turning for 7.5 miles to

SUMMIT—a station on the dividing ridge between Pine Valley on the west and Diamond Valley on the east. The face of the country is not very inviting, except for those "narrow-gauge mules." Near the summit the old overland stage road crosses from Jacob's Wells on the east to Austin on the west.

From the summit the road enters Diamond Valley, and follows it up to Eureka, the road making, between Summit and

GARDNER PASS—six miles from Summit —a great horse-shoe curve, and fairly doubling back upon itself to get around the projecting spurs that shoot out from the range of high bluffs on each side.

Continuing along up the narrow valley nine miles, we reach

DIAMOND—an unimportant station, and another run of twelve miles brings our train to the end of the road at

Eureka—This city is 90 miles south of Palisade, and contains, with the near surroundings, a population of 6,000, nearly all of whom are engaged in mining and dependent pursuits. Besides the usual number of stores, hotels and small shops, there are two 30-stamp mills, seven smelting works and 16 furnaces, with a capacity of 50 tons of ore each, daily. These extensive establishments, running night and day, make business pretty lively, and will account for the quantities of base bullion hauled over the railroad to Palisade, as above noticed. Of the hotels, the Jackson and the Parker are the principal ones. There are two daily papers, the *Sentinel* and the *Republican*.

The Ruby Hill railroad, really an extention of the Eureka & Palisade, runs from the depot at Eureka around the various smelting and refining works and mines of the different mining companies, and around Eureka, delivering freight and handling ores. This road is about six miles in length. The most prominent mines at Eureka are Eureka Consolidated, Richmond Consolidated, the K. K., the Jackson, Hamburg, Matamoras and Atlas.

Stages connect at Eureka, carrying passengers, mails and express to the various mining towns and camps in the adjoining country; to Hamilton, 40 miles, daily, which runs through the Ward and Pioche districts to Ward, 100 miles; Pioche, 190 miles; to Tybo, 100 miles; Austin, 80 miles; tri-weekly stage to Belmont, 100 miles.

The freighting business to Pioche and all intermediate towns and camps is very extensive, most of which is hauled by the Railroad Company's teams, as previously stated.

THE WHITE PINE COUNTRY, is situated to the southeast from Eureka, the principal city of which is

HAMILTON—This city contains a population of about 800, all of whom are engaged in the mining business. Milling and smelting are the only occupations, there being two smelters and six mills. An English company is now engaged running a tunnel under Treasure Hill, to strike the great mineral deposit known to be there. This tunnel, when completed, will be 6,000 feet long—7x9 feet, double track, "T" rail —and will tap the mines at a depth of 1,600 feet. It is now completed about 3,000 feet. Hamilton had one weekly newspaper—the *News*.

WHITE PINE—is nearly due east of Virginia City, where the first silver mining excitement occurred on the Pacific slope, and by many is supposed to be on the same range which produced the Comstock and other famous lodes. Possibly such is the case, though "ranges" have been terribly shaken about in this section of our country. The Eberhardt mine, which first attracted attention to this locality, was discovered in 1866, but the great stampede of miners and speculators to that quarter did not take place until the winter and spring of 1869. As far as prospected, the veins, in a majority of cases, are not regular, being broken and turned in every direction. Some are flat, others dip at a regular angle and have solid walls. The Base Metal Range in this vicinity is very extensive, and a number of furnaces have been erected to reduce the ores into base bullion for shipment. For items of interest see ANNEX No. 30.

We will now return to Palisade, and resume our place in the C. P. cars. Passing down the canyon, winding and twisting along around a succession of projecting spurs, we pass the "Devil's Peak," on the opposite side of the river, a perpendicular rock, probably 500 feet high, rising from the water's edge. On, past the towering bluffs and castellated rocks—which, at first view, look like some old brown castle, forsaken by its founders and left to ruin, desolation and decay—we cross the river on a fine Howe truss bridge; and from this point we shall keep on the southern side of the stream until we near Humboldt Lake, when we cross it again, and for the last time.

Cluro—is a flag station, 10.4 miles west of Palisade, reached just after emerging from the canyon. We now enter a more open country, with strips of meadow along the river's brink. Near this point is where the powder magazine of the railroad company exploded in 1868, while the road was building through the canyon.

North of the river, at the point on the opposite side, can be seen a peculiar formation, not seen elsewhere in the canyon. Where the road is cut through these points, they consist of gravel, sand and cement, having all the appearance of gold-bearing gravel-beds. It is an unmistakable water-wash, and not caused by volcanic wear—fine layers of sand, from one to five feet thick, and interspersed through the gravel, showing where the water rested and the sediment settled.

GRAVELLY FORD—one of the most noted points on the Humboldt River in early days, is near Cluro. Then the canyon through which we have just passed was impassable. The long lines of emigrant wagons could not pass through the mighty chasm, but were obliged to turn and toil over the mountains until they could descend into the valley again. Coming to this point on the south side of the river, they crossed and followed up a slope of the opposite hills, thence along the table-land, and from thence to the valley above. A few would leave the river lower down and bear away to the south, but the road was long and rough before they reached the valley above the canyon. There were and now are other fords on the river, lower down, but none were as safe as this. With sloping gravelly banks and a hard gravel bottom, it offered superior advantages to the emigrant. Hence it became a noted place—the point to which the westward-bound emigrants looked forward with great interest. Here was excellent grazing for their travel-worn teams. Owing to these considerations, large bodies of emigrants were often encamped here for weeks. At times the river would be too high, and they would wait for the torrent to subside. The Indians—Shoshones—knew this also, and many a skirmish took place between them and their white brothers, caused by mistaken ideas regarding the ownership of the emigrant's stock.

Connected with this place is an incident which, for the honor of the men who performed the Christian act, we will relate:

In the early times spoken of, a party of emigrants were encamped here, waiting for the water to subside. Among these emigrants were many women and children. While here, an estimable young lady of 18 years fell sick, and despite the watchful care and loving tenderness of friends and kindred, her pure spirit floated into that unknown mist, dividing the real from the ideal, the mortal from the immortal. Her friends reared an humble head-board to her memory, and, in course of time—amid the new life opening to them on the Pacific slope—the young girl's fate and grave were alike forgotten by all but her immediate relatives. When the advance guard of the Central railroad—the graders and culvert men—came to Gravelly Ford, they found the lone grave and fast-decaying head-board. The site awoke the finer feelings of their nature and aroused their sympathies, for they were *men*, these brown, toil-stained laborers.

The "culvert men" (masons) concluded that it was not consistent with Christian usage to leave a grave exposed and undefended from the incursion of beasts of prey. With such men, to think was to act, and in a few days the lone grave was enclosed with a solid wall, and a cross—the sacred emblem of immortality—took the place of the old head-board. In the day when the final reckoning between these men and the recording angel is adjusted, we think they will find a credit for that deed which will offset many little debits in the ledger of good and evil. Perhaps a fair spirit above may smile a blessing on their lives in recompense of the noble deed. The grave is on the south side of the road, upon a low bluff, about five miles west of Cluro. In October, 1871, the Superintendent of the Division erected over it a fine large cross.

"CORRAL," (Spanish) a pen made of posts set on end in the ground close together, and fastened with rawhide thongs, or by wagons drawn in a circle forming an inclosure.

TELEGRAPH OPERATORS are called "lightning shovers."

OUTFIT—Necessary supplies for a journey.

Upon one side is inscribed "THE MAIDEN'S GRAVE," on the other, her name, "Lucinda Duncan."

Passing on, we cross narrow patches of meadow land, and wind around the base of low hills until we reach a broad valley. Across the river to the northward can be seen the long, unbroken slopes which stretch away until they are lost in that cold blue line—the Idaho Mountains—which rises against the northern sky. Behind that gray old peak, which is barely discernible, the head waters of the North Fork of the Humboldt break away when starting on their journey for the main river. Farther to the left, and nearer, from among that darker clump of hills Maggie's Creek finds its source.

Be-o-wa-we—is reached at a narrow point called Copper Canyon, 8.7 miles from Cluro.

The Cortez mines and mills are situated about 25 miles south of this station, with which they are connected by a good road. At this point the Red Range throws a spur nearly across the valley, cutting it in two. It looks as though the spur extended clear across, at one time, damming up the waters of the river, as at the Palisades. The water-wash far up the hillside is in evidence of the theory that such was once the case, and that the waters cut this narrow gorge, through which they speed along unmindful of the mighty work done in former years, when the resistless current "forced a highway to the sea," and drained a mighty lake, leaving in its place green meadows.

Here, on this red ridge, is the dividing line between the Shoshones and the Piutes, two tribes of Indians who seemed to be created for the express purpose of worrying emigrants, stealing stock, eating grasshoppers, and preying on themselves and everybody else. The Shoshones are very degraded Indians, and until recently, were like the Ishmaelites or Pariahs of old—their hand was against every man, and every man's hand was compelled, in self-defense, to be against them until they became almost unable to commit depredations, and could make more by begging than they could by stealing. The term Be-o-wa-we signifies gate, and it is literal in its significance.

After leaving Be-o-wa-we, we pass through the gate, and wind along by the hillside, over the low meadows, which here are very narrow. The "bottom" is broad, but is covered with willows. with the exception of the narrow meadows spoken of. Amid these willows the stream winds and twists about through innumerable sloughs and creeks, as though undecided whether to leave this shady retreat for the barren plains below. Perhaps the traveler will see a flock of pelicans disporting in the waters on their return from their daily fishing excursion to Humboldt Lake. These birds, at certain seasons of the year, are to be found here and there along the river for about 20 miles below, in great numbers. They build their nests in these willow islands and rear their young undisturbed, for even an Indian cannot penetrate this swampy, treacherous fastness. Every morning the old birds can be seen taking their flight to Humboldt Lake, where, in its shallow waters, they load themselves with fish, returning towards night to feed their young and ramble about the bottom.

Soon after leaving the station, Hot Spring Valley comes in on the left—south—and by looking away to the south eight miles, can be seen columns of steam, from one of the many "hot springs" which abound in the "Great Basin."

If you do not behold the steam—for the springs are not always in active operation—you will behold a long, yellowish, red line, stretching for a full half-mile around a barren hill-side. From this line boiling, muddy water and sulphuric wash descends the hill-side, desolating everything in its course, its waters escaping through the bogs of the valley.

Sometimes for hours these springs are inactive, then come little puffs of steam, then long and frequent jets, which often shoot 30 feet high. The waters are very hot. Woe to the unlucky hombre who gets near and to the windward of one of these springs, when it sends forth a column of spray, steam and muddy sulphur water from 20 to 30 feet in height. He will need a change of clothes, some simple cerate, a few days' rest, and the prayers of his friends—as well as of the congregation. There are over 100 of these spurting, bubbling, sulphuric curiosities around the hills in this vicinity. The general character of all are about the same.

There are a great many theories regarding these springs—what causes the heat, etc. Some contend that the water escapes from the regions of eternal fires, which are supposed to be ever burning in the center of the globe. Others assert that it is mineral in solution with the water which causes the heat. Again, irreverent persons suggest that this part of the country is but the roof of a peculiar place to which they may well fear their wicked deeds may doom them in the future.

Shoshone—is ten miles west of Be-o-wa-we; elevation 4,636 feet. Across the river to the right is Battle Mountain, which rises up clear and sharp from the river's brink. It seems near, but between us and its southern base is a wide bottom land and the river, which here really "spreads itself." We saw the same point when emerging from Be-o-wa-we, or "the gate," and it will continue in sight for many miles.

This mountain derives its name from an Indian fight, the particulars of which will be related hereafter. There are several ranges near by, all bearing the same general name. This range being the most prominent, deserves a passing notice. It lies north of the river, between the Owyhee Range on the north and the Reese River Mountains on the south. Its base is washed by the river its entire length—from 50 to 75 miles. It presents an almost unbroken surface and even altitude the entire distance. In places it rises in bold bluffs, in others it slopes away from base to summit, but in each case the same altitude is reached. It is about 1,500 feet high, the top or summit appearing to be table-land. Silver and copper mines have been prospected with good results.

Behind this range are wide valleys, which slope away to the river at either end of the range, leaving it comparatively isolated.

Opposite to Shoshone, Rock Creek empties its waters into the Humboldt. It rises about 40 miles to the northward, and is bordered by a beautiful valley about four miles wide. The stream is well stocked with fish, among which are the mountain trout. In the country around the headwaters of the stream is found plenty of game of various kinds, including deer and bear.

Copper mines of vast size and great richness are found in the valley of Rock Creek, and among the adjoining hills. Whenever the copper interest becomes of sufficient importance to warrant the opening of these mines, this section will prove one of great importance.

Leaving Shoshone, we pursue our way down the river, the road leading back from the meadow land and passing along an upland, covered with sage-brush. The hills on our left are smooth and covered with a good coat of bunch-grass, affording most excellent pasturage for stock, summer and winter.

Argenta—is 11.1 miles further west. This was formerly a regular eating station and the distributing point for Austin and the Reese River country; but is now a simple side-track. Paradise Valley lies on the north side of the river, nearly opposite this station. It is about 60 miles long by eight wide, very fertile and thickly settled. Eden Valley, the northern part of Paradise Valley, is about 20 miles long and five wide. In general features it resembles the lower portion, the whole, comprising one of the richest farming sections in the State. Camp Scott and Santa Rosa are situated in the head of the valley, and other small towns have sprung up at other points.

Paradise Creek is a clear, cold mountain stream, upon which are a number of grist and saw mills. It rises in the Owyhee Mountains and flows through these valleys to the Humboldt River. Salmon trout of enormous size are found in the stream and its tributaries. Bears, deer, silver-gray foxes, and other game, abound on the hills which border the valley.

These valleys—the Humboldt for 50 miles east and west, and the adjoining mountains—are the stock-raisers' paradise.

Tens of thousands of cattle are now roaming along the Humboldt and adjoining valleys, and surrounding hills. It is computed that there are not less than 350,000 head between Promontory Mountain and the Sierra Nevada Mountains. One firm near this station has over 40,000 head, and one range fenced of 28,000 acres.

A few miles after leaving Argenta, Reese River Valley joins the Humboldt—coming in from the south. It is very diversified in feature, being very wide at some points—from seven to ten miles—and then dwindling down to narrow strips of meadow or barren sand. Some portions of the valley are susceptible of cultivation, and possess an excellent soil. Other portions are barren sand and gravel wastes, on which only the sage-brush flourishes. This valley is also known by old emigrants as "Whirlwind Valley," and passengers will frequently see columns of dust ascending skywards. Reese River, which flows through this valley, rises to the south, 180 to 200 miles distant. It has many tributaries, which find their source in the mountain ranges that extend on either side of the river its entire length. It sinks in the valley about 20 or 30 miles before reaching the Humboldt. During the winter and spring floods, the waters reach the Humboldt, but only in very wet times.

Near where Reese River sinks in the valley was fought the celebrated battle between the Whites and Indians—settlers and emigrants, 30 years ago—which gave the general name of Battle Mountain to these ranges. A party of marauding Shoshone Indians had stolen a lot of stock from the emigrants and settlers, who banded themselves together and gave chase. They overtook them at this point, and the fight commenced. From point to point, from rock to rock, down to the water's edge they drove the red skins, who, finding themselves surrounded, fought with the stubbornness of despair. When night closed in, the settlers found themselves in possession of their stock and a hard-fought field. How many Indians emigrated to the Happy Hunting Grounds of the spirits no one knew, but from this time forward the power of the tribe was broken.

From Argenta, it is 11.8 miles to

Battle Mountain—This is a dinner station for passenger trains from both the East and West, where trains stop 30 minutes. The waiters are Chinese, and very lively while serving a good meal.

Water for the little fountain in front of the Battle Mountain House, the railroad, and the town, is conducted in pipes from a big spring in the side of the mountain, three miles to the south.

Battle mountain is the distributing point for a great number of mining districts, towns and camps, both north and south of the road.

Stages and fast freight lines leave daily for the northward: To Tuscarora, 68 miles; Rock Creek, 80 miles; Cornucopia, 100 miles. The shipments from Battle Mountain Station average over 500 tons per month, and is increasing.

The surrounding country is alive with herds of cattle, particularly on the north side of the river, and this place has become quite a point for cattle buyers from California to congregate.

Nevada Central Railway.
General Offices, Battle Mountain.
S. H. H. CLARK, (of the U. P. R. R.).... President.

This is a three-foot narrow gauge, organized September 2, d 1879, completed the December following, and sold to Union Pacific parties in the Summer of 1881. Its general course is to the southward, up the valley of Reese River, 93 miles to Austin. The grades are easy, and the country tributary rich in mines, agricultural lands, cattle and sheep. The principal stations and distances between, are: Galena, 11 miles; Mound Springs, 10 miles; Bridges', 22 miles; Walters', 13 miles; Hallsvale, 10 miles; Caton's, 10 miles; Ledlies', 10 miles; Austin, 7 miles. There are twenty mining districts tributary to this road, among which are Battle Mountain, Galena, Austin, Lewis, Reese River, Washington, Kinsley, Dun Glen, Cortez, Diamond, Humboldt, Grass Valley and Belmont. In all these districts rich mines are being worked; stamp mills and smelting furnaces are numerous. Railroads and low freights have resulted in vigorous development and better machinery.

Stage connections are made at several points on the road, and at Austin for all points in Central and Southern Nevada.

AUSTIN—is situated near the summit of the Toiyabe Range, on the ground where the *first* silver ore was discovered in this district, in May, 1862. The discoverer, W.

M. Talcott, located the vein and named it Poney.

As soon as it became known, prospectors flocked in, and the country was pretty thoroughly prospected during 1862 and 1863. Many veins were located, some of them proving very valuable. Mills were erected at different points, and from that time forward the district has been in a prosperous condition.

Austin contains a population of about 2,000, nearly all of whom are engaged in mining operations. The town has some extensive stores and does a very large business in the way of furnishing supplies for the mining camps surrounding it, for from 50 to 100 miles. The *Reese River Reveille* is a live daily published here. To the south of Battle Mountain Station, about 20 miles, are several hot springs, strongly impregnated with sulphur and other minerals, but they attract no particular attention, being too common to excite curiosity.

Leaving the station we skirt the base of the mountains to the left, leaving the river far to our right over against the base of Battle Mountain. We are now in the widest part of the valley, about opposite the Big Bend of the Humboldt.

After passing the Palisades the river inclines to the south for about 30 miles, when it sweeps away to the north, along the base of Battle Mountain, for 30 miles further; then turning nearly due south, it follows that direction until it discharges its waters in Humboldt Lake, about 50 miles by the river course from the great elbow, forming a vast semi-circle, washed by its waters for three-fourths of the circumference. This vast area of land, or most of it, comprising many thousand acres of level upland, bordered by green meadows, is susceptible of cultivation when irrigated. The sage-brush grows luxuriantly, and where the alkali beds do not appear, the soil produces a good crop of bunch-grass. The road

TRUCKEE RIVER.

takes the short side of the semi-circle keeping close to the foot of the isolated Humboldt Spur. On the opposite side of the river, behind the Battle Mountain Range, are several valleys, watered by the mountain streams, and affording a large area of first-class farming land. Chief among these is QUINN'S VALLEY, watered by the river of that name. The arable portion of the valley is about 75 miles long, ranging in width from three to seven miles. It is a fine body of valley land, capable of producing luxuriant crops of grain, grass or vegetables. The hills which enclose it afford excellent pasturage. Timber of various qualities—spruce and pine predominating—is found in the gulches and ravines of the mountains.

QUINN'S RIVER, which flows through this valley, is a large stream rising in the St. Rosa Hills of the Owyhee range, about 150 miles distant. From its source the general course of the river is due south for about 80 miles, when it turns and runs due west until it reaches Mud Lake. During the summer but little, if any, of its waters reach that place, being absorbed by the barren plain which lies between the foothills and the Humboldt River. Near the head-waters of Quinn's River, the CROOKED CREEK, or Antelope, rises and flows due north for about 50 miles, when it empties its waters into the Owyhee River. The head-waters of the streams which run from the southern slops of the Owyhee Mountains are well supplied with salmon and trout, and other varieties of fish. Quail, grouse, and four-footed game are abundant in the valleys and timbered mountains.

Piute—is 4.9 miles west of Battle Mountain Station. Here passenger trains from the east and west meet.

Coin—a flag station, is 7.8 miles west of Piute.

Stone House—is 7.1 miles further. This place was once an old trading post, strongly fortified against Indian attacks. The Stone House stood at the foot of an abrupt hill, by the side of a spring of excellent water, but is now a mass of ruins. To the south of this station are more of the many hot springs that abound in the Nevada Basin.

We cross a broad sage-brush bottom, the soil of which in places is sandy and in others alkaline, and then wind along around the base of a mountain spur that shoots away to the northward, and come to

Iron Point—a small side-track, 12.4 miles from Stone House. Here are located a few cattle-yards and shutes for loading cattle. At this station the bluffs draw close and high on each side, with the river and a narrow strip of meadow land on the right. After passing around the point and through numerous cuts for two miles, the canyon widens into a valley for several miles, then closes in, and the train passes around another rocky point into another valley, and stops at

Golconda—a station 11.4 miles from Iron Point. This is a small station with a few good buildings. Large herds of cattle range near by in the surrounding valleys, and on the bluffs. Rich mines of gold and silver are located both to the north and south; one, the Golconda mine and mill, only three miles distant to the south. Close to the west of the station, under the edge of the bluff on the right, are located some hot springs. Here some of the settlers—as at Springville, Utah—use the hot water for their advantage—one for milling the other for stimulating the soil.

Continuing our journey, we pass over a broad sage-brush plain, with wide meadows beyond, for 10.9 miles to

Tule—an unimportant station. Passing on down the valley we skirt the hills on our left, drawing still closer, in some places the spurs reaching to the track. On our left is an opening in the hills, from whence a canyon opens out near the roadside. It is about five miles long, containing living springs. Here were discovered the first mines in this part of Nevada. In the spring of 1860, Mr. Barbeau, who was herding stock for Copernig, discovered the silver ore, and from this beginning, the prospecting was carried on with vigor, which resulted in locating many very valuable bodies of ore.

From Tule it is 5.8 miles to the end of the Humboldt division, at Winnemucca.

Winnemucca—is the commencement of the Truckee Division. The station was named after a chief of the Piute Indians who formerly resided here. Elevation 4,331 feet. Distance from Omaha 1,451 miles; from San Francisco 463 miles.

Winnemucca is the county seat of Humboldt county, and is composed of what is known as the old and new towns, which, together, contain about 800 inhabitants. The old town is situated on the low land directly fronting the station, about 300 yards distant. Though so near, it is hid from sight until you approach the bank and look over. The town contains about 150 buildings of all sorts, among which are a fine new court-house, stamp mill, smelting works, flouring mill, and a good hotel, the Central Pacific.

The buildings with few exceptions, are of wood, new, and like most of the railroad towns, more useful than ornamental. The company have located here the usual division work and repair shops, including a 16-stall round-house. They are built of wood in the most substantial manner, as are all the shops along the road.

There is considerable mining going on around and near Winnemucca, and quite a number of mills and furnaces are in operation, all of which are said to be doing well. In the Winnemucca Range, many lodes of silver-bearing ore have been located which promise a fair return for working.

Stages leave here daily for Camp McDermott, 80 miles, fare $15; Paradise, 40 miles, fare $5; Silver City, Idaho, 200 miles, fare $40; Boise City, 255 miles, fare $40. Fast freight trains run from this station to all the above towns, and to the mining camps in the adjoining country.

HUMBOLDT HOUSE —See Next Page.

The *Silver State*, a weekly newspaper, is published here. Winnemucca is the great distributing point for a number of mining districts to the north and south, and does an extensive freighting business. Many herds of cattle and sheep range the adjacent country, and large numbers are brought here for shipment to California.

Mud Lake—is about 50 miles west of this station, across the Humboldt, which here turns to the south, and is one of those peculiar lakes found in the great basin of Nevada. The lake receives the waters of Quinn's River and several smaller tributaries during the wet season. It has no outlet, unless its connection with Pyramid and Winnemucca lakes could be so designated. It is about 50 miles long by 20 wide, in high water; in summer it dwindles down to a marshy tract of land and a large stagnant pool. At the head of the lake is Black Rock, a noted landmark in this part of the country. It is a bold, rocky headland, rising about 1,800 feet above the lake, bleak, bare, and extending for several miles. It is an isolated peak in this desert waste, keeping solitary guard amid the surrounding desolation.

Pyramid Lake—is about 20 miles south of Mud Lake, which receives the waters of Truckee River. It is about 30 miles long by 20 wide during the wet seasons. The quality of the water is superior to that of Mud Lake, though the water of all these lakes is more or less brackish.

Winnemucca Lake — a few miles east of Pyramid Lake, is another stagnant pond, about fifteen miles long by ten wide. This lake is connected with Pyramid Lake by a small stream, and that in turn with Mud

No. 42 Annex. **Sierra Nevada Mountains.**—The large illustration, No. 14, of the Sierras, is from a photograph, and affords a beautiful view of the highest point of the Sierra Nevada Mountains, passed over by the Central Pacific railroad. There are to be seen a succession of tunnels and snow-sheds, which extend without a break for 28 miles; below is the "Gem of the Sierras," Donner Lake. (For description, see pages 135 and 156 of this book.)

Lake, but only during high water, when the streams flowing into them cause them to spread far over the low, sandy waste around them.

Returning to Winnemucca, we resume our journey. The road bears away to the southward, skirting the low hills which extend from the Winnemucca Mountain toward Humboldt Lake. The general aspect of the country, is sage and alkali on the bottoms, and sage and bunch-grass on the bluffs.

Rose Creek—comes next, 8.88 miles, and 10.2 more,

Raspberry Creek—Both the last named are unimportant stations where passenger trains seldom stop. They are each named after creeks near the stations, but *why* one should be Rose Creek and the other Raspberry Creek, we never could learn. We saw no indication of roses or raspberries at either creek. But they *do* have queer names for things in this country. Where they call a Jack rabbit a "narrow gauge mule," we are prepared to hear sage-brush called roses, and greasewood raspberries.

Mill City—is 7.49 miles from Raspberry, and has some good buildings, among which is a fine hotel, close to the track on the right—and large freight warehouses; also cattle pens and shutes for shipping cattle, great numbers of which roam over the bottoms and adjacent bluffs. Stages leave this station on arrival of the cars for Unionville, a thrifty and promising silver mining town, 18 miles distant to the southward.

Humboldt—11.7 miles from Mill City, is a regular eating station where trains for the West stop 30 minutes for supper, and those for the East the same time for breakfast. The meals are the *best* on the road.

Here will be found the clearest, coldest mountain spring water along the road, and viewing it as it shoots up from the fountain in front of the station, one quite forgets the look of desolation observable on every side, and that this station is on the edge of the great Nevada Desert.

It is worth the while of any tourist who wishes to examine the wonders of nature to stop here and remain for a few days at least—for one day will not suffice—although to the careless passer-by the country appears devoid of interest. Those who wish to delve into nature's mysteries can here find pleasant and profitable employment. The whole sum of man's existence does not consist in mines, mills, merchandise and money. There are other ways of employing the mind besides bending its energies to the accumulation of wealth; there is still another God, mightier than Mammon, worshiped by the *few*. Among the works of His hands—these barren plains, brown hills and curious lakes—the seeker after knowledge can find ample opportunities to gratify his taste. The singular formation of the soil, the lava deposits of a by-gone age, the fossil remains and marine evidences of past submersion, and, above all else, the grand and unsolved problem by which the waters that are continually pouring into this great basin are prevented from overflowing the low land around them, are objects worthy of the close attention and investigation of the scholar and philosopher. From this station, the noted points of the country are easy of access.

Here one can observe the effects of irrigation on this sandy, sage-brush country. The garden at the station produces vegetables, corn and fruit trees luxuriantly, and yet but a short time has elapsed since it was covered with a rank growth of sage-brush.

About seven miles to the northeast may be seen Star Peak, the highest point in the Humboldt Range, on which the snow continues to hold its icy sway the whole year round. Two and one-half miles southeast are the Humboldt mines—five in number—rich in gold and silver. The discovery of a borax mine near the station has been recently reported. Five miles to the northwest are the Lanson Meadows, on which are cut immense quantities of as good grass as can be found in the country. Thirty miles north are the new sulphur mines, where that *suspicious* mineral has been found in an almost pure state, and so hard that it requires to be blasted before it can be got out of the mine, and in quantities sufficient to enable those operating the mine to ship from 20 to 30 car-loads a week to San Francisco. Leaving Humboldt, about one mile distant, on the right near the road, is another sulphur mine—but it is undeveloped.

Rye Patch—is 11.23 miles from Humboldt, named for a species of wild rye that grows luxuriantly on the moist ground near the station.

To the left of the road, against the hillside, is another hot spring, over whose surface a cloud of **vapor** is generally floating.

The medicinal qualities of the water are highly spoken of by those who never tried them, but we could learn of no reliable analysis of its properties. To the right of the track is located a 10-stamp quartz mill, the ores for which come from the mountains on the left. The Rye Patch and Eldorado mining Districts are to the left, from five to fifteen miles distant, for which most of the supplies are hauled from this station.

Oreana—is reached after passing over a rough, uneven country for 10..2 miles. To the southeastward are located a number of mining districts, in which are located a number of stamp mills and smelting works.

Leaving the station to the west, the long gray line of the desert is seen cheerless and desolate. We draw near the river again and catch occasional glimpses of narrow, green meadows, with here and there a farm-house by the river-side; pass a smelting furnace and stamp mill—on the Humboldt River—to the right, which has been dammed near by to afford water power. Five miles from the station we cross a Howe truss bridge over the river, which here winds away on our left until it reaches the lake a few miles beyond. The current and volume of the river has been materially reduced since we left it at the head of the Big Bend.

Lovelocks—11.86 miles from Oreana, derives its name from an old meadow ranche which is situated near, upon which, during the summer, large quantities of hay are cut and baled for market. Some attempt has been made at farming near by, but little of the country is adapted to the purpose. Cattle and sheep raising is the principal occupation of the people—but few herds will be seen hereafter on our route. Reliable authority places the number of head of stock now along the Humboldt River and adjacent valleys—1878—cattle, 352,000 head; sheep, 30,000 head. During the year there were shipped from the same section to Chicago 20 car-loads of cattle, and 486 car-loads to San Francisco.

Passing on over alkali beds, sand-hills and sage-brush knobs, the meadow-lands along the bottom get narrower, and finally fade from sight altogether, and we find ourselves fairly out on the

GREAT NEVADA DESERT—This desert occupies the largest portion of the Nevada Basin. In this section, to the northward, is Mud Lake, Pyramid Lake, Humboldt, Winnemucca and Carson lakes, which receive the waters of several large rivers and numerous small creeks. As we have before stated, they form a portion of that vast desert belt which constitutes the central area of the Nevada Basin. The desert consists of barren plains destitute of wood or water, and low, broken hills, which afford but little wood, water or grass. It is a part of that belt which can be traced through the whole length of the State, from Oregon to Arizona, and far into the interior of that Territory. The Forty Mile Desert, and the barren country east of Walker's Lake, are part of this great division which extends southward, continued by those desolate plains, to the east of Silver Peak, on which the unfortunate Buel party suffered so terribly in their attempt to reach the Colorado River. Throughout this vast extent of territory the same characteristics are found—evidences of recent volcanic action—alkaline flats, bassalt rocks, hot springs and sandy wastes abounding in all portions of this great belt.

Although this desert is generally spoken of as a sandy waste, sand does not predominate. Sand hills and flats occur at intervals, but the main bed of the desert is lava and clay combined—one as destitute of the power of creating or supporting vegetable life as the other. The action of the elements has covered these clay and lava deposits with a coarse dust, resembling sand, which is blown about and deposited in curious drifts and knolls by the wind. Where more of sand than clay is found, the sage-brush occasionally appears to have obtained a faint hold of life, and bravely tries to retain it.

Granite Point—a flag station, is 8.33 miles from Lovelock's. Passing on, an occasional glimpse of Humboldt Lake, which lies to the left of the road, can be obtained, and in full view 7.65 miles further, at

Brown's—This station is situated about midway of the northern shore, directly opposite

HUMBOLDT LAKE—This body of water is about 35 miles long by ten wide, and is in reality a widening of the Humboldt River, which after coursing through 350 miles of country, empties its waters into this basin. Through this basin the water flows to the plains beyond by an outlet at the lower end of the lake, uniting with the waters of the sink of Carson Lake which lies about ten miles distant. During the wet season, when the swollen rivers have overflowed the low lands around the lakes and united them, they form a very respectable sheet of water, about eighty miles or more in length, with a large river emptying its waters into each end; and for this vast volume of water there is no visible outlet.

Across the outlet of Humboldt Lake a dam has been erected, which has raised the water about six feet, completely obliterating the old emigrant road which passed close to the southern shore. The necessities of mining have at length utilized the waters of the lake, and now they are employed in turning the machinery of a quartz mill. In the lower end of the lake is an island—a long narrow strip of land—which extends up the lake and near the northern shore. Before the dam was put in the outlet, this island was part of the main land. There are several varieties of fish in the lake, and an abundance of water-fowl during portions of the year.

Leaving Brown's, and passing along the shore of the lake for a few miles, an intervening sand ridge hides the lake from our sight, and about eight miles west we obtain a fine view of the Sink of Carson Lake, which is a small body of water lying a few miles north of the main Carson Lake, and connected with that and the Humboldt during the wet season.

CARSON LAKE lies directly south of Humboldt Lake, and is from 20 to 25 miles long, with a width of ten miles. In the winter its waters cover considerable more area, the Sink and lake being one.

The Carson River empties into the southern end of the lake, discharging a large volume of water. What becomes of the vast body of water continually pouring into these lakes, is the problem yet unsolved. Some claim the existence of underground channels, and terrible stories are told of unfortunate people who have been drawn down and disappeared forever. These stories must be taken with much allowance. If underground channels exist, why is it that the lakes, which are 10 to 15 miles apart in low water are united during the winter floods? And how is it, that when the waters have subsided from these alkaline plains, that no openings for these channels are visible? The only rational theory for the escape of the water is by evaporation. Examine each little stream bed that you meet with; you find no water there in the summer, nor sink ho'es, yet in the winter their beds are full until they reach the main river. The sun is so powerful on these lava plains in summer that the water evaporates as soon as it escapes from the cooling shadows of the hills. By acutal experiment it has been demonstrated that at Carson and Humboldt lakes the evaporation of water is equal, in the summer, to six inches every 24 hours. In the winter, when the atmosphere is more humid, evaporation is less, consequently the waters spread over a larger area.

CARSON RIVER, which gives its name to the lake, rises in the eastern slope of the Sierra Nevada Mountains, south of Lake Tahoe and opposite the head waters of the American River. From its source to its mouth is about 150 to 200 miles by the river's course. From its source its course is about due north for about 75 miles, when it turns to the east, and follows that direction until it enters the lake.

Under the general name of Carson Valley, the land bordering the river has long been celebrated as being one of the best farming sections in the State. The thriving towns of Carson City and Genoa are situated in the valley, though that portion around Carson City is frequently designated as Eagle Valley. The upper portion, from Carson to the foot hills, is very fertile, and yields handsome crops of vegetables, though irrigation is necessary to insure a good yield. In some portions the small grains are successfully cultivated, and on the low lands an abundant crop of grass is produced. The valley is thickly settled, the arable land being mostly occupied. South and west of the head waters of Carson River, the head waters of Walker's River find their source. The west fork of Walker's River rises within a few miles of the eastern branches of the Carson. The east fork of Walker's River runs due north until joined by the west fork, when the course of the river is east for about forty miles, when it turns to the south, following that direction until it reaches Walker's Lake, about forty miles south of the sink of the Carson, having traversed in its tortuous course about 140 miles. In the valleys, which are found at intervals along the rivers, occasional spots of arable land are found, but as an agricultural country the valley of Walker's River is not a success.

WALKER LAKE is about 45 miles long by 20 miles wide. Like all the lakes in the basin, it has no outlet. The water is brackish and strongly impregnated with alkali. The general characteristics of the other lakes in the great basin belong to this also; the description of one embracing all points belonging to the others.

White Plains—is 12.17 miles west of Brown's. This station is the lowest elevation on the Central Pacific railroad east of the Sierra Nevada Mountains. As indicated by the name, the plains immediately around the station are white with alkali, solid beds of which slope away to the sinks of Carson and Humboldt lakes. No vegetation meets the eye when gazing on the vast expanse of dirty white alkali. The sun's rays seem to fall perpendicularly down on this barren scene, burning and withering, as though they would crush out any attempt which nature might make to introduce vegetable life.

The water to fill the big tank at the station is pumped from the "Sink" by means of a stationary engine, which is situated about midway between the station and the Sink.

Mirage—is 7.96 miles from White Plains. This station is named for that curious phenomenon, the mirage (meerazh) which is often witnessed on the desert. In early days the toil-worn emigrant, when urging his weary team across the cheerless desert, has often had his heart lightened by the sight of clear, running streams, waving trees and broad, green meadows, which appeared to be but a little distance away. Often has the unwary traveler turned aside from his true course and followed the vision for weary miles, only to learn that he had followed a phantom, a will-o'-the-wisp.

What causes these optical delusions no one can tell, at least we never heard of a satisfactory reason being given for the appearance of the phenomenon. We have seen the green fields, the leafy trees and the running waters; we have seen them all near by, as bright and beautiful as though they really existed, where they appeared too, in the midst of desolation, and we have seen them vanish at our approach. Who knows how many luckless travelers have followed these visions, until, overcome with thirst and heat, they laid down to die on the burning sands, far from the cooling shade of the trees they might never reach; far from the music of running waters, which they might hear no more.

Onward we go, reclining on the soft cushions of the elegant palace car, thirty miles an hour; rolling over the alkali and gray lava beds, scarcely giving a thought to those who, in early days, suffered so fearfully while crossing these plains, and, perchance, left their bones to bleach and whiten upon these barren sands.

Hot Springs—is 6.57 miles west of Mirage. Here, to the right of the road, can be seen more of these bubbling, spurting curiosities—these escape pipes, or safety valves for the discharge of the super-abundant steam inside the globe, which are scattered over the great basin. Extensive salt works are located at this station, from which a car-load or more of salt is shipped daily. The salt springs are about four miles west of the station.

The Saxon American Borax Co. have erected works here which cost about $200,-000. They are situated a half-mile south of the station, in plain view.

Passing on, we find no change to note, unless it be that the beds of alkali are occasionally intermixed with brown patches of lava and sand. A few bunches of stunted sage-brush occasionally break the monotony of the scene. It is worthy of notice that this hardy shrub is never found growing singly and alone. The reason for it is evident. No single shrub could ever maintain an existence here. It must have help; consequently we find it in clumps for mutual aid and protection.

Desert—is 11.7 miles from Hot Springs. This is, indeed, a desert. In the next 5.97 miles, we gain about 100 feet altitude, pass Two Mile Station, descend 82 feet in the next 2.37 miles, and arrive at

Wadsworth—This town is situated on the east bank of the Truckee River and the western border of the desert, and contains some good buildings, and a population of about 600.

The division workshops are located here, and consist of a round-house of 20 stalls, car, machine and blacksmith shops. Adjoining the workshops, a piece of land has been fenced in, set out with trees, a fountain erected, and a sward formed, by sowing grass-seed and irrigating it—making a beautiful little oasis. Considerable freight is shipped from this station to mining camps to the south.

Pine Grove Copper Mines lies six miles south of the town. They attract little attention, that mineral not being much sought after. Ten miles south are the Desert mines, which consist of gold-bear-

SNOW SHEDS—SEE PAGE 156

The Piute Indians have two reservations; one is situated eighteen miles northward, and another to the southeast, at Walker Lake.

Leaving Wadsworth, we cross the Truckee River, on a Howe truss bridge, our course being to the southwest. This stream rises in Lakes Tahoe and Donner, which lie at the eastern base of the Sierras, about 80 miles distant. From its source in Lake Tahoe, the branch runs north for about twelve miles, when—near Truckee City—it unites with Little Truckee, the outlet of Donner Lake, and turns to the east, following that course until it reaches this place, where it turns north about 25 miles, branches, and one portion enters Pyramid and the other Winnemucca Lake.

The level lands bordering the Truckee ing quartz lodes. Some of the mines there are considered very rich. Ninety miles south, at Columbus, are located the famous Borax mines of Nevada, said to be very rich. consist mostly of gravelly upland covered with sage-brush. It is claimed that they might be rendered productive by irrigation, and the experiment has been tried in a small way, but with no flattering result.

The upper portions of the valley, especially that which borders on Lake Tahoe, is excellent farming land. Between these two points—the meadows and the lake—but little meadow land is found, the valley being reduced to narrow strips of low land in the canyons and narrows, and broad, gravelly uplands in the more open country.

Salvia—a small side-track, comes next after Wadsworth, 7.25 miles distant. Soon we pass around a lava bluff, called Red Rock, on the right, and 7.55 miles brings us to

Clark's—in a round valley, surrounded by fenced fields, where good crops of vegetables are raised for market in mining towns to the south. From Clark's, it is 11.96 miles to

Vista—a small station situated on the northern edge of what is known as the Truckee Meadows. In early days these meadows were a noted rendezvous of the emigrants, who camped here for days to recruit their teams after crossing the desert. They have an extent of about twelve miles in length by about two miles in width, inclosing considerable excellent grass land. Vegetables and small grains are successfully cultivated on portions of the moist land.

Reno—is 7.64 miles from Vista; is the county seat of Washo county, and contains a population of about 1,500. It was named in honor of General Reno, who was killed at the battle of South Mountain. This city has rapidly improved within the last six years, and now contains five church edifices, two banks, a fine court-house, a number of good business blocks, a steam fire department, several small factories, two daily newspapers, the *Journal* and the *Gazette*, and is the distributing point for an enormous freighting business to the north, as well as the south. Some good agricultural land surrounds the town, as well as many herds of cattle and sheep. The State Agricultural grounds are located here, in which is a very fine race track. The Lake House is the principal hotel. Stages leave daily for Susanville, 90 miles.

The English works are near the town, affording excellent means by which to test the ores discovered in the neighborhood.

The greatest mining region in the world is reached via Reno. Virginia City, located over the mountain to the southeast, from this station is *only* 21 miles distant, by the old wagon road, but by rail it is 52 miles.

Before the completion of this road, Virginia City was reached by stage, over a fearfully steep zig-zag mountain road, but the difference between the "old and the new" is more than made up in the comfort of the passage if not in time.

At the time when these stages were running to convey passengers, a fast "Pony Express" was run for the purpose of carrying Wells, Fargo & Co.'s letter bags. This pony express was once a great institution. Approaching Reno, the traveler could have observed that the mail express bags were thrown from the cars before the train had ceased its motion. By watching the proceedings still further he would see that they are transferred to the backs of stout horses, already bestrode by light, wiry riders. In a moment all is ready, and away they dash under whip and spur to the next station, when, changing horses, they are off again. Three relays of horses were used, and some "good time" was often made by these riders.

Let us take a run up and see this
Huge Bonanza Country.

Virginia & Truckee Railroad.

Principal office, Carson, Nevada.

D. O. MILLS.. ......*Pres't*........*San Faancisco.*
H. M. YERINGTON....*Gen'l Sup't*....*Carson, Nev.*
D. A. BENDER......*Gen'l T. A.* ...   "   "

This road was commenced at Carson City, March 19th, 1869, completed to Virginia City in the following November, and to Reno in 1871. The length is 52.2 miles; the grade in places is 115 feet to the mile, and there are six tunnels, of the aggregate length of 3,000 feet; the shortest curve is 19 degrees—between Gold Hill and Virginia City.

The train for Virginia stands on the opposite side of the station building from the C. P. Let us step on board. From Reno, our course is east of south, crossing a portion of the Truckee Meadows, a few well-cultivated fields and greater quantities of sage and grease-wood. The first station on the hills is 8.5 miles from Reno, called

ANDERSON'S—but we do not stop. Crossing the river, we pass the first of a series of V-shaped flumes, which are constructed to float down wood and lumber from the mountains. The one we are now passing is said to be 15 miles in length.

HUFFAKER'S — comes next—after 3.6 miles, where another flume is passed, both of which are on the right, and land their

COLUMBIA RIVER, OREGON. (See Annex No. 38.)

freight—wood—close to the track of our road. Along here we find some broad meadows on the left, but sage on the right. Passing over 1.9 miles from the last station we arrive at

BROWN'S—Here is the end of another flume, and 2.4 miles further, and after curving around to the right, up a broad valley, arrive at the

STEAMBOAT SPRINGS—which are eleven miles south of Reno. There are several of these curious springs within a short distance of the road. They are near each other, all having a common source, though different outlets, apparently. They are situated to the right of the road, just before reaching the station, a short distance above the track; are strongly impregnated with sulphur, and are very hot, though the temperature varies in different springs.

They are said to possess excellent medicinal qualities. At times they are quite active, emitting jets of water and clouds of steam, which at a distance resemble the blowing off of steam from a large boiler. The ground around them is soft and treacherous in places, as though it had been thrown up by the springs, and had not yet cooled or hardened. It is related that once upon a time, when a party of emigrants, who were toiling across the plains, arrived near these springs about camping time, they sent a man ahead—a Dutchman—to look out for a suitable place for camping— one where water and grass could be obtained. In his search the Dutchman discovered these springs, which happened to be quiet at the time, and knelt down to take a drink of the clear, nice-looking water. Just at that instant a jet of spray was thrown out and over the astonished Dutchman. Springing to his feet, he dashed away to the train, shouting at the top of his voice, "Drive on! drive on! h—l is not five miles from this place!" Guess the innocent fellow firmly believed what he uttered.

The traveler will find the springs sufficiently interesting to repay him for the trouble of pausing here awhile and taking a look around. At the station will be found a comfortable hotel, ample bath accommodations, and about a half-dozen residences.

Leaving the springs, our course is south, in a narrow valley, in which is some good farming land, with high bluffs on each side; cross and re-cross Steamboat Creek, curve to the right through a narrow canyon where there are many evidences of placer mining; twist and climb, between high projecting cliffs, and suddenly emerge into a great valley, and stop at

WASHOE CITY—Ah! here is a child of the past. In its palmy days Washoe was as lively a city, or camp, as could be found in the whole mining region. Where thousands of people once toiled, there are now only a few dozen, and most of those are engaged in other pursuits than mining. On the right is another flume for floating wood from the mountains on the westward.

The valley near this place is from half to a mile in width, surrounded by high mountains, the highest peak of which is Mt. Rose, at the south end of the valley, over 8,000 feet in height. The mountains on the east are bare, with some sage and bunch-grass, while those on the west are covered, the greater part, with pine and spruce timber.

Leaving Washoe, we pass, on the left, the Old Ophir Mill, a stone building—now in ruins—which once gave employment to about 150 men, besides a $30,000 a year superintendent.

FRANKTOWN—is 4.7 miles from Washoe, a growing station in the midst of Washoe valley; population about 150. A "V" flume comes down on the right. There are some good farming lands along here, but the greater portion is only adapted for grazing purposes.

WASHOE LAKE, on the left, is about four miles long and one mile wide. On the east side of the lake is Bower's Hotel, a great resort in the summer for pic-nic parties from the cities to the southward. From Franktown it is 2.6 miles to

MILL STATION—near the site of an old mill, where another "V" flume comes down from the mountains on the right, making six since leaving Reno.

Proceeding south, the valley narrows and is soon crowded out completely, and we rise up onto the southern rim; and then, a look back will take in the whole valley and lake from end to end, and a beautiful view it is. At this narrow gorge the railroad track crosses the great

WATER SYPHON, through which the water is conducted from the Sierra Nevada Mountains, on the west, across this narrow gorge, for supplying Virginia City, Gold Hill and Silver City. It is an achievement which finds no parallel in the history of hydraulic engineering. The total length of the pipe used is but little less than *seven* miles.

At the point where the water is taken from Dall's Creek, up in the Sierras, it is brought in an 18-inch flume, four miles long, to the point of a spur on the west side of Washoe Valley, the height of which is 2,100 feet above the railroad track. At the point where the water in the flume reaches the spur it is received in an iron pipe, which, after running along the crest, descending, crossing and ascending twelve steep cañons on its route, finally descends into this gorge, crosses it from the west, and ascends the cliff on the east side to a height of 1,540 feet, where it is taken by another flume and conducted to a reservoir on the Divide between Virginia City and Gold Hill. The pipe has an orifice twelve inches in diameter, and where the pressure is the greatest, is five-sixteenths of an inch in thickness, riveted with five-eighth inch rivets in double rows. Where the pressure lessens, the thickness of the material gradually decreases.

The amount of rolled iron used in constructing the pipe was 1,150,000 lbs. One million rivets and 52,000 lbs of lead were used on the pipe. Before being used each length of pipe—26 feet long each—was heated to a temperature of 380 degrees, and submerged in a bath of asphaltum and coal tar, to prevent corroding. At the bottom of each depression there is a blow-off cock, for removing any sediment that might accumulate, and at each elevation is an air-cock to let out the air when the water is first introduced into the pipes. Where the water pipe runs under the railroad track, it is surrounded by a massive iron sleeve, 12 feet long, to protect it from the jar of passing trains. This pipe is capable of furnishing 2,000,000 gallons daily. The whole cost of construction was seven hundred and fifty thousand dollars. A movement is now on foot to lay another and larger

Giant Geyser, Yellowstone Park.

No. 24 Annex. Hauling Ores in Hides—On a recent visit to Little Cottonwood Cañon, Utah, we saw a very novel contrivance being used by the Emma Mining Company, whose mine is situated about 1,000 feet above the railroad depot, on the side of the mountains. The snow was several feet deep, and the ore was being hauled down to the depot in drags. The drags consisted of a green ox hide. The ore is first sewed up in sacks of 100 lbs. each, then placed on the hide, which has loops around the edge, and when 15 of these sacks are in position, a rope is run through the loops in the hide, the edges drawn together, then a mule or horse is hitched to the head portion of the hide—with the hair outwards—and a brake to the tail. The brake is of iron, shaped like a horse-shoe, with teeth that drag through the snow, holding back.

In coming down the hill the driver stands on the hide, but when it becomes necessary on account of heavy grade to "*down brakes,*" the driver changes his position and stands on the horse-shoe instead of the hide.

THE first half of a wagon train is called the "right wing," the other half the "left wing." In forming a corral, the wagons of the "right wing" form a half circle on the right hand side of the road, hauled close together, teams on the outside; the "left wing" forms on the left side in the same manner, leaving a passage way open at the front and rear ends of each "wing," called "gaps."

pipe near the present one.

From Mill station it is 2.5 miles to LAKE VIEW—situated just south of the southern rim of the Washoe Valley, in the gorge above alluded to. Soon after leaving the station, we pass into a tunnel, through a projecting cliff, which shoots out from the right, and comes out on the side of the mountain overlooking the beautiful Eagle or Carson Valley. Away in the distance, four miles away, can be seen Carson City, a little further, Carson River, and beyond both, the mountains, just beyond which is Walker's River, and then Walker's Lake.

Winding and descending around the side of the mountain, through numerous rocky cuts, a distance of 4.6 miles from Lake View, we arrive at

CARSON CITY—the capital of the State of Nevada. It is situated in Eagle Valley, on the Carson River, at the foot of the eastern base of the Sierras, and contains about 4,000 population; is 31.1 miles south from Reno, and 21.1 miles southwest from Virginia City. It is the oldest town in the State, and has a good many fine private and public buildings. The town is tastefully decorated with shade trees, and has an abundance of good water. The United States Branch Mint of Nevada is located here. The capital is located in the center of a Plaza, and is surrounded by an iron fence. It is two story and basement, made of cut stone.

Carson is a busy city, has some good blocks of buildings, several good hotels, chief of which is the Ormsby; four churches, five schools, two daily newspapers—the *Morning Appeal* and the *Nevada Tribune*. Here are located the machine shops of the Railroad Company, and several manufactories. Carson City is in the center of the best farming land on Carson River, and the best in this part of the State, and is the distributing point for a vast amount of freight, destined for the southern mines.

To the south of the city, comes down the large "V" flume from the Sierras, via Clear Creek Canyon, owned by the Railroad Company through which thousands of cords of wood and millions of feet of lumber are landed at Carson weekly. Four and six horse coaches leave Carson daily, carrying passengers, mails and express. From Carson to Monitor, the distance is 46 miles, and to Silver Mountain, in Alpine county, Cal., 54 miles; to Bishop's Creek, 192 miles;

Benton, 150 miles in Mono county, Cal., Sweetwater, 73 miles, Aurora, 105 miles, Bodie, 119 miles, Mariette, 145 miles, Bellville, 155 miles, Candelaria, 165 miles, Columbus, 173 miles and Silver Peak, 228 miles in Esmeralda county, Nevada,—To Independence, is 234 miles; Lone Pine, 252 miles, and Cerro Gordo, 274 miles, in Inyo county, Cal. The fare to these places averages about 15 cents per mile. A stage also runs to Genoa and Markleville, and in the summer to Lake Tahoe, at Glenbrook, 15 miles. This line connects at Tahoe City, with stages for Truckee, the Summit, and also with the new line over the mountains to the Calavaras Grove. Leaving Carson, our course is to the northeast, across a broad bottom. To the right, about two miles distant, beside a round butte, is a large building—a huge boarding-house—conducted by the State. The guests are numerous, and are not inmates of their own *free will*, but by due course of law, and when the law is satisfied, it is hoped they will leave this STATES PRISON and become better citizens.

Near the prison are the Carson Warm Springs, where are ample accommodations for bathing.

LOOKOUT—is the first station from Carson, 1.1 miles distant, but our cars lookout not to stop, and 1.3 miles further, brings

EMPIRE.—This is a town of about 1,000 population, situated on the north bank of Carson River. Here are located the big Spanish or Mexican mill, on the right, then the Morgan or Yellow Jacket mill, and then the Empire. Passing on, down the bank of the Carson, we curve around the point of a bluff, pass the Brunswick mill on the left, near the station of the same name, 1.3 miles from Empire. Soon the valley is crowded out, and we enter a canyon, with the river to our right, just below, as we are now climbing up a heavy grade. To our right, but far below, is the Vivian, and the Merrimac mills, nearly one mile from the Brunswick. Continuing on up, still upward, we come to the Santiago mill, 1.8 miles further. This mill is situated about 500 feet below the road, on the right, and almost under it. Shutes run from the track above to the mill below, for dumping ore or coal. The road is now far up on the side of the mountain, much of the way blasted out from the solid rock, and very crooked. The canyon on the Carson River is far below, on the right, and soon will be lost to view.

EUREKA—is half a mile from Santiago, with a narrow-gauge track on our right, away down the river. Near the track on the right, is the dump-shute of the Eureka mill. Ascending rapidly and tortuously it is two miles to

MOUND HOUSE.—Here connections are made with the

Carson & Colorado Railroad.

General Office, Carson.

H. M. YERINGTON......*President and Gen'l Supt.*
D. A. BENDER............*Gen'l F. & P. Agent.*

This railroad—a three-foot narrow gauge—was completed and opened for business April 18, 1881, to Hawthorne, a distance of 100 miles from Mound House, and trains run regularly in connection with the V. & T. R. R. Stages leave Hawthorne daily for: Aurora, 26 miles; Bodie, 37 miles; Candelaria, 50 miles; Columbus, 55, miles; and connecting for Belmont, Silver Peak, Montezuma, Gold Mountain, Benton and Independence.

SUTRO TUNNEL.—This tunnel is one of the most important enterprises ever inaugurated in mining operations in this or any other country. The object sought is ventilation, drainage, and a cheap means of working the mines, or bringing the ores to the surface. The tunnel commences in the valley of the Carson River; is 14 feet wide at the bottom, 13 feet at the top, and 10 feet high.

The main tunnel is 20,018 feet in length, and the cross tunnels will be about 12,000 feet more. The tunnel strikes the Comstock ledge at a depth 2,000 feet below the point of the croppings. The work of extending the cross tunnels is being pushed ahead vigorously. Distance from Mound House to Sutro, five miles.

Near Mound House is a gypsum mine of good quality, large amounts of which are shipped to San Francisco. A track branches off near the station, to the right, for Silver City, situated about two miles to the eastward, in a narrow cañon, in plain view, where are located a number of quartz mills.

SILVER—is the next station, 3.3 miles from Mound House. Here ore is dumped down a shute to the right, and taken to the mills below. The best view of Silver City—a place of 1,000 population, all of whom are engaged in mining, having one newspaper, the *Reporter*—can now be had on the right; curving around to the left, we come to the American Flat tunnel, 900 feet long. It was at this tunnel where a thrilling incident occurred, October 17, 1872. (See ANNEX No. 34.) The fire alluded to in the annex cost the Railroad Company $500,000. It took two months to replace the timbering, during which time all passengers, freight, mails and express, had to be transferred by teams.

Passing through the tunnel, Mt. Davidson looms up directly ahead, 7,827 feet above sea-level; to the right is Gold Hill—far away, in a narrow canyon. The train runs around the side of the mountain, describing a great curve to the north and eastward, passing numerous mills, among which are the Rock Island, down on the right; the Baltimore, a track to the Overman, the Knickerbocker, Belcher, Baltic, and many other mills, both on the right and left, and finally cross over a huge mill, and one of the principal streets of the city of Gold Hill, which extends to the left up a narrow canyon, and stops at the depot in the city of

GOLD HILL—This is a flourishing mining city, 19 miles from Carson and two from Virginia; population, 6,000. It consists mostly of one main street, built along a steep ravine. The city has some good buildings, among which are one good hotel, the Vesey House; and one daily newspaper, the *Gold Hill News*. The city is surrounded with mills of all sorts, sizes and kinds, and all is noise and *business* night and day. The street between Gold Hill and Virginia is so generally built up that one cannot tell where the dividing line is between them. An omnibus line plies between the two cities, running every fifteen minutes.

Passing on from Gold Hill two miles, around sharp curves, through three short tunnels in quick succession, with mills to the right, mills to the left, and mills all around us, we arrive at

Virginia City—This city is on the southeastern slope of Mt. Davidson, at an elevation of 6,200 feet, with the mountain rising 1,627 feet above it. The city is built along the side of the mountain—one main street, with numerous steep cross-streets—and contains some very fine business blocks. Virginia is 21 miles from Carson, and 52 miles by rail, from Reno, and contains a population of about 16,000, a great proportion of whom are engaged in mining operations immediately *under the city*, from 1,500 to 2,200 feet.

The *Enterprise* and the *Chronicle*—both daily and weekly papers—are published here.

The religious and educational interests are well represented by churches and schools.

There are a number of hotels in the city, at which the traveler will find good accommodations. The INTERNATIONAL is the principal one.

At both Virginia and Gold Hill, are located great numbers of smelting furnaces, reduction works and stamp mills, all thundering away, night and day. The fires from these works, at night, light up with a lurid glare all surrounding objects. There are no two cities in the world more cosmopolitan; here, meet and jostle, the people from every land and clime; the rich and the poor; the miser and the spendthrift; the morose and jolly. *Self* here predominates. "Rich to-day, poor to-morrow," is the rule. *All* gamble in mining stocks, from the boot-black or servant girl to the Rev. Mr. Whitetie, or the Bonanza-Nabob. The whole people are alive to each movement of the "stock indicators," as much as the "Snowballs" were in Baltimore twenty-five years ago on the lottery and policy business; 4-11-44 was their god; the *tick* of the "indicator" is the god of *this* people. The saying that "It is not birth, nor wealth, nor State—its git up and git that makes men great," has been thoroughly demonstrated by many of the citizens of Virginia City and Gold Hill.

EARLY HISTORY—The first gold mines were discovered in 1857, by Joe Kirby, and some others, who commenced mining in Gold Canyon (Gold Hill), and continued working the place with indifferent success until 1859. The first quartz claim was located by James Finney, better known as "Old Virginia," on the 22d of February, 1858, in the Virginia mining district and on the "Virginia Croppings." The old prospector gave his name to the city, croppings and district. In June, 1859, rich deposits of silver ore were discovered by Peter O'Reilly and Patrick McLaughlin, on what is now the ground of the Ophir Mining Company. They were engaged in gold washing, and uncovered a rich vein of sulphuret of silver, when engaged in excavating a place wherein to catch a supply of water for their rockers. The discovery was made on ground claimed by Kirby and others. A Mr. Comstock was employed to purchase the claims of Kirby and those holding with him, hence, Comstock's name was given to the lode.

THE COMSTOCK LODE—is about four miles in length, the out-croppings extending in a broad belt along the mountain side. It extends under Virginia City and Gold Hill; the ground on which these cities are built being all "honey-combed" or undermined; in fact, the whole mountain is a series of shafts, tunnels and caverns from which the ore has been taken. The vein is broken and irregular at intervals along its length as far as traced, owing to the formation of the mountain. It is also very irregular in thickness. In some places the fissure ranges from 30 to as high as 200 feet in width, while at other points the walls come close together. The greatest variation in width occurs at a depth of from 400 to 600 feet from the surface. The principal silver ores of this lode are stephanite, vitreous silver ore, native silver and very rich galena. Pyrargyrite, or ruby silver, horn silver and polybasite, are found in small quantities, together with iron and copper pyrites, zinc-blende, carbonate of lead, pyromorphite and native gold.

The number of mills in and around Gold Hill and Virginia, and at other points, which work on ore from this lode, is between 75 and 80. They are scattered around through several counties, including Storey (where the lode lies), Lyon, Washoe and Ormsby, from 30 to 40 in number being in Storey county. The product of the Comstock lode has been beyond that of any silver vein of which we have any record; furnishing the largest portion of bullion produced on the Pacific Slope.

But let us return to Reno before we get to watching the "indicator;" and start once more for the West.

Leaving Reno, our course is south of west, up Truckee River. The hills are

NOTE.—**Stop over Privileges.** Passengers traveling on unlimited First-Class Tickets, who are desirous of securing a full daylight view of the *wonderful scenery* on the Sierra Nevada Mountains, can now do so. The "Pacific Express" arrives at Reno [see page 144] for supper, at 8:20 p. m. Passengers can stop over night at the hotel, and take the "Reno & San Francisco Express" next morning, [Sundays excepted] at 6:15; reach Truckee at 8:10; Summit, 9:00; Cape Horn, 11:20; Sacramento, 2:10 p. m.; Benicia, 4:55, and San Francisco at 7:05 p. m. Only 7½ hours later than the regular Pacific Express.

loftier, and were—at the time the railroad was built—covered with d nse pine forests; now, only stumps and rocks appear, with very little undergrowth. As we enter the Truckee Canyon, we seem to l ave entered a cooler, pleasanter, and more invigorating atmosphere. The aroma of the spruce and pine, which comes with the mountain breeze, is pleasant when compared with that of the alkali plains.

Verdi—is the first station from Reno, 10.48 miles. Passing on, up, 1.23 miles brings us to a new side-track station called

Essex—which we pass; continuing along the river, with its foaming current now on our left, first on one side, then on the other, runs this beautiful stream un'il we lose sight of it altogether. The road crosses and re-crosses it on fine Howe truss bridges, running as straight as the course of the mountains will permit. The mountains tower up on either hand, in places sloping and covered in places with timber from base to summit, in others precipitous, and covered with masses of black, broken rock. 'Tis a rough country, the canyon of the Truckee, possessing many grand and imposing features.

On the road up we pass a new station called

Mystic—5.11 miles from Essex, and four miles further

Bronco—another side-track. Occasional strips of meadow land are seen close to the river's edge, but too small and rocky to be of any use, only as grazing land. Now we cross the dividing line, and shout

as we enter California, a few miles east of

Boca—a small station 5.7 miles from Bronco. The lumber interest is well represented here, huge piles of ties, boards and timber lining the roadside. The river seems to be the means of transportation for the saw logs, immense numbers of them being scattered up and down the stream, with here and there a party of lumbermen working them down to the mills. A great quantity of ice is cut and housed here, and an extensive beer brewery erected.

The Truckee River, from Reno to its mountain source, is a very rapid stream, and affords dam-sites and mill-sites innumerable; yet, it is related that some years ago, before the completion of the Pacific railroad, a certain Indian agent, who is now an Ex-U. S. Senator, charged up to the Government an "item" of $40,000, as being the purchase-money for a mill-site on the Truckee, near a dam site.

Some hungry aspirant for official position, who had a hankering after the "loaves and fishes," exposed the "item," and a committee was sent out from Washington to investigate the matter. This committee, went out by "Overland Stage," had a good time, traversed the country in every direction, explored the river thoroughly, from the Desert to Lake Tahoe, and reported that they could find numberless dam-sites by mill-sites, but could not find a mill by a dam-site.

From Boca it is 1.57 miles to

Prosser Creek—Here is a long "V" flume for the accommodation of the lumbermen, and where large quantities of ice is cut and stored for market. Another run of 4.1 miles and the train passes

Proctors—and 2.76 miles more and our train comes to the end of the Truckee division, at Truckee City.

Truckee City—This place is situated on the north bank of the Truckee River, in the midst of what was once a heavily timbered region, much of which has been cut off. The principal business of the place is lumbering, though an extensive freighting business is carried on with other points in the mountains. One can hardly get around the town for the piles of lumber, ties and wood, which cover the ground in every direction. Some fine stores and a good hotel are the only build ngs which can lay claim to size and finish corresponding with the growth and business of the place. The town is composed of wooden buildings, mostly on the north side of the railroad.

The very sharp roofs of the buildings point out the fact that the snow falls deep and moist here, sufficiently so to crush in the roofs—unless they are very sharp and strong. The town contains about 2,000 inhabitants, nearly all of whom are directly or indirectly connected with the lumber trade.

The educational interests have been provided for, Nevada county, in which Truckee City is situated, being justly celebrated for her public schools.

The Truckee House is the headquarters of the tourists who stop over to visit objects of interest in this locality. This station is the end of the Truckee and the commencement of the Sacramento division.

The company have a 24-stall round-house and the usual machine and repair shops of a division located here.

OVERLAND PONY EXPRESS—*See Footnote.*

A line of stages leaves Truckee daily for Donner Lake, two miles; Lake Tahoe, 12 miles; Sierraville, 30 miles. A good wagon-road connects Sierra City with Truckee, *via* the Henness Pass and Donner Lake.

Freight is reshipped here for Donner and Tahoe Lakes, Sierraville and the various towns in the Sierra Valley. There are some wholesale and retail houses in Truckee, which do a large business.

LAKE TAHOE, or Bigler, as it is called on some of the official maps, is located 12 miles south of Truckee. Tahoe is an Indian name, signifying "big water," and is pronounced by the Indians "Tah-oo," while the "pale-faces" pronounce it "Ta-hoe." From Truckee a splendid road af-

---

No. 41 ANNEX. The Pony Express—was an enterprise started in 1860, by Majors, Russell & Co., of Leavenworth, Kan., to meet the pressing business wants of the Pacific Coast. It will be remembered that the usual time made on the mail service, by steamer, between New York and San Francisco, was about twenty-six days. The first *Overland* mail—which arrived in San Francisco Oct. 10th, 1858—carried it from St. Louis, Mo., via Los Angeles, in twenty-three days, twenty-one hours. The *Pony Express*—which left St. Joseph, Mo., and San Francisco, simultaneously, April 3d, 1860—succeeded in transporting it through safely, on its first trip, in ten days; on its second, in 14 days; third, nine days; fourth, ten days; fifth nine days; sixth, nine days;—a distance of 1,996 miles. This rapid transmission of business correspondence was of incalculable value to businessmen in those days.

This service, we can readily see, required courage and endurance, as well as enterprise and the expenditure of large sums of money. The moment the ferry boat touched land on the opposite shore, the Pony Expressman mounted his horse; and by day or by night, in starlight or darkness; whether sun-dried or soaked, snow-covered or frozen; among friends or through foes; be he lonely or merry—onward he hasteneth, until, at the thrice-welcomed station, he leaped from his saddle to rest. Here another was ready, whose horse, like himself, had been waiting, perhaps, without shelter; and with a cheery "Good-night, boys," he galloped off, and was soon lost in the distance. He rides on alone, over prairies and mountains; whether up hill or down; on rough ground or smooth, until he descries in the distance the goal of his hopes, and the station is reached. To tell of the losses in men from the Indians, and of horses and other property, both from volunteers as well as Indians, with the many thrilling adventures of those who participated in this daring enterprise, however interesting, would make too long a recital for these pages.

FIRST WELLS, FARGO & CO.'S EXPRESS OVER THE MOUNTAINS.

fords one of the best and most pleasant drives to be found in the State. The road follows the river bank, under the shade of waving pines, or across green meadows until it reaches Tahoe City, at the foot of the lake. Here are excellent accommodations for travelers—a good hotel, boats, and a well-stocked stable.

According to the survey of the State line, Lake Tahoe lies in two States and five counties. The line between California and Nevada runs north and south through the lake, until it reaches a certain point therein, when it changes to a course 17 degs. east of south. Thus the counties of El Dorado and Placer, in California, and Washoe, Ormsby and Douglas, in Nevada, all share in the waters of the Tahoe. Where the line was surveyed through the lake it is 1,700 feet deep.

There are three steamboats on the lake, but only one, the "Stanford," takes excursionists. The trip on this steamer is very fine, but for our *personal* use, not the way we like to travel for sight-seeing, at *this*, the loveliest of all drives in the world. Our choice is a good saddle animal, or a good team of horses, an agreeable companion, and start around the western shore. Six miles from Tahoe, over a beautiful road, we reach Sugar Pine Point, a spur of mountains covered with a splendid forest of sugar pine, the most valuable lumber, for all uses, found on the Pacific coast. There are fine streams running into the lake on each side of the point. We now arrive at EMERALD BAY, a beautiful, placid inlet, two miles long, which seems to hide itself among the pine-clad hills. It is not over 400 yards wide at its mouth, but widens to two miles inland, forming one of the prettiest land-locked harbors in the world. It is owned by Ben Holiday. At the south end of Tahoe, near the site of the Old Lake House, near Tallac Point, Lake Valley Creek enters the lake, having wound among the hills for many miles since it left the springs and snows which feed it. The valley of Lake Creek is one of the loveliest to be found among the Sierras. The whole distance, from the mountain slope to the lake, is one continual series of verdant meadows, dotted with milk ranches, where the choicest butter and cheese are manufactured. The next object of interest met with is a relic of the palmy days of staging:

FRIDAY STATION, an old stage station, established by Burke in 1859, on the Placerville and Tahoe stage road. Ten miles further on we come to the Glenbrook House a favorite resort for tourists. From Glenbrook House there is a fine road to Carson City, between which ply regular stages. This is a lovely place, and a *business* place too, as a half-dozen saw mills are located here, which turn out a million and a half feet of lumber weekly.

Four miles further we come to

THE CAVE, a cavern in the hillside fully 100 feet above and overhanging the lake.

Following around to the north end of the lake, and but a short distance away, are the celebrated HOT SPRINGS, lying just across the State line, in Nevada. Near them is a splendid spring of clear, cold water, totally devoid of mineral taste. The next object which attracts our attention is CORNELIAN BAY, a beautiful indenture in the coast, with fine gravel bottom. Thus far there has been scarcely a point from which the descent to the water's edge is not smooth and easy.

Passing on around to the west side we return to TAHOE CITY. Around the lake the land is generally level for some distance back, and covered with pine, fir and balsam timber, embracing at least 300 sections of as fine timbered land as the State affords. It is easy of access and handy to market, the logs being rafted down the lake to the Truckee, and thence down to any point on the railroad above Reno. So much for the general appearance of Lake Tahoe. To understand its beauties, one must go there and spend a short time. When once there, sailing on the beautiful lake, gazing far down its shining, pebbly bottom, hooking the sparkling trout that make the pole sway and bend in the hand like a willow wand, few will have a desire to hurry away. If one tires of the line and of strolling along the beach, or sailing over the lake, a tramp into the hills with a gun will be rewarded by the *sight* of quail, grouse, deer and possibly a bear.

We have now circled the lake and can judge of its dimensions, which are 22 miles in length and ten in width.

While on a recent visit to San Francisco, we learned, on good authority, that a movement was on foot, urged by several capitalists in that city, to build a large hotel at Tallac Point during the year, from which a stage line will convey passengers over the High Sierras, via Hope Valley and Blue Lake, to the Calavera Big Trees; distance 65 miles; fare, about $20. This would certainly be a lovely trip, passing as it does, through the grandest of the High Sierra range, and to the noted Blue Lake, so long talked about as the great reservoir from which the City of San Francisco is to be supplied with water in the future. For scenery, variety of game, trout, etc., this route will be found very attractive.

We will now return to Truckee.

DONNER LAKE—a lovely little lakelet, the "Gem of the Sierras," lies two and a half miles northwest of Truckee. It is about three and a half miles long, with an average width of one mile, and at the deepest point sounded, is about 200 feet. This and Lake Tahoe are, by some, thought to be the craters of old volcanoes, the mountains around them presenting unmistakable evidences of volcanic formation. The waters of both lakes are cold and clear as crystal, the bottom showing every pebble with great distinctness under water 50 feet deep. It is surrounded on three sides by towering mountains, covered with a heavy growth of fir, spruce and pine trees of immense size. Were it not for the occasional rattling of the cars, away up the mountain side, as they toil upward to the "Summit," and the few cabins scattered here and there along the shore, one would fancy that he was in one of nature's secret retreats, where man had never ventured before. A small stream, which tumbles down the mountain side, winds its way through the dense wood, and empties its ice-cold flood in the upper end or head of the lake, which rests against the foot of "Summit" Mountain. From the Lake House, situated as it is on a low, gravelly flat, shaded by giant pines, a very fine view of the railroad can be obtained. Within sight are four tunnels and several miles of snow-sheds, while behind and seemingly overhanging the road, the mountains—bald, bleak, bare, massive piles of granite—tower far above their precipitous sides, seeming to bid defiance to the ravages of time. A fine road has been graded along the right-hand shore, from the station, forming a splendid drive. The "old emigrant road" skirts the foot of the lake (where the Donner party perished, see ANNEX No. 28), and following up the stage road, climbs the "Summit" just beyond the long tunnel. Originally, it struck the Divide at Summit Valley; from thence it followed the valley down for several miles, then struck across the crest-spur, and followed the Divide down from Emigrant Gap.

The business of lumbering is carried on quite extensively at the lower end of the lake. The logs are slid down the mountain sides in "shoots," or troughs made of large trees, into the lake, and then rafted down to the mill. On the west side of the lake the timber has not been disturbed, but sweeps down from the railroad to the water's edge in one dense unbroken forest. The lower end of the lake is bordered with green meadows, covering an extent of several hundred acres of fine grazing land.

From the foot of the lake issues a beautiful creek, which, after uniting with Coldstream, forms the Little Truckee River.

COLDSTREAM—is a clear, cold mountain stream, about fifteen miles long. It rises in the "Summit" Mountain, opposite Summit Valley. Some excellent grazing land borders the creek after it leaves the mountain gorge.

FISHING AND HUNTING—In Donner and Tahoe lakes is found the silver trout, which attains the weight of 20 pounds. There are many varieties of fish in these lakes, but this is most prized and most sought after by the angler. It is rare sport to bring to the water's edge one of these sleek-hided, sharp-biting fellows—to handle him delicately and daintily until he is safely landed; and then, when fried, baked, or broiled brown, the employment of the jaws to masticate the crisp, juicy morsels—it's not bad *jawing*. The water near the lake shore is fairly alive with white fish, dace, rock-fish, and several other varieties—the trout keeping in deeper water. There is no more favorite resort for the angler and hunter than these lakes and the surrounding mountains, where quail, grouse, deer, and bear abound.

These lakes were once a favorite resort for the "San Francisco schoolmarms," who annually visit this locality during the summer vacation. The Railroad Company generally passed them over the route, and they had a happy week—romping, scrambling and wandering over the mountains, and along the lake shore, giving new life and animation to the scene. The gray old hills and mighty forests re-echo with their merry laughter, as they stroll around the lake, gathering flowers and mosses, or, perhaps, essaying their skill as anglers, to the great slaughter of the finny inhabitants of the lake.

SIERRA VALLEY—lies about 30 miles from Truckee City, among the Sierras. It is about 40 miles long, with a width of from five to seven miles. It is fertile, thickly settled, and taken in connection with some other mountain valleys, might be termed the Orange county of California —from the quantity and quality of butter and cheese manufactured there. In the mountain valleys and on the table-lands the best butter and cheese found in the State are manufactured—the low valleys being too warm, and the grasses and water not so good as found here. In Sierra, and many other mountain valleys, good crops of grain and vegetables are grown in favorable seasons, but the surest and most profitable business is dairying. The flourishing town of Royalton is situated in this valley.

HONEY LAKE—an almost circular sheet of water, about ten miles in diameter, lies about 50 miles north of Truckee City. Willow Creek and Susan Creek enter it at the north, while Long Valley Creek empties its waters into the southern portion of the lake. Some fine meadow and grazing land is found in the valleys bordering these streams, which has been occupied by settlers, and converted into flourishing farms.

Susanville, the principal town in the valley, is situated north of the lake. It is connected by stage with Reno, Nevada, and Oroville, California.

We now take leave of Truckee City and its surroundings, and prepare to cross the "Summit of the Sierras," 14 miles distant. With two locomotives leading, we cross the North Fork or Little Truckee on a single-span Howe truss bridge, and make directly across the broken land bordering the lake meadows, for the foot of the Sierras. Then skirting along the hill-side, through long snow sheds, with the sparkling Coldstream on our right, winding through the grassey valley and among waving pines, for 6.52 miles, we pass

Strong's Canyon---and bend, around the southern end of the valley, which borders Donner Lake, then crossing Coldstream, commence the ascent of the mountains. Soon after passing this side-track, our train enters a snow-shed, which—with a number of tunnels,—is *continuous for twenty-eight* miles, with but a few "peek-holes," through which to get a glimpse at the beautiful scenery, along this part of the route—yet, we shall describe it, all the same. As the train skirts the eastern base, rising higher and higher, Donner Lake is far below, looking like a lake of silver set in the shadows of green forests and brown mountains. Up still, the long, black line of the road bending around and seemingly stealing away in the same direction in which we are moving, though far below us, points out the winding course we have followed.

Up, still up, higher and higher toils the train, through the long line of snow-sheds leading to the first tunnel, while the locomotives are snorting an angry defiance as they enter the gloomy, rock-bound chamber.

Summit---is 14.31 miles west of Truckee, the highest point on the Sierra Nevada Mountains, passed over by the Central Pacific railroad, 7,017 feet above the level of the sea. Distance from Omaha, 1,669 miles; from San Francisco, 245 miles. This is not the highest land of the Sierra Nevada Mountains, by any means, for bleak and bare of verdure, rise the granite peaks around us, to an altitude of over 10,000 feet. Piles of granite—their weather-stained and moss-clad sides glistening in the morning sun—rise between us and the "western shore," hiding from our sight the vast expanse of plain that we know lies between us and the golden shores of the Pacific Ocean. Scattering groups of hardy fir and spruce, line the mountain gorges, where rest the everlasting snows that have rested in the deep shady gulches, near the summit of these towering old mountains— who can tell how long? They have lain, evidently, since Adam was a very small boy, or the tree sprouted from which our apple-loving ancestor, Eve, plucked that bedeviled fruit.

We are on the dividing ridges which separate the head-waters of several mountain rivers, which, by different and tortuous courses, find at last the same common receptacle for their snow-fed waters - the Sacramento River. Close to our right, far down in that fir-clad gorge, the waters of the South Yuba leap and dance along, amid dense and gloomy forests, and over almost countless rapids, cascades and waterfalls. This stream heads against and far up the Summit, one branch crossing the road at the next station, Cascade. After passing Cisco, the head waters of Bear River can be seen lying between the Divide and the Yuba, which winds away beyond, out of sight, behind another mountain ridge. Farther on still, and we find the American River on our left. These streams reach the same ending the Sacramento River but are far apart, where they mingle with that stream. There is no grander scenery in the Sierras, of towering mountains, deep gorges, lofty precipices, sparkling waterfalls and crystal lakes, than abound within an easy distance of this place. The tourist can find scenes of the deepest interest and grandest beauty; the scholar and philosopher, objects of rare value for scientific investigation; the hunter and the angler can find an almost unlimited field for his amusement; the former in the gorges of the mountains, where the timid deer and fierce grizzly bear make their homes; the latter among the mountain lakes and streams, where the speckled trout leaps in its joyous freedom, while around all, is the music of snow-fed mountain torrent and mountain breeze, and over all is the clear blue sky of a sunny clime, tempered and softened by the shadows of the everlasting hills.

TUNNELS AND SNOW-SHEDS—From the time the road enters the crests of the "Summit," it passes through a succession of tunnels and snow-sheds so closely connected that the traveler can hardly tell when the cars enter or leave a tunnel. The Summit tunnel, the longest of the number, is 1,659 feet long, the others ranging from 100 to 870 feet in length.

The snow-sheds are solid structures, built of sawed and round timber, completely roofing in the road for many miles (see illustration, pp. 72-67-143). When the road was completed, there were 23 miles of shed built, at an actual cost of $10,000 per mile. With the additions since made, the line reaches about 45 miles, which includes the whole length of the deep snow line on the dividing ridge. When we consider that along the summit the snow falls from 16 to 20 feet deep during a wet winter, we can imagine the necessity and importance of these structures. By this means the track is as clear from snow in the winter as are the valleys. The mighty avalanches which sweep down the mountain sides in spring, bearing everything before them, pass over the sloping roofs of the sheds and plunge into the chasms below, while beneath the rushing mass the cars glide smoothly along, the passengers hardly knowing but that they are in the midst of an enormous tunnel.

Where the road lies clear on the divide or level land, the sheds have sharp roofs, like those of any building calculated to withstand a great weight of snow. But where the road is built against the side of these bare peaks, the roof of the shed can have but one slope, and that must reach the mountain side, to enable the " snow-slides" to cross the road without doing harm to that or the passing trains. (See illustration, pages, 67 and 143.)

Fires sometimes cause damage to sheds and road, but seldom any delay to the trains, as the company have materials of all kinds on hand for any emergency, and, with their swarm of men, can replace everything almost as quick as it is destroyed; but, to further protect the snowsheds and **bridges from fire**, and the more effectually to extinguish them, the Railroad Company have stationed the locomotive Grey Eagle at the Summit (with steam always up and ready to answer a summons), with a force pump of large capacity, supplied with steam from the engine. Attached to the locomotive are eight water cars, the tanks on which are connected with each other and with the tender of the engine, so that the supply of water will always be sufficient to check any ordinary fire.

The Summit House, located at the station, is one of the best hotels on the road and can furnish tourists with every accommodation required, while spending a few days or weeks exploring this very interesting region.

Passengers from the West, desiring to visit Lake Tahoe, can take a stage at the Summit House, which will afford them a fine view of Donner Lake, while rolling down the mountain and around to the north and east side of it, en route to Tahoe. Returning, those who choose, can take the cars for the East, at Truckee, without returning to the Summit. Fare for the "round trip," $6.00.

Leaving the Summit, we pass on through the long shed, and tunnels alternately, around the base of towering peaks anon over high, bare ridges, then through grand old forests, for 5.77 miles to

CASCADE—Here we cross one of the branches of the Yuba, which goes leaping down the rocks in a shower of spray during the summer, but in the winter the chasm shows naught but a bed of snow and ice.

Summit valley, one of the loftiest of the Sierra valleys, lies to the west, a broad, grassy meadow, dotted with trees and lying between two lofty mountains, about two miles long by one mile wide. It is covered with a luxuriant growth of grass, affording pasturage for large bands of cattle, during the summer. It is all occupied by dairymen and stock raisers, at whose comfortable dwellings the tourist will find a hearty welcome. It is a delightful summer retreat; a favorite resort for those who prefer the mountains, with their cool breezes and pure water. The valley is watered by many springs and snow-fed rivulets, whose waters flow to the American River.

This valley is becoming noted in a business point of view, as well as being a place of summer resort. It is becoming celebrated as a meat packing station, it having been demonstrated that pork and beef can be successfully cured here during any portion of the year.

SODA SPRINGS—are situated near the foot of Summit Valley, their waters uniting with others, forming the head waters of the American River. The springs are very large and numerous, and the water is pronounced to be the best medicinal water in the State. It is a delightful drink, cool and sparkling, possessing the taste of the best quality of manufactured soda water. The larger of the springs have been improved, and great quantities of the water are now bottled and shipped to all parts of the State. Near the Soda Springs are others, the waters of which are devoid of mineral or acidous taste, and boiling hot.

In the summer these springs are much resorted to by people from the "Bay." There is a comfortable hotel at the Springs which is reached from the Summit by stage, and sometimes at a side-track, called "Soda

Spring Station," midway between Summit and Cascade stations.

Tamerack—is the next station, 4.2 miles from Cascade, and 3.51 miles from

Cisco—At one time this was quite an important place, being the "terminus" during the time occupied in tunneling through the summit; *then*, it was a place of 500 inhabitants, *now*, a score or so make up the town.

From this station we pass along rapidly and easily, without the help of the locomotive. To the right, occasional glimpses of the Bear and Yuba Rivers can be seen far below us.

Emigrant Gap—is 8 5 miles west of Cisco, at the place where the old emigrant road crossed the Divide, and followed down the ridges to the valley of the Sacramento. The emigrants passed *over* the "gap," we pass *under* it, making a slight difference in elevation between the two roads, as well as a difference in the mode of traveling. We have seen the last of the old emigrant road that we have followed so far. No more will the weary emigrant toil over the long and weary journey. Space is annihilated, and the tireless iron horse will henceforth haul an iron wagon over an iron road, landing the tourist and emigrant fresh and hearty, after a week's ride, from the far eastern shores of our country to the far western—from ocean to ocean.

Passing on amid the grand old pines, leaving the summit peaks behind, we turn up Blue Canyon, the road-bed on the opposite bank apparently running parallel with the one we are traversing. Swinging around the head of the canyon, past sawmills and lumber side-tracks, 5.2 miles, we reach

Blue Canyon—a freight and lumber station, where immense quantities of lumber are shipped from mills in the vicinity. Before the railroad reached these mountains, the lumber interest of this section was of little value, there being only a local demand, which hardly paid for building mills and keeping teams. The mines were then the only market—the cost of freight to the valleys forbidding competition with the Puget Sound lumber trade, or with mills situated so much nearer the agricultural districts. Now the lumber can be sent to the valleys, and sold as cheaply as any, in a market rarely overstocked; for the one item of lumber forms one of the staple market articles, ruling at more regular prices, and being in better demand than any other article of trade, on the coast, if we except wheat.

Leaving Blue Canyon, we speed along around the hill-sides, past

China Ranch—a side-track, about two miles west. The passenger should now watch the scenery on the left.

Shady Run—is 4.72 miles west of Blue Canyon, but passenger trains seldom stop. On the left, south side, can be seen one of the grandest gorges in the Sierra Nevada Mountains, "The Great American Canyon." (See illustration, page 130). At this point the American River is compressed between two walls, 2,000 feet high, and so nearly perpendicular that we can stand on the brink of the cliff and look directly down on the foaming waters below. The canyon is about two miles long, and so precipitous are its sides, which are washed by the torrent, that it has been found impossible to ascend the stream through the gorge, even on foot. This is a beautiful view—one of nature's most magnificent panoramas. But we soon lose sight of it, as our train turns to the right, up a side canyon, 4.84 miles from Shady Run, and stops at

Alta—Alta looks old and weather-beaten, and its half-dozen board houses, with sharp roofs, look as though there was little less than a century between the present and the time when they were ushered into existence—like its namesake in San Francisco, after which it was named.

Dutch Flat—is 1.87 miles from Alta; old settlers call it German Level. The town of Dutch Flat is situated in a hollow, near by and to the right of the road, a portion of it being in plain view. The town contains many good buildings, churches, schools, and hotels. The *Farmer*, a weekly newspaper, is a new institution at Dutch Flat. Population, about 2,000. One feature of this town is worth noting, and worthy of commendation—the beautiful gardens and fine orchards which ornament almost every house. In almost all of the mountain towns—in fact in all of the older mining towns—the scene is reproduced, while many of the valley towns are bare of vines, flowers or fruit trees; the miner's cabin has its garden and fruit trees attached, if water can be had for irrigation, while half of the farm-houses have neither fruit trees, shrubs, flowers nor gardens around them.

Stages leave this station daily for Little York, You Bet and Red Dog. Freight teams leave here for all the above named towns and mining camps in this vicinity.

LITTLE YORK—a mining town, three miles northwest of Dutch Flat, contains about 500 inhabitants.

YOU BET—is six miles from Little York, also a mining town, about the same size.

RED DOG—seven and a-half miles from You Bet, is still another small mining town.

These towns are situated on what is called the Blue Lode, the best large placer mining district in the State. The traveler will see the evidences of the vast labor performed here, while standing on the platform of the cars at Alta, Dutch Flat or Go'd Run stations. The Blue Lode extends from below Gold Run, through the length of Nevada, on, into and through a portion of Sierra county. It is supposed to be the bed of some ancient river, which was much larger than any of the existing mountain s'reams. The course of this old river was nearly at right angles with that followed by the Yuba and other streams, which run across it. The channel is from one to five miles wide in places—at least the gravel hills, which are supposed to cover the bed, extend for that distance across the range. Many of these gravel hills are from 100 to 500 feet high, covered with pine trees from two to six feet in diameter. Petrified trees, oak and pine, and other woods, such as manzanita, mountain mahogany and maple, are found in the bed of the river, showing that the same varieties of wood existed when this great change was wrought, as are now growing on the adjacent hill-sides.

HYDRAULIC MINING—The traveler will observe by the road-side, mining ditches and flumes, carrying a large and rapid stream of water. These ditches extend for many miles, tapping the rivers near their sources—near the regions of perpetual snow. By this means the water is conveyed over the tops of the hills, whence it is carried to any claim below it. The long, high and narrow flume, called a "telegraph," carries the water from the ditch, as nearly level as possible, over the c aim to be worked. To the "telegraph" is attached a hose with an iron pipe, or nozzle, through which the water rushes with great velocity. When directed against a gravel bank, it cuts and tears it down, washing the dirt thoroughly, at a rate astonishing to those unacquainted with hydraulic mining. (See accompanying illustration.) The water carries rocks, dirt and sand through the tail race, and into the long flumes, where the riffles for collecting the gold are placed. Miles and miles of the flumes have been built, at an enormous expense, to save the gold carried away in the tailings.

Around Little York and You Bet, the lode is mixed too much with cement to mine in this manner with profit, hence mills have been erected where the cement is worked in the same manner as quartz rock—crushed and then amalgamated.

Gold Run—is 2.13 miles beyond Dutch Flat, and is a small mining town, containing about 200 inhabitants. Around it you

HYDRAULIC MINING.

can see, on every hand, the miner's work. Long flume beds, which carry off the washed gravel and retain the gold; long and large ditches full of ice-cold water, which, directed by skillful hands, are fast tearing down the mountains and sending the washed debris to fill the river-beds in the plains below. There are a set of "pipes" busily playing against the hill-side, which often comes down in acres. All is life, energy and activity. We don't see many children peeping out of those cabins, for they are not so plentiful in the mining districts as in Salt Lake. But we do see nearly all of the cabins surrounded with little gardens and orchards, which produce the finest of fruits.

Descending the mountain rapidly, amid mining claims, by the side of large ditches, through the deep gravel cuts, and along the grassy hill-sides, until, on the left, a glimpse of the North Fork of the American River can be had, foaming and dashing along in a narrow gorge full 1,500 feet beneath us. Farther on we see the North Fork of the North Fork, dashing down the steep mountain at right angles with the other, leaping from waterfall to waterfall, its sparkling current resembling an airy chain of dancing sunbeams, as it hastens on to unite with the main stream. Now we lose sight of it, while it passes through one of those grand canyons only to be met with in these mountains.

C. H. Mills—a station where trains seldom stop, is 5.96 miles from Gold Run. The passenger should be on the lookout, and look to the left—south—as the scene changes with every revolution of the wheels. A few moments ago we left the canyon behind—now, behold, it breaks on our view again, and this time right under us, as it were, but much farther down. It seems as though we could jump from the platform into the river, so close are we to the brink of the precipice; steadily on goes the long train, while far below us the waters dance along, the river looking like a winding thread of silver laid in the bottom of the chasm, 2,500 feet below us. This is Cape Horn, one of the grandest scenes on the American Continent, if not in the world. Timid ladies will draw back with a shudder—one look into the awful chasm being sufficient to unsettle their nerves, and deprive them of the wish to linger near the grandest scene on the whole line of the trans-continental railroad.

Now look farther down the river and behold that black speck spanning the silver line. That is the turnpike bridge on the road to Iowa Hill, though it looks no larger than a foot plank. Now we turn sharp around to our right, where the towering masses of rock have been cut down, affording a road-bed, where a few years ago the savage could not make a foot trail. Far above us they rear their black crests, towering away, as it were, to the clouds, their long shadows falling far across the lovely little val'ey now lying on our left, and a thousand feet below us still. We have lost sight of the river, and are following the mountain side, looking for a place to cross this valley and

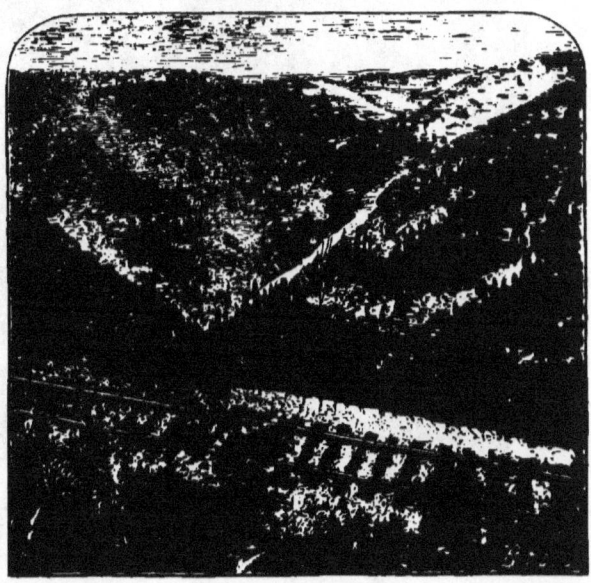

ROUNDING CAPE HORN

LOOKING UP AT CAPE HORN—See next page.

WOOD-HAULING IN

ADA. (See Annex No. 39.)

reach the road-bed on the opposite side, which we can see ruts parallel with us. Soon it is found, and turning to our left, we cross the valley—Rice's Ravine—on a trestle bridge 113 feet high and 878 feet long, under which can be seen the track of the narrow gauge railroad, from Colfax to Nevada. Gradually the height grows less, until it is reduced, at the end of 600 feet, enough to admit of an embankment being raised to meet it. On, over the embankment, which cu ves around to the left, and now we are on the solid hill-side, and running along opposite the road by which we passed up the valley. We now have our last and best look at the bold bluff.

The best view of this noted place is obtained when going east, or from the river below. Viewed from the river, the passing train looks like some huge monster winding around the bluff, bold point, puffing and blowing with its herculean labors, or screaming angry notes of defiance, or perhaps of ultimate triumph at the obstacles overcome (see page 160)

When the road was in course of construction, the groups of Chinese laborers on the bluffs looked almost like swarms of ants, when viewed from the river. Years ago, the cunning savage could find only a very roundabout trail by which to ascend the point, where now the genius and energy of the pale-face has laid a broad and safe road, whereon the iron steed carries its living freight swiftly and safely on their way to and from ocean to ocean.

When the road-bed was constructed around this point, the men who broke the first standing ground were held by ropes until firm foot-holds could be excavated in the rocky sides of the precipitous bluffs.

Colfax—is 4.5 miles from C. H. Mills, two miles west of the high bridge, trains until recently stopped for meals, they now stop at Sacramento.

The company have a large depot here, this being the distributing point for freight bound for Grass Valley, Nevada, and a large scope of mining country. The town is named in honor of Schuyler Colfax, one of the warmest friends and earliest supporters of the road.

Colfax is a substantial railroad town. It contains about 1,000 inhabitants, is well watered, and has an air of general thrift about it, which marks all the permanent towns along the road. The school and church accommodations are ample; the climate is invigorating and healthy, and the inhabitants a thrifty, driving, enterprising people; the greater number, natives of the State of Illinois, who emigrated to this country in early days—1849—50.

ILLINOIS TOWN— is a half-mile west, once a noted freighting point for the surrounding mines, now the only business is raising fruit, apples, peaches and pears.

IOWA HILL—is a mining town, 12 miles south of Colfax. A good toll-road crosses the American River on the bridge which we saw when rounding Cape Horn, and follows up the mountain to the town, which contains about 250 inhabitants. Formerly stages ran *daily* to Iowa Hill and the mining camps to the southward, but for some reason, they now run only semi-occasionally. Private conveyance can always be secured at Colfax at reasonable charge.

As our trip is for pleasure, and to see all that is worth seeing, we will need to take a trip to the old mining towns of Grass Valley and Nevada.

Nevada County Narrow Gauge Railroad.

General Offices are at Grass Valley.

J. C. COLEMAN..................*President*.
JOHN F. KIDDER.........*General Superintendent*.
GEO. FLETCHER......................*Secretary*.

This road is a three-foot narrow gauge; commenced in January, 1875, and completed May 22, 1876; length, 22½ miles. This is a very crooked road has 15 trestle bridges, aggregating 5,176 feet, two truss bridges, each 165 feet long, and 97 and 85 feet high, respectively; and two tunnels, aggregating 800 feet. As for the scenery—well, it is *immense*—the rapid and astonishing changes remind one of the *kaleidoscope*, and its wondrous changes. Here are to be seen every variety of mountain scenery, as though a choice morsel of each of the grand beauties of nature had slipped from the hand of the great Architect while distributing them, giving such a variety of magnificent views as are seldom, if ever, found in the same distance traveled.

On the route we shall pass through fearful chasms, and tortuous canyons; *under* and over lofty bridges, through forests, beside green fields and towering mountains; tall pines and diminutive manzanitas; huge furnaces, and thundering quartz mills; long flumes and beautiful cascades; large rivers, and tiny sparkling creeks; dark and gloomy gorges, and fruit-laden orchards; old placer diggings, new diggings, and immense quartz mines. But come along, and take a look. The train stands just on the south side of the depot at Colfax, and leaves on the arrival of the overland train.

Passing along to the eastward, we gradually descend the canyon with the track of the C. P. road away above us on the left, and again to the right, where it curves around Cape Horn, a beautiful view of which is obtained. Following on up, we pass *under* the high bridge of the Central, one and a-half miles from Colfax, and reach the Divide, where the waters run to the north, to Bear River—which we soon reach and cross on a bridge 750 feet long, and 97 feet high; amid towering pine and spruce trees and the most romantic scenery—then, 4.5 miles from Colfax, we come to the side-track station of

You Bet—the town of which is four miles to the east—heretofore described. We now come to the Greenhorn. Following it up through a 350-foot tunnel, we cross that creek on a trestle and bridge 700 feet long; on, up and over another 450 feet trestle, along the side of the mountain, overlooking the Greenhorn, around the great "S" curve, on a grade of 105 feet to the mile; through heavy rock cuts, almost doubling back on our route.

Storms—another side-track, by a great saw-mill, is four miles further, but the trains stop only on signal. The mountains on the route up to this station are covered with pines, spruce and oaks. The chasms are fearfully grand in places on the left.

Buena Vista—another side-track, in the Noonday Valley, is four miles from Storms, from whence, continuing on up two miles, we reach

Kress Summit—with an altitude of 2,851 feet. From the summit the descent is rapid, 151 feet to the mile; the mountains are here covered with small pines and manzanitas, the big timber that once covered them having long since been cut off, and used to a great extent in the mines at Grass Valley. On the road down, we pass many evidences of placer mining, and, doubtless, will see some Chinamen working over the old placers near Union Hill. On the left are several old mills, and just before reaching Grass Valley, away to the right, across a low place in the ridge, can

be seen two great Q and Eureka. Tho on our *right*, will b pass the next station from track to track is three miles.

Grass Valley— a thriving mining habitants. It is s Colfax, 35 miles c five and a-half mil the sides of the h with comfortable li about the nooks a ously. It contain blocks, and some The private dwelli closed in fine orcha give them an air of beauty. The town from the quartz m No town in the equal amount of none has added n State at large.

In September, 18 a piece of gold-be Hill. From this, p and soon several opened. In 1851, t erected in Boston most populous port Grass Valley nc mills, agregating three large de-sul city is illuminated hotels—the Exchan one daily newspa *Union*, and the *Fw* Of the quartz mills, note—the Idaho. this mill had nev dend for 100 month from $5 to $25 per shares of a par v shares have sold as total receipts for ni 255; dividends pai

Stages leave Gr ville, west 35 mile which is on the sou the town, we turn mill and the old E doned), on our left, a section of countr and many signs of as old washed ou some orchards of land, cross Wolff C digging over the old claims, note the young pines, and the long "V" flume which brings lumber from the mountains to the eastward twelve miles, and climb up to the Town-Talk Tunnel, 450 feet long; elevation 2,774 feet; and then descend, passing old mills and new mills, a portion of the city of Nevada, away across on the opposite side of the mountain, peacefully reposing—probably all unconscious of our near approach. On the descent to the city, we pass the New England mill on the left, and the Pittsburgh mill on the right; then cross a bridge 450 feet long over Gold Run Creek, where considerable placer mining is still being done, and after a run of five and a half miles from Grass Valley, arrive at

Nevada—This city is the county seat of Nevada county, situated on Deer Creek, a rapid stream with rugged canyon walls, and contains a population of about 4,300. There are here seven stamp mills, agregating 110 stamps, two de-sulphurizing works, and, when sufficient water can be had for the purpose, an extensive business is carried on in hydraulic mining.

The place is rather irregularly laid out, owing to the formation of the land and the creek which runs through a portion of the the town. There are some good business blocks, good county buildings, several hotels, of which the principal is the Union; one daily newspaper, the *Transcript;* and one weekly, the *Gazette.* There are some very nice private residences, surrounded with orchards, fruit and beautiful shrubbery, which contrast strikingly with the bare, brown, or red old hillsides.

The first mining in Nevada was placer, creek and gulch-washing. The mines were very rich, and lasted several years. During this time the famous hill "diggings," a part of the "old river bed," were discovered and opened. They, too, proved a source of great wealth, though many miners became "dead broke" before the right system—hydraulic mining with long flumes—was inaugurated. These mines proved very extensive and lasting, and yet form one of the chief sources of the city's wealth. Of late years the attention of the people has been directed to cement and quartz mining, and several very valuable quartz veins have been opened, and fine mills erected on them. The quartz interest is now a decided feature in the business of the city.

Stages leave Nevada daily for North San Juan, 14 miles; Comptonville, 22 miles; Forest City, 45 miles, and Downieville, 50 miles.

NORTH SAN JUAN—is a lively mining town of 1,500 inhabitants, most of whom are engaged in hydraulic or other mining. The yield of the Milton Company's mill for 1877, was $233,000; the Manzanita mine, $155,713, for the same year. Orchards and vineyards are numerous, also some fine private residences.

COMPTONVILLE—is another small mining town, of about 500 inhabitants, most of whom are dependent on placer mining, and they have a portion of the " old channel " or hill mines in the immediate vicinity.

FOREST CITY—is a place of about 400 inhabitants, also a mining town, working " drift diggings."

DOWNIEVILLE — the largest town in Sierra county, is situated on Yuba River, with a population of about 1,000.

BLOOMFIELD—is twelve miles from Nevada, sometimes called "Humbug," but the yield of the North Bloomfield Co.'s mine for 1877, $291,125, was *not* much of a humbug.

With this hasty glance at a country where the material for a big book lays around loose, we return to the Overland road, and again to the westward.

Leaving Colfax, we follow down Auburn Ravine, at times near its bed and anon winding in and out among the hills, which are here and there covered with small oaks and an occasional large oak and pine, together with the Manzanita, a peculiar shrub, resembling the thorn of the Eastern States, which sheds its *bark* instead of its leaves. (See page 164.)

N. E. Mills—is the first station after Colfax, 5.6 miles distant, but trains stop only on signal. The country is very rough and broken, and 3.31 miles more brings our train to

Applegate—another side-track near some lime kilns. Continuing along with numerous cuts, fills, bridges and one tunnel near the next station, 700 feet long, for 2.97 miles, we arrive at

Clipper Gap—an unimportant station. Again onward, we leave the ravine and keep along the foot hills, to hold the grade — passing through many an old washed placer mine, in which, only a few short years ago, could be seen thousands of men digging and washing, washing and digging, from morning till night, seeking what is said to be "the root of all evil"—GOLD—and a *huge root it is*; they all point to it; we are hunting it; have hunted it for forty years; struck the trail several times, but it soon got cold; and it has been difficult for some time to find a "color."

Auburn—The county seat of Placer county—is 7.71 miles west of Clipper Gap, and contains about 1,000 inhabitants. Gardens, vineyards, and orchards abound, and everything betokens quiet, home-comforts and ease. It has excellent schools and fine churches, and is one of the neatest looking towns in the county. The public buildings, court-house etc., are good, and the grounds well kept. The greater portion of the dwellings stand a little distance from the road. The American, Orleans, and Railroad House, are the principal hotels. The *Placer Herald* and the *Argus*, both weekly newspapers, are published here.

Stages run daily from Auburn to Pilot Hill, eleven miles; Greenwood, 14 miles; Georgetown, 20 miles; Coloma, 22 miles; Forest Hill, 23 miles; Michigan Bluffs, 30 miles; Placerville, 30 miles.

We are now in the foot-hills:

After leaving Auburn, we pass through Bloomer Cut, (see illustration, p. 168) then near the next station we pass over the New Castle Gap Bridge, which, before it was filled up with earth, was 528 feet long and 60 feet high. All trestle bridges and trestle works on both the Union and Central Pacific roads, have all been filled in with rock, earth or iron, within the last five years.

New Castle—is a small place of about 200 inhabitants, about 4.89 miles from Auburn.

We pass on through little valleys.

No. 6 ANNEX. The Madrone Tree.—This peculiar tree can be seen in many parts of California, particularly on excursions, Nos. 4 and 5. It sheds its *bark* in the fall of the year, much the same as other trees their leaves. The tree, after shedding its bark, has a bright salmon color, then turns gradually darker, until, at the shedding time the following year, the bark is quite dark.

The Manzanita, which means in Spanish "little apple," a small shrub, also sheds its bark. It is found along the foot-hill ranges of California. The root is very tough, fine grained and polishes very beautifully. Many fine boxes, and handles for canes, umbrellas and parasols are made from the root of the Manzanita.

In Utah a man is rated according to the number of wives he has, thus: A man with two wives is a "2-ply" Mormon; one with three, a "3-ply." Each wife adds a "ply."

Virtue and honor are very nice for Sunday wear, but too rare for every day use.

and among low hills, with evidences of past and a little present mining.

Off to the right are the old-time mining camps of Ophir, Virginia City, Gold Hill, and several others, where yet considerable placer mining is indulged in by the old settlers who are good for nothing else.

There are several stone quarries near the station, where a very good article of granite is procured.

Just after leaving Newcastle, we catch the first glimpse of the beautiful valley of the Sacramento, from the windows on the right-hand side of the cars. There has been several points above, where the valley could be seen for a moment, but very indistinctly. Passing on by several valuable stone quarries, for 6.06 miles, we come to

Pino—We are rapidly descending, but among the low hills, covered with chaparral, manzanita and grease-wood, the road winds onward for 3.04 miles further, passing several valuable quarries, to the right and left, when we arrive at

Rocklin—Here the company have a machine shop and round-house of 28 stalls, built in the most substantial manner, of granite obtained near by. The celebrated Rocklin Granite Quarries are close to the station, on the left-hand side of the road. The granite obtained here is of excellent quality, and does not stain on exposure to the weather. The stone for the State Capitol and for many of the best buildings in San Francisco was quarried here.

Leaving Rocklin and the foot-hills—the country now opening out into the plains, or the valley bordering the American River —we have no more hills to encounter; yet the country is somewhat uneven, and after winding around, on a regular grade, for 3.91 miles further, we reach the

Junction—Roseville Junction—This place is 18.24 miles from Sacramento. Here are several stores, a hotel, and one of the best places on the coast to establish a flouring mill. Using the wheat that is raised near, and finding a ready market in the mines to the east, would have a decided advantage in point of location and freights over any other mill on the coast. At this junction branches off the Oregon division of the Central Pacific, north for Marysville and Oregon. The road is completed to Redding, 169 miles, and passengers can change cars here if they choose, or go on to Sacramento, as the trains for the Oregon division are made up at that city, and start on the arrival of the morning train from San Francisco, about 3 P. M. We shall take a run over this division, starting from Sacramento. (See page 166) See also map on page 120, and description of depot buildings, page 173.

Antelope—is 3.9 miles west. The country is more level, and dotted here and there with varieties of oaks. Passenger trains do not stop, but pass on, and 6.42 miles further come to

Arcade—a mere side-track.

Rolling down 4.56 miles from Arcade, the train slowly crosses a long stretch of elevated road, and then on to the American River Bridge, 3.36 miles east of Sacramento—which spans the main stream of the American River—and pass along by the orchards and gardens which fringe the suburbs of the capitol of California, the dome of which can be seen on the left, also the State Agricultural Fair Grounds. The long line of machine shops belonging to the Railroad Company, on the left, are passed, and then we come to the Sacramento River, on the right, with its crowded whaives, and stop at the city of

Sacramento—Until the spring of 1870, this was the western terminus of the Grand Trans-Continental railroad. But upon the completion of the Western Pacific, from Sacramento to San Francisco, the two roads were consolidated under the name of the Central Pacific railroad of California, making one unbroken line from San Francisco to Ogden, 833 miles long. The distance from Sacramento to Omaha is 1,776.18 miles; Stockton, 50 miles; San Francisco via Livermore, 138 miles; via Benicie 89 miles; Vallejo, 60 miles; Marysville, 52 miles.

The city is situated on the east bank of the Sacramento River, south of the American, which unites with the Sacramento at this point. It is mostly built of brick; the streets are broad, well paved, and bordered with shade trees throughout a large portion of the city. It contains numerous elegant public and private buildings, including the State Capitol and county buildings. Population, 23,000. Churches, of all denominations, are numerous, as well as public and private schools. There are two orphan asylums; one Catholic, by the Sisters of St. Joseph, and the other Protestant. Secret orders are well represented, and newspapers are also plentiful, chief of which are the *Record Union* and the *Bee*, both daily; the *Journal*—German—is a tri-weekly; the *Leader*, the *Agriculturalist* and the *Rescue*, weeklies. The city is lighted with gas and supplied with water by two huge pumps in a building just north of the depot—with a capacity of 90,000 gallons per hour.

Hotels are numerous, but the principal ones are the Eagle, Arcade, Orleans and Western. Free "Buses" convey passengers from the depot to any of them, or, they can ride past them all on the street cars. In or near the city are located four flouring mills, six iron works, two potteries, smelting works, distilleries, plow works, planing mills, and many other small factories. The Capital Woolen Mills are located here, and consist of main building, 216 by 60 feet, with extention 40 by 60 ft.; total number of spindles, 1,440; employ about 65 hands, and use 1,000 lbs. of wool daily.

The Johnson & Brady Wine Co. work up 400 tons of grapes annually. The Sacramento Beet Sugar Factory is near the city—capacity, from 80 to 100 tons per day; main building, 150 by 63 feet. The factory grounds produce about 700 tons of beets annually. The company employ, when making sugar, 200 whites and 300 Chinese.

The principal machine shops of the Central Pacific railroad are situated, as we have seen, on the north side of the city, and with the tracks, yards, etc., cover about 20 acres. The buildings first erected are of wood, still standing and in use. The new buildings are of brick, comprising a machine, car, paint and blacksmith shops, round-house, and several other buildings. Nearly all the cars used by the company are manufactured here. It is a noted fact that the cars on both C. P. and U. P. R. R., are far superior in size, style and finish to those on the majority of the Eastern roads, and for strength and completeness of the arrangements for comfort in riding, they have no superior on any road.

The hospital belonging to the Railroad Co., a large, airy and comfortable building, is located near the shops, where their men are taken care of when sick or disabled It is well conducted, a credit to the company, and of inca'culable benefit to those unfortunates who are obliged to seek its shelter.

As for the mercantile business, let a few "figures talk":

During the year 1881 the aggregate sales of all kinds of merchandise and manufactured wares, exclusive of local in-

BLOOMER CUT,
85 feet deep and 800 feet long. See page 164.

agricultural land in the State. All kinds of grain and produce are raised in abundance. The vineyards are numerous, producing large quantities of wine and brandy annually. Raisins are produced in large quantities, and an immense amount of peanuts are gathered for market every year Stock-raising is also an important feature. Wool is a staple export of the county. Schools and churches are in a flourishing condition—a sure evidence of a people's prosperity. Stages leave Oroville regularly for La Porte, 2 miles; Susanville, 105 miles; as well as to most of the adjoining towns.

Returning to Marysville, we will now take a run to

YUBA CITY—situated about one mile west of Marysville, containing a population of about 1,000, and located on the eastern bank of the Feather River, just above its junction with the Yuba It is the county seat of Sutter county—first settled in 1849. The county was named after General Sutter, the old pioneer, at whose mill race at Coloma, El Dorado county, on the south fork of the American River, January 19th, 1848, the first gold was discovered in California. The county has a population of about 6,000, mostly engaged in agriculture. The soil is very fertile, and produces large crops of wheat, oats, and barley; there are also some very fine vineyards, producing a superior quality of fruit, from which many thousand gallons of wine and brandy are made annually.

The city has one newspaper—the *Sutter Banner*—and several hotels. It is at the head of steamboat navigation, and is connected with San Francisco and the world by the road over which we came—the Oregon division of the C. P., via. Marysville, between which cities a hack runs regularly.

THE "BUTTES"—called the "Marysville Buttes," are a noted land-mark to the westward, about ten miles. They consist of a series of peaks that rise from the crest of an isolated mountain range, which stands bold and clear among the plains, 1,000 feet in height. From appearances, one would be led to suppose that this ridge crossed the valley at one time, when this was an inland sea; and when the waters escaped from the lower valley, those confined above cut a portion of the ridge down level with the plain, and escaping, left a beautiful valley above.

From the summits of their bald peaks a fine view can be had of a great portion of the Sacramento Valley, including MOUNT SHASTA, away to the northward, 220 miles distant, in latitude 41 deg. 30 min., an isolated and lofty volcanic mountain, full 14,440 feet high. It is covered with perpetual snow, and is the head and source of the Sacramento River. (See ANNEX No. 43, page 214. To the northwest, in the Coast Range, can be distinctly seen Mts. LINN, ST. JOHN and RIPLEY. On the south, Mt. DIABLO, in the Contra Costa range, while on the east, from north to south, is the long range of the Sierra Nevadas, as far as the eye can reach.

Returning to Marysville, we again start on our journey. One mile north of the city we cross the track of the Oroville railroad, pass several cemeteries on the right, also a race track; then, bear away to the left—northwest—and cross the Feather River on a long trestle bridge. Along this river, live oaks and sycamore trees abound by the million.

LOMO—a flag station, comes first from Marysville, 6.8 miles. Here are wheat lands which continue, with an occasional clump of trees, 3.9 miles, to

LIVE OAK—another side track, in the centre of some thousands of acres of young live oaks, and manzanita shrubs. Passing along with the broad valley of the Sacramento on the left, which stretches away as far as the eye can reach, and the Feather River Valley on the right, beyond which are the lofty Sierras, we reach

GRIDLEY—which is 6.5 miles further north. This station has several hotels and stores, a dozen residences, and a large grain warehouse, and one flouring mill, as, be it remembered, we are now in one of the great wheat sections of the State. The station was named for a Mr. Gridley—maybe it's "Old Bob Gridley"—who knows?—who owns somewhere about 35,000 acres of land adjoining the station, much of which he has worked by farmers on a division of crops. Live oaks, big ones are numerous all the way for 3.4 miles to

BIGGS—This is a lively town of about 1,200 population, in the midst of fine wheat lands, with extensive warehouses for storing and handling wheat in nearly all seasons. This cereal is a large and sure crop. Biggs has a weekly paper—the *Register*—several hotels, chief of which is the Planters, and a stage line to Oroville, twelve miles east; fare, $1.00.

The place was named for a Mr. Biggs, who, like the Mr. Gridley, is troubled with about 30,000 acres of this wheat land, much of which yields, when properly farmed, 50 bushels to the acre. Leaving Biggs we cross the big canal of the Cherokee Company, which is 18 miles long and 400 feet wide; the water is used for hydraulic mining, and then for irrigating purposes. After a run of ten miles, all the way through wheat fields, we reach

NELSON—composed of about a dozen buildings, surrounded with wheat, wheat, all wheat. These fields extend far away in every direction.

Passing along, we cross Butte Creek, and 6.6 miles from Nelson, come to

DUNHAM—Here is about a dozen buildings, in the midst of a broad plain studded with occasional oaks. A flouring mill and large warehouses are near the station. Continuing on 6.1 miles, and we stop at the beautiful town of

CHICO—It is 43 miles from Marysville; 25 miles northwest from Oroville, and five miles east of the Sacramento River, situated in the Chico Valley, Butte county, in the midst of as rich a farming section as the State affords; population 5,000. The city is lighted with gas, has ample water-works situated near the depot, and has several banks and hotels, chief of which are the Chico House and the Union; one daily paper, the *Record*, and one weekly, the *Enterprise*. To the eastward looms up the Sierra Nevada Mountains, covered with a dense forest of timber, in which are many sawmills, the lumber from which is floated down to within three miles of the city, in a "V" flume, 35 miles in length. The streets are lined with shade trees, groves of oaks, and orchards and gardens are on every hand. Near the town, General Bidwell, the old pioneer, has an extensive ranche—or farm, as it would be called in the Eastern States—which is in a very high state of cultivation, producing abundantly all kinds of fruits and plants of the temperate and semi-tropical climes.

Stage routes from Chico are numerous. Six-horse coaches, in summer, leave for Oroville, 25 miles; Butteville, Plumas Co., 63 miles; fare, 10 cents per mile. Stages leave for Diamondville, eleven miles; Butte Creek, 12 miles; and Helltown, 14 miles; also, for Dayton, six miles; Jacinto, 14 miles; Germantown, 13 miles; Willows, 56 miles; Colusa, 40 miles; Williams, 40 miles; Allen, 55 miles; and Bartlette Springs, 58 miles. Stages run Mondays, Wednesdays and Fridays, to St. John, ten miles; Orland, 23 miles; O.impo, 30 miles; Coast Range, 35 miles; Newville, 40 miles. The average fare to all these places is ten cents per mile.

Leaving Chico, our course is more westward for seven miles to

NORD—a small station about three miles east of the Sacramento River. Next comes a side-track, 2.3 miles further, called

ANITA—and 2.4 miles further

CANA—This place has a population of about 100, most of whom are farmers, as wheat fields are still the rule. On, 2.8 miles further comes

SOTO—near Deer Creek, and 4.3 miles from

VINA—a small station near the Sacramento River, in the center of a very fertile region and a great point for grain shipments.

Continuing on 7.5 miles further, crossing several small creeks, passing some oaks and willows along the creeks, we come to

SESMA—a side-track, on the east bank of Sacramento River, which we cross, and stop at

TEHAMA—on the west bank of Sacramento River, where boats often land, being a thriving town of about 700 population. The *Tocsin* heralds the news daily in clarion tones, that all may learn that Tehama has a live daily newspaper. The place was, in early days, known as "Hall Crossing." Agriculture is the principal feature of the place, although the lumber business is an important item. A "V" flume brings the lumber down from the mills in the Sierras on the northeast, a distance of forty miles, with a capacity of 40,000 feet per day. The country is very fertile. Live oaks are numerous.

Lassen's Peak, to the northeast, is a prominent feature of the landscape, as it rises 10,578 feet above sea level, which would be about 10,000 feet higher than Tehama. Continuing along 12.1 miles further, crossing several small creeks, we come to

RED BLUFFS—the county seat of Tehama county, at the head of navigation on the Sacramento River, with a population of about 2,000. It is situated in the midst of rich agricultural and grazing land, with many thriving vineyards and several hotels, chief of which are the Tremont and Red Bluffs Hotel; also two weekly newspapers, the *Sentinel* and the *People's Cause*. Lumber is an important industry, and the manufacture of doors, sash and blinds is carried on to a great extent. Mt. Shasta, to the north, is a prominent object, rising up out of the valley.

Continuing along, crossing several little creeks, bearing away more to the westward, 10.3 miles, we come to

HOOKER—a signal station, 4.9 miles from

BUCKEYE—another signal station, located 1.8 miles from

COTTONWOOD—This is a small village in Shasta county, of about 300 population, situated on Cottonwood Creek, about five miles west of the Sacramento River.

Turning more to the northeast, a short run of 7.6 miles brings us to

ANDERSON'S—a village of about 100 population, 6.3 miles from

CLEAR CREEK—a small station on a creek of that name, which comes in from the west, and after 4.8 miles further, we come to the end of the road at

REDDING—This place was named for the land commissioner of the railroad company. This is the terminus of the road, having a population of about 500.

Redding is 308 miles north of San Francisco, 2.5 miles south of Roseburg, Oregon, and 475 miles from Portland.

Stages leave Redding daily, with passenger, mails and express, for Roseburg, and all intermediate places, distance 275 miles. At Roseburg connections are made with the Oregon & California railroad, for Portland, 200 miles further. Fare to Roseburg from Redding, $41.25.

This region is fertile in subject matter for our book, but we are limited as to space, and with this hasty glance at the most important features of the country, now return to the city of Sacramento, and again start out on another route.

Up the Sacramento River.

Stepping on board a light draft steamboat, at the wharf in Sacramento, we are off for a trip by water.

About the first thing to attract the attention of the traveler after leaving the city, will be the

"TULES," which is the native name given to the rushes which cover the low lands and bays of California. They are of the bulrush family, probably the father of all rushes. They grow from six to ten feet high, and so thick on the ground that it is extremely difficult to pass among them. The lands on which they grow are subject to annual overflows. During the prevalence of the floods, miles and miles of these lands are under water, presenting the appearance of one vast lake or inland sea. In the fall and early winter, when the tuiles are dry, they are often set on fire, forming a grand and terrible spectacle, especially during the night. When once the fire attains headway, nothing can quench its fury until the tuiles are swept away to the bank of some water-course, which bars its further progress.

The soil composing the land is *adobe*, of a purely vegetable mold. Wherever it has been reclaimed, it produces grain and vegetables in almost fabulous quantities. It is claimed by many, that, with proper appliances, these lands could be converted into magnificent rice fields; the advocates of this measure asserting that they possess every requisite of soil, clime, and adaptability to irrigation. The State has provided for a system of levees, by which a large amount of land, heretofore known as tuile land, has been successfully reclaimed, and is now counted the most valuable in the State.

The country, after leaving Sacramento, is level for a vast distance on either hand; the "tuiles" are disappearing, and, before we reach Knight's Landing, the left-hand shore is more bold, and the wheat-fields and gardens have taken the place of "tuiles" along the river bank.

We have already been to Marysville by railroad, now let us go via FEATHER RIVER, a beautiful stream, its clear waters contrasting advantageously with the muddy waters of the river we have left. We pass through a fine country with wheat farms on the higher lands, and reach Nicholas, a dull, quiet town of about 300 inhabitants, situated at the junction of Bear River with the Feather. Proceeding up the Feather, we pass HOCK FARM, the home of the venerable pioneer of California, General Sutter. It is a lovely place—the old farm-house and iron fort standing on the bank of the stream. Enormous fig trees line the bank, while behind them can be seen the fine orchards and vineyards planted by the General over 50 years ago. General Sutter settled in California under a grant from the Russian Government, which conveyed to him large tracts of land around Sacramento City, including the city site; also a large tract, of which Hock Farm is a part. Sharpers and swindlers deprived the old pioneer of most of his property, leaving him penniless, and a pensioner on the State.

Passing on by the junction of the Yuba and Feather rivers, we soon reach Yuba near Marysville, 65 miles from Sacramento, by water. (See description on page 167) Returning to the Sacramento, the right-hand bank of the river appears low and swampy, covered with "tuiles" for a great distance inland. Passing on, we soon arrive at

KNIGHT'S LANDING—a small place—46 miles from Sacramento. Population about 200. It is quite a shipping point for Yolo county, and is on the line of the California Pacific railroad. This road, in 1873-4, extended to Marysville, crossing the river at this point, but the floods washed away the road-bed on the east side of the river. The road passed into the hands of the "Central" Company, who had a road to Marysville via the Roseville Junction. So it never was repaired, but the bridge turned to allow a free passage of the river boats, and has been so, for the last four years, and probably will so remain until it rots down.

For a long distance above Knight's Landing, the low marshy plains continue on our right, the higher land covered with wheat on our left, with no towns of any importance to note until we arrive at

COLUSA—This is a point of considerable trade—125 miles from Sacramento. It is the county seat of Colusa county, situated on the west bank of the Sacramento River, and contains about 1,500 inhabitants. The Colusa *Sun* is published here.

The town was laid out in 1850, by Colonel Semple, the owner of the "Colusa Grant"—containing two Spanish leagues. It is now the center of a very large farming and grazing country. Schools and churches are well represented. Stages run daily between Colusa and Marysville—29 miles; also to Williams, nine miles west.

Passing on up the river, the country seems to gradually change to a grazing, instead of a grain country, more especially on the west.

About 200 miles further and we reach the Landing of Chico—but we have already described the town on page 186, so we will pass on. On the right-hand side, the shores are low and sedgy most of the way, fit only for grazing when the floods have subsided; yet we pass intervals of grain fields till we arrive at Red Bluffs—270 miles from Sacramento, at the head of navigation. See description of the town on page 170.

We will now return to the city of Sacramento, and there take another route.

Down the Sacramento River.

Stepping on board of a large passenger and freight boat, we start down the river towards San Francisco. The plains stretch away on either hand, and there is little to be seen except the gardens and farms along the banks on the higher ground, the wide waste of "tules," and the plains and mountains beyond. On the left—away in the dim distance, the hills succeed the plains, the mountains the hills, until the vast pile towers among the clouds.

Winding around curves, where the stern of the boat is swept by the willows on the shore, we glide down the river, past sloughs, creeks, and tule swamps, until we pass FREEPORT, 12 miles from the city, a little hamlet of half a dozen dwellings.

Floating along between the low banks, covered with willow and shrub, we pass MISSISSIPPI BEND—24 miles from Sacramento. Here the river makes one of its numerous curves, almost doubling back on itself.

To the left is the little town of RICHLAND, containing a half-dozen dwellings. Now the Nevada Mountains fall behind, and we have one vast plain around us. We pass the outlet of Sutter's Slough, and then the Hog's Back—a long sand-bar, which stretches diagonally across the river. The water here is very shoal. A wing dam has been built from the western shore, half way across the channel, which throws the water into a narrow compass, giving greater depth to the bar. Next comes Cache Creek Slough, on which large quantities of grain are shipped to San Francisco via Sacramento River, from Yolo and Solano counties. Now we are passing along by the Rio Vista hills, which come close to the water's edge on the right-hand shore. These hills are the first we have seen near the river since leaving the city. They consist of one long, low ridge, broken into hillocks on its crest. These hills are excellent wheat land, yielding an abundant harvest. The land is very valuable, though but a few years have passed since it was sold for 25 cents per acre. The town of RIO VISTA is situated on the slope of the foot-hills, and contains about 300 inhabitants. Formerly the town stood on the low ground, near the river bank, but the flood of '62 washed it away, carrying from 40 to 60 houses down the river. The people fled to the high lands, where they remained until the passing steamers took them away. For days the little steamer Rescue was plying up and down the river, running far out over the submerged plains, picking up the "stragglers," who were surrounded by the waters. Some were found on the house-roofs, with the flood far up the sides of their dwellings, and others were rescued from the branches of trees, which afforded them the only resting-place above the waters. The flood of '62 will long be remembered by those who then dwelt on the banks of the Sacramento.

We next pass COLLINGSVILLE, a long wharf on the right-hand side of the river, with a house or two standing close by. It is a point of shipment of considerable freight, for the country, and grain for the city. A little below this point, the San Joaquin River unites with the Sacramento, entering from the left, forming *Suisun Bay* (pronounced Soo-e-soon).

ANTIOCH—just across the bay—is in Contra Costa county; population, 500. Three miles south, by railroad, are the Mount Diablo coal mines, which yield large quantities of coal.

There are several manufactories of pottery in the town—the clay in the vicinity being a very superior article. The *Ledger*, a weekly paper, is published here. Attention has lately been attracted to the silk culture, and many thousand mulberry trees have been planted.

Passing on down the bay, we enter the Straits of Carquinez, when a long, low wharf on the right attracts our attention. It is fronting the old town of Benicia, of which more hereafter.

Passing on down the straits we have a fine view of Vallejo, which lies to our right, near where we enter San Pablo Bay. Turning to the left, 23 miles more brings us to San Francisco. But we must return to Sacramento and

Take the old Overland Route, Via Stockton and Livermore Pass. Leaving Sacramento, the route is along the east bank of the Sacramento River, through well-cultivated gardens, orchards and vineyards 5.7 miles to

Brighton.—Here we learn that the cars of the Sacramento Valley railroad, managed by the Central Pacific Company, run down on the same track as the "Central" to this station, where they branch off. Let us step into them, and see where they go. Patterson's is five miles; Salsbury's six miles; Alder Creek, three miles, and three more to

FOLSOM—twenty-five miles east from Sacramento, in Sacramento county, on the south bank of the American River; population about 2,000. Vine culture is an important industry. Some of the finest vineyards in the State are located here, including the Natoma, which is celebrated for its fine quality of raisins and wine. To the north and east of the town placer mining is the principal business; to the south and west, farming and grazing.

There are extensive granite quarries in the vicinity. From the bed of the river, near this point, large quantities of cobble-stones have been obtained, for paving the streets of Sacramento and San Francisco.

Folsom is ornamented with fruit and shade trees, and has many fine public and private buildings, with magnificent scenery. Regular stages leave for Coloma, daily, via Mormon Island, Salmon Falls and Greenwood Valley, twenty-four miles distant.

The Indians call the telegraph the "whispering spirit."

Emigrants, on the plains, are called, by the older settlers, "pilgrims."

No. 15 ANNEX. State Capitol of California.—This is one of the first objects which meets the eye when approaching Sacramento from the east. It is a conspicuous landmark. The building occupies the center of four blocks, bounded by 10th and 12th, and by L and N streets. The grounds form three terraces, slightly elevated above each other, and connected by easy flights of steps. They are regularly laid out, and covered with a beautiful sward, closely shaven by the lawn-cutter. They are interplanted with shrubs and evergreen trees. The outer border of the lowest terrace is studded with flowers. Its front is toward 10th street, and is 350 feet long. Approaching it from this point you may regard it as a great central building, from which rises the lofty dome, and having on each side a large wing. A flight of granite steps, 25 feet high by 80 feet in width, leads to a front portico of ten columns, through which, and a large hall, the rotunda of 72 feet diameter is found in the center; and from this, in each story, halls, elegantly arched, extend through the front and wings, the State offices being on either side. Five female figures ornament the front above the columns. The central one is standing, the remaining four are in sitting postures. They represent war, science, agriculture and mining. The wings forming the flanks of the building are 164 feet above the first or basement story. The north and south flanks of the building form, respectively, the Assembly and Senate chambers, the former being 82x72, and the latter 72x62. In the rear center, a circular projection of 60 feet diameter forms the State Library. The first story of 25 feet is of white granite, from neighboring quarries, and is surmounted by a cornice of the same. Above this the body of the main dome is surrounded by an open balcony, which is supported by 24 fluted Corinthian columns and an equal number of pilasters. Above this balcony the body of the dome is supported by an equal number of ornamental pilasters. From these rises the great metallic dome. From the top of this dome in turn rise 12 fluted Corinthian pillars, which support the final or small dome, and this is surmounted by the statue of California.

The whole interior is one solid mass of iron and masonry. The dome of the interior rotunda, which is of iron ornaments and brick work, is exceedingly handsome. The panels and pedestals under the windows are of the beautiful laurel, well known in California for its susceptibility to receive a high polish. All the first floor doors are of walnut, with laurel panels, as are also the sashes throughout the building. The stories are, respectively, 21 feet 6 inches, 20 feet, and 18 feet in height. It covers, with its angles, nearly 60,000 surface feet of ground, and measures over 1,200 lineal feet round in all the angles. See large illustration, No. 16.

No. 58 ANNEX.—New Sacramento Depot.—From page 165.—The Central Company have completed a depot at Sacramento, that is the largest, finest and most commodious on the Pacific Coast. It is constructed of the best material and in the most substantial manner. It is situated about midway between the bridge over the Sacramento river and the company's shops, fronts north, on ground filled in and specially prepared for that purpose. The main building is 416 feet long, and 70 feet 6 inches wide, two story. The front has four large arches in the center, and eight smaller ones on each side. Three tracks run through the building, and a platform 22 feet wide. In the rear is an annex, 160 feet long, and 35 feet wide, one story, in which is a dining-room, 40x55 feet, 14 feet high, two waiting-rooms, 26x35 feet. On the first floor are ticket, sleeping-car, and telegraph offices, lunch counter and baggage-room, news-room, etc. The second story is occupied by the offices of the Sacramento Valley Railroad, Supt. of Division of the C. P., Train Dispatchers, Conductors, Rooms for Storage, Stationery, etc.

Sacramento is now a regular eating station.

Passing on seven miles to White Rocks, eight to Latrobe, and eleven to Shingle Springs, brings us to the end of the railroad, 49 miles from Sacramento.

PLACERVILLE.—is twelve miles from Shingle Springs, with which it is connected with daily stages. It is the county seat of El Dorado county, 60 miles east of Sacramento, at an altitude of 1,880 feet above tide—present population, about 2,000.

Who has not heard of Placerville, El Dorado county ? It was in this county, at Coloma, eight miles northeast of the city, where the

FIRST GOLD DISCOVERY—was made January, 19th, 1848—by J. W. Marshall, in the mill race of General Sutter. The announcement of this discovery caused the *wildest gold fever excitement ever experienced* not only in America, but in every part of the civilized world.

The news of these rich discoveries sped with the wings of the wind, and thousands, yes, tens of thousands, in the Atlantic States left homes, friends, and all they held dear, to make their fortunes in this, the new El Dorado. With many the excitement became intense, ships, steamers, barks, brigs, and all manner of sailing vessels were chartered or purchased for a trip "around the Horn;" and no sacrifice was thought to be too much to make to procure the necessary outfit for the expedition. Again there were thousands who, choosing the land, boldly struck out toward the setting sun to cross the *then* almost unknown trackless deserts, and pathless mountains. Horses, mules and cattle were pressed into service, as well as all kinds of conveyances, while many started with hand-carts, propelling them themselves, upon which they packed their tools and provisions for the trip. Again, others started on foot, with only what they could pack on their backs, "*trusting to luck.*" Very few, if any, had a thought of the privations to be endured, or the obstacles to be overcome, so anxious were they to arrive at the Land of Gold.

Those who came by water, passed in at the Golden Gate, and up the Sacramento, while those by land came pouring over the Sierra Nevada Mountains, by natural passes, down, down into this beautiful valley, where a city of many thousands suddenly sprang into existence. From a "little unpleasantness" the place was first known as "Hangtown," but in 1852 it was changed to Placerville, which indicated at that time, the nature of the mining done in the vicinity. Of the many thousands who started across the plains and mountains, hundreds died by the wayside, and were buried by their companions, while the greater number were "lost" by the hand of the *friendly* Indian or the *hostile* Mormon.

It has been estimated, and we think correctly, could the bones of these emigrants be collected, and those of their animals, together with their wagons and carts, in one continuous line, between the Missouri river and the Pacific coast, since the rush commenced in 1848, they would be *more numerous* and *closer together* than the telegraph poles on the line of the Pacific railroad across the Continent.

The early mining done about Placerville was done by hand, the pan, rocker, and long Tom; these have long since given place to the quartz mills—there are 32 in the county—and the hydraulic process, by which nearly all the mining is now done.

Vine culture and fruit culture are now the most important occupations of the people of the county. Fresh and dried fruits are shipped by the hundreds of tons, while the annual crop of wine and brandy produced is over 300,000 gallons.

Placerville contains a goodly number of schools, and churches of almost every denomination, including a "Joss House." The different secret orders are well represented, and two newspapers, the *Democrat* and the *Republican*, make their appearance every week.

Placerville is situated in what is known as the FOOT-HILLS as the chain of broken land is called, which lies between the Sierra Mountains and the plains, extending from Fresno county on the south, through Tuolumne, Calaveras, Amador, El Dorado, Placer, Nevada, Yuba, Butte to Tehama, on the north, comprising nearly one-fourth of the arable land of the State. The soil is altogether different from that of the valleys, being generally of a red, gravelly clay and sandy loam. In the little valleys which are found among these hills, the soil is generally a black loam—the product of the mountain washings. Experiments, however, have decided the fact that these foot-hills are the natural vineyards of California. In El Dorado and Placer counties, on these sandy foot-hills are now the finest vineyards in the State, from which are manufactured fine wines and raisins.

Here among these hills are as cozy homes as one could wish to have, where grain, vegetables and all kinds of fruit are raised in abundance, while thousands of acres are lying vacant awaiting the emigrant.

The mulbery tree and the silkworm are cultivated to some extent in the foot-hills, and this branch of industry is lately receiving considerable attention.

Again we return to the TRANS-CONTINENTAL, which we left at Brighton.

Florin—is four miles from Brighton. The many new buildings the well-cultivated fields, the raisin grape vineyards, all denote a spirit of progress in the settlers, that would seem to say, "*We have come here to stay.*"

The traveler has probably noticed several windmills along the road, before arriving at this station. The CALIFORNIA WINDMILL is a great institution in its way. It seems to have been brought to a greater state of perfection on this coast than anywhere else. From this place we will find their numbers to increase until we get to the "Windmill City," as Stockton is often called, where they can be seen in great numbers, in every direction. Many times the water is pumped into reservoirs built on the tops of the houses, resembling a cupola, from which pipes take the water to the different rooms throughout the house and grounds; the waste water is conducted into the gardens and fields for irrigating purposes. These mills are numerous in San Francisco and throughout the State. From Florin it is 6.2 miles to

Elk Grove—Here, too, many recent improvements are noticeable. In a clear day the snow-capped Sierras, on the left can be plainly seen, and away to the south 60 miles distant,

MOUNT DIABLO rising clear and grand from out the plains, an unerring pilot to those who wandered across these once trackless plains that now are teeming with life and industry. It is situated in the Contra Costa range of mountains, and is the meridian point in the land surveys of the State. Elevation 3,876 feet. The view from the summit includes the country and towns around San Francisco, San Pablo and Suisun bays, and the valleys of the Sacramento and San Joaquin Rivers. It is reached by rail or steamboats from San Francisco, Stockton or Sacramento.

The beautiful valley through which the road passes is now spreading out before us until we begin to realize that nature has done much for this "sunset land."

McConnell's—is a small station 8.5 miles from Elk Grove. Near are large fields, where cattle and sheep are fed and fattened for market.

Before reaching the next station, we cross Cosumnes River, which rises in the mountains to the northeast. The bottom lands are very wide, and covered with both white and live oaks, and near the stream with willows. The water gets very high in the spring-time, and very low in the summer.

Galt—a station with a few dozen buildings, large warehouses, big cattle pens, and shutes for loading cattle and sheep—is 7.7 miles from McConnell's. At this station we find the

Amadore Branch Railroad,

Let us "change cars," and note the result. Leaving the station, our course is north a short distance, and then northeast towards the mountains, over a section of country devoted pretty generally to stock-raising—sheep principally.

CICERO—is the first station, 8.8 miles distant, a small place of about 150 inhabitants. Passing on 11.7 miles further, we are at

CARBONDALE—Here are extensive coal mines, operated by the Ione Coal Co., who load on an average fifteen cars per day—150 tons.

At Michigan Bar, eight miles north, large quantities of the best pottery are manufactured, which finds its market over this road. From Carbondale it is 6.7 miles to

IONE—the end of the road, 27.2 miles from Galt. This place is in Amadore county, in a section devoted to mining and agriculture.

The coal mines located here have yielded about 200 tons per day during the year 1878. A new vein of coal, struck towards the last of the year 1877, opens up an almost unlimited deposit. Placer mining is carried on to some extent on Sutter Creek.

The *News*, a weekly paper, is published at Ione, which is also a point from which fourteen mining towns, large and small, draw their supplies. Some of these are reached by stage as follows: Volcano, population, 500; West Point, 300; Jackson, San Andreas, Mokelumne Hill, Sutter, Amador, Drytown, Plymouth, and Fiddletown. These places are from ten to fifteen miles distant. From Ione it is fifteen miles southeast to Mokelumne Hill, (pronounced Mokel-m-ne,) county seat of Calaveras county. This is one of the early mining towns of the State. Placer mines were worked as early as 1848, and are worked to some extent at the present time; but quartz mining and agriculture are the principal occupation of the people. It is a pretty little town; the streets are ornamented with shade trees on each side, and has some beautiful gardens and private residences, with good schools and churches, several good hotels, and one weekly newspaper—the *Chronicle*—the oldest paper in the State. Population, 1,200.

We will now return to Galt, and start once more south.

Acampo—is a small station where trains seldom stop, 5.4 miles from Galt, and 2.9 miles from

Lodi—The country along here has been settled up very much in the la-t four years; the fields are pretty generally fenced and well cultivated, and some fine vineyards of the raisin grape can be seen. Many new buildings attest the thrift of the people. Stages leave this station daily—except Sunday—for Mokelumne Hill, 35 miles east.

Castle—is six miles from Lodi. Our train rolls along through fine broad bottom lands, dotted here and there with white-oak trees, which, at a distance, appear like an old New England apple-tree.

Six miles further, just before reaching the next station—on the right, that large building is the STATE INSANE ASYLUM. The grounds devoted to the use of the asylum occupy 100 acres. The first building in view is the male department; the second, the female. We are now in the suburbs of

Stockton—the county seat of San Joaquin county. Population, 13,000. Elevation, 23 feet. The city was named in honor of the old naval commodore of that name, who engaged in the conquest of California. It is situated on a small bay, of the San Joaquin River, at the head of navigation; navigable for crafts of 200 tons; yet steamboats of light draft ascend the river (San Joaquin) 150 miles farther. Stockton is situated in the midst of level plains, celebrated for their great yield of grain. It is the center of an immense grain trade. In early times, the only trade depended upon for the support of the city was derived directly from the work-ing of the mines to the eastward. Some of this trade is still retained; but, compared with the tremendous grain trade which has sprung into existence within the last six years, it sinks to a unit. The city has many beautiful public and private buildings, thirteen churches, fourteen public and many private schools; is lighted with gas, and well supplied with water, the latter from an artesian well 1,002 feet deep, which discharges 360,000 gallons per day—the water rising ten feet above the city grade. There are several good hotels; the Yo-Semite and Grand are the principal ones. The *Independent* and the *Herald* are daily papers, published here. The city is embowered in trees and shrubbery; most of the private residences and gardens of the citizens are certainly very tastefully ornamented with all kinds of vines, shrubbery, and flowers.

The soil around Stockton is "adobe," a vegetable mold, black and very slippery, and soft during the rainy season. This extends southward to the Contra Costas, and west about five miles, where the sand commences and extends to the river.

Stockton, for several years after the completion of the Overland road, was the starting point—in stages—for Yo Semite Valley and *all* the big trees. But, by the building of the Visalia Division of the "Central," most, if not all, the travel for Yo Semite, Mariposa and the Tuolumna big tree groves, has taken that route, the distance by stage being much shorter. (See map of route, page 120.) Again, by the building of the Stockton & Visalia, and Stockton & Copperopolis railroads, all the travel for the Calaveras Big Tree Grove, goes by that route. (See map.) For a description of the route to Yo-Semite Valley and the Mariposa and Tuolumna big trees, see under "Towards Sunrise," page 209.

To CALAVERAS BIG TREES—we will simply note the route. Take S. & C. R. R.—a branch of the Central Pacific—at the same depot, and the route is east, six miles to Charleston, three to Walthall, two to Holden and four more to Peters, 15 miles from Stockton. From Peters, a line branches off to the south-east. On that branch, it is five miles to Farmington, three to Grigo, five to Clyde, four to Burnett's and two more to Oak Dale, the end of the road, 34.4 miles from Stockton.

Returning to Peters, it is 7 miles to

MIRROR LAKE AND REFLECTIONS.

1ITE VALLEY, CAL. (See Annex No. 40.)

LIVERMORE PASS TUNNEL.—See page 179

Waverly and eight more to Milton, the end of the road, 30 miles, from Stockton. Some travel leaves Milton for Yo Semite Valley, Chinese Camp, Big Oak Flat and the Tuolumne Big Trees; distance from Milton, 87 miles.

Stages leave Milton regularly for Murphys, 30 miles; Sonora, 36 miles; Chinese Camp, 28 miles and all mining towns of note to the north and east. From Milton it is 45 miles to the Calaveras Big Tree Grove, via Murphys.

BIG TREES—There has been, up to the present time, ten "Big Tree Groves" discovered on the western slope of the Sierra Nevada Mountains, numbering from 92 to 1,000 trees each, and ranging in height from 250 to 321 feet, with a circumference, at the ground, of from 60 to 95 feet each.

The largest ever discovered is called the "Father of the Forest"—now prostrate—and measures 435 feet in length and 110 feet in circumference. It is in the Calaveras grove. The elevation of this grove above tide is 4,735 feet. The trees number 92, ranging from 150 to 321 feet in height. The most notable are the "Father of the Forest," as above stated; the "Mother of the Forest," 321 feet high, 90 feet in circumference; "Hercules," 320 feet high, 95 feet in circumference; "Hermit," 318 feet high, 60 feet in circumference; "Pride of the Forest," 276 feet high, 60 feet in circumference; "Three Graces," 295 feet high, 92 feet in circumference; "Husband & Wife," 252 feet high, 60 feet in circumference; "Burnt Tree"—prostrate—330 feet long, 97 feet in circumference; "The Old Maid," "Old Bachelor," "Siamese Twins," "Mother & Sons," the "Two Guardsmen," and many others range from 261 to 300 feet in height and from 59 to 92 feet in circumference. Of over 350 big trees in the Mariposa grove, 125 are from 250 to 350 feet in height and 40 feet in circumference. The "Rambler" is 250 feet high, and 102 feet in circumference, at the ground.

Hotel accommodations at the different groves, and in Yo Semite Valley will be found ample. Returning to Stockton, we once more take the Overland train for San Francisco. This is a great country for rapid changes. Where to-day there are only stages, to-morrow there may be palace cars; so that it is almost impossible for us to keep up with the times. When our guide was *first* published, in 1869, *not one foot* of all the roads we have come over, from the Missouri to this place, or will go over, or that are, was built or hardly thought of. How fast we live! At the same proportional increase for the next ten years, where will we be?

Lathrop—is 8.9 miles south of Stockton, at the junction of the "Visalia Division" of the Central Pacific railroad. Here the R. R. Co. have erected a fine, large hotel, before which trains stop 30 minutes, to afford passengers an opportunity to take a meal, for which the moderate charge of 50 cents is made. (For a description of the country to the south, including Yo Semite Valley and the big trees of Mariposa and Tuolumna groves, See page 211.) Here passengers for Fort Yuma, Los Angels and intermediate country can "change cars," if they choose, without going to San Francisco.

From Lathrop, it is 3.6 miles to the bridge over the San Joaquin River. Here the cars come to a full stop before crossing, to be *sure* to guard against accidents—as the bridge has a "draw" for the accommodation of the river boats. This company has a rule for all their employes, and a "GOLDEN" ONE IT IS, that "*In case of uncertainty, always take the safe side.*" This rule is *well observed;* few "accidents" take place on the roads operated by this company, for the reason that the road is constructed of *good* materials, and in the most *substantial manner*, with all its equipments of the *first class*. The officers are thoroughly practical men, who never discharge an employe, on any consideration, who has proved to be a *competent man* for his position, simply to make room for a *favorite*, or a *worthless* "*cousin*."

Crossing the bridge, the long range of the Contra Costa Mountains looms up in the distance, directly ahead, and extends a long distance to the right and left, on either hand, as though to effectually stop our progress. We cannot see any place to get through or over them, yet we are *sure* San Francisco is on the other side.

Banta—is 1-4 miles from the bridge, and is reached after passing over a broad bottom, the soil of which is a rich, deep sandy loam and very productive. This station consists of a postoffice, a half-dozen stores, several large freight warehouses, with a surrounding country, well settled, most of which is under cultivation. Stages leave this station on arrival of trains for Graysonville, 20 miles; Mahoney, 35 miles; and Hill's Ferry, 44 miles.

After leaving the station, we have witnessed, on several occasions, by looking away to the right, that curious phenomenon, the mirage, which is often seen on the desert. (See page 142).

AT TRACY—three miles west of Banta we come to the junction of the new San Pablo & Tulare Railroad, a division of the Central, opened in the summer of 1878, and runs via Martinez, along San Pablo Bay via Berkley to Oakland, distance 83 miles to San Francisco, with *no grades*, being 12 miles longer than the route via Livermore Pass. We shall take this ROUTE *first*, and then return.

Tracy—is situated on a broad plain, with the Contra Costa range of mountains to the south and westward, and the San Joaquin River on the east and northward, the broad bottom lands of which extend for over thirty miles north, where they unite with, and from thence become the great Sacramento Valley, extending farther than the eye can reach. The station buildings are located between the old and the new tracks, and are very commodious.

From Tracy, our course is north of west, direct for the junction of the Sacramento and San Joaquin Rivers. The country is principally settled by agriculturists and small stock raisers. The lands where cultivated, are fenced, and can always be depended upon for raising good crops; some without irrigation, but *all* with it—the facilities for which are readily obtained, and with little expense.

Bethany—is the first station after leaving Tracy, trains stop on signal.

To the left five miles, are the Contra Costa mountains, grass covered, with timber in the ravines, and some trees on the higher peaks. Near the station, on the left, is a grove of eucalyptus trees.

Continuing on nine miles, we reach another signal station, named BYRON, from which it is five miles to the signal station of BRENTWOOD. Oak trees are numerous for the last ten miles, with some chaparral in places on the right. The soil is a rich sandy loam, warm, and easily worked. The mountains to the left, are closer and covered with trees. Five miles further and we stop at

Antioch Station—The town is one mile to the right, on the banks of the San Joaquin River, above its junction with the Sacramento. See page 189.

Leaving Antioch, we pass under three railroad tracks which come down from the coal mines, two miles to the left, and run to shipping wharfs one mile to the right. These mines are extensive, but the quality is quite ordinary.

Our road for the next twenty miles is cut through many narrow rocky or sandy spurs, from 50 to 100 feet in height that shoot down from the mountains on our left, to the water's edge on our right, between which, are as many little creeks, or sloughs, over which the road is built, sometimes on piles, and with tule lands on each side.

Cornwall—is five miles from Antioch, opposite Suisun Bay, just below the junction of the Sacramento and San Joaquin Rivers, which are one mile to the right.

At this station we get a first view of the town of Benicia, five miles ahead to the right, on the opposite side of the waters. Those large buildings on the higher ground are the U. S. arsenal and barracks. (See description page 187.)

To the left of our train, all along beside the mountains, are many well-fenced and cultivated farms, with neat cottages surrounded by orchards, vineyards, fruit, and flowers, together with evergreen shade trees in great variety, and in great abundance.

Passing BAY POINT in eight miles, AVON in three, four more brings us to

Martinez—the county seat of Contra Costa county, a small town of about 600 inhabitants. The country is principally devoted to agriculture and vine culture, the vineyards being numerous

Continuing along, at times on the river bank and through several rocky cuts, 3½ miles brings us to the Short Line Overland Route. (See page 183).

Ellis—is 5.2 miles from Bantas, west, situated in the midst of a beautiful valley, which is rapidly settling up. The coal mines of Corral Hollow are fourteen miles distant from this station to the southeast, connected a portion of the way by rail-track. The "Central" Co. use large quantities of this coal—besides transporting it to San Francisco, and other cities and towns. Since leaving the last station we have gained altitude, this station being 76 feet elevation. Another engine will be attached here, as the grade increases rapidly after leaving this station until we get to the summit of the mountain.

Midway—formerly called "Zink House," is 5.7 miles from Ellis; elevation, 357 feet. Soon after leaving the station, we enter the bluffs, pass through deep cuts and over high fills, our two iron horses puffing and blowing furiously as they labor up the heavy grade. These bluffs are heavy sand, and almost destitute of vegetation. To our right can be seen the old wagon road, but now almost deserted. Still upward and onward, the long train thundering around this jutting point, and over that high embankment, twisting and turning, first to the right, and then to the left, like some huge serpent, while the bluffs seem to increase in height, and the canyon is narrower and darker at every turn, until, at last, we are plunged into *total darkness*, and the tunnel of Livermore Pass; elevation, 740 feet. This tunnel is the only one on the road from Sacramento to San Francisco; is 1,116 feet long, supported by heavy timbers. (See illustration, page 177

Passing through the tunnel, our descent is rapid, through a narrow canyon, down into one of the loveliest little valleys in the whole country, and 7.9 miles from Midway our train stops at

Altamont—This is a small station at the foot of the mountain. Speeding to the westward 8.1 miles, brings us to

Livermore—a thrifty town of about 1,200 inhabitants, most of whom are engaged in agricultural pursuits. The town is the center of Livermore Valley—one of the most fertile in the State—is about 20 miles long and fifteen miles wide, surrounded by the Contra Costa Mountains, and their numerous spurs. To the north, away above the head of the valley, rises Mt. Diablo, the meridian center of the Pacific coast.

At the station are several very large

warehouses for storing grain—as this is a productive grain valley—several large hotels, some stores and many fine, costly residences. Here are particularly noticeable the eucalyptus, or Australian blue gum tree; we have seen it in a few places before, but from this time forward they will be found numerous, in some places comprising immense groves. These trees are planted along the sides of the streets, around public buildings, in the grounds of private residences, and by the Railroad Company, in immense quantities. The latter had 300,000 of these trees growing beside their road and around their stations in the year in 1877, and we understand 500,000 more are to be set out as soon as they can be procured. One peculiarity of this tree, besides its being an evergreen and unusually thrifty, is, that it will grow on the most sandy, alkaline, dry and barren soil, and it is said to be a sure preventive against chills and fever, where it is grown in profusion. Some claim that it is fire-proof, and that shingles or plank sawed from these trees will not burn, and for that reason they are very much esteemed in Australia—its native country—and from which the first on this coast were imported. There are 125 known species of the eucalypti, about 50 of which are to be found in California.

Leaving Livermore, the ground is covered very thickly in places with white, water-worn pebbles, from the size of a mustard seed to that of a bird's egg; when the ground is bare of grass or grain, they show very plainly.

Pleasanton—is reached 6.1 miles west of Livermore, after crossing a long bridge over Alameda Creek. The town contains about 600 population and is beautifully situated on the western edge of the valley, and is a thrifty, substantial town.

Leaving the station, the mountain again looms up directly ahead, and it looks to be impossible this time to get through it; but soon the train passes around, or through several mountain spurs, and emerges into a narrow canyon, down which ripples the sparkling Alameda Creek. The bluffs on each side are steep, and covered with scrub oaks, wild oats, and bunch grass. Sycamore trees are to be seen, also white and live oaks, some with long, drooping moss-covered boughs—some very large—growing on the banks of the creek, presenting at a distance the appearance of an apple-tree laden with fruit. On we go, down,

down, first on one side of the creek, then on the other, the bluffs drawing in close on both sides, through deep cuts, over high bridges, with rapidly changing scenery on either hand. Soon we enter a little valley where once was located the San Jose Junction at

Sunol—the road bed of which can be seen on the left. Then the canyon narrows to a gorge, and on, on we go past the old "Vallejo Mill," the track curving to the westward, and 11.6 miles from Pleasanton the train arrives at

Niles—elevation 86 feet. From Livermore Pass we have descended 654 feet, and are now in the valley, which continues to San Francisco Bay.

Niles is situated in the thickest settled portion of Alameda Valley, surrounded by the finest lands in the State of California, and will, at no distant day, be a place of considerable importance. Seven miles to the south, by rail, are the noted warm springs of Alameda county.

Niles is now the junction of the San Jose branch, which runs around the head of Alameda Valley and San Francisco Bay. As the train for San Jose is ready, let us step on board and take a look at the country. Four miles brings us to WASHINGTON. About two miles to the east is located the old mission San Jose, in a delightful nook in the mountains, just such a beautiful site as all the old Padres were sure to select. Three miles further is the WARM SPRINGS, where the traveler will find ample accommodations for a pleasant sojourn. These springs are situated a short distance from the station, in a quiet little valley among the foot-hills, rather retired, surrounded by attractive scenery. The waters are impregnated with sulphur, and are highly spoken of for their medicinal qualities. Near these springs is one of several country residences owned by the Hon. Leland Stanford, President of this road. It is now occupied by his brother, Josiah, who overlooks many orchards of choice fruit, besides a vineyard containing 100,000 vines. From the Springs it is four miles to MILPETAS, and seven more to SAN JOSE, at which place the "Central" connects with the Southern Pacific railroad for the north and south. (See description of San Jose on page 206.)

Returning to Niles, we continue our journey towards "Frisco."

On a clear day, the city of San Francisco 26 miles distant—can be distinctly seen a little to the left, ahead of the train across the bay.

Decota—is three miles from Niles, through beautiful fields on our left and high bluffs on the right. This is a new town—one preparing for the future, and promises at this time to be one of unusual importance as a suburban residence for the merchant princes of San Francisco.

The lots are very large, with wide avenues, beside which are planted long rows of trees—mostly eucalyptus—to the number of from 40,000 to 50,000. The water comes from living springs, which flow abundantly a few miles to the east in the mountains. To the left the valley stretches away ten miles to San Francisco Bay, dotted here and there with comfortable farm-houses, and on all sides extensive and well-cultivated fields.

Passing along, many young orchards and groves of trees will be noticeable, also some of the beautiful country residences of San Francisco's merchants.

Hayward's Station—is 6.3 miles from Decota. The town is one mile to the east, nestling in beside the mountains, and a lovely hamlet it is, completely embowered in ornamental trees, among which are the Japanese persimmon. Near the town are two groves of eucalyptus trees, of about 150 acres, aggregating 250,000 trees, 200,000 of which are raising for the Railroad Company. The town has a population of about 1,000, many of whom do business in San Francisco, going and returning daily, 22 miles.

Along this valley for many miles, the Railroad Company have planted, beside their road, double rows of the "gum-tree," as the eucalyptus is called here, and we understand, should the experiment prove satisfactory they will continue the planting until their whole 2,000 miles of road and branches from Ogden, in Utah, to Yuma, in Arizona, will all be lined with these peculiar trees.

From Hayward's it is 2.7 miles to

Lorenzo—This is a small village surrounded by wealth of all kinds. Away to the right, beside the mountain, that large building is the County Poor House. Some of the pioneers of this country, and others that were once rich, are in that building. This is a country where the "ups and downs" are *very rapid;* one is rich to-day, with gold to throw away, then poor to-morrow, without a farthing in the world.

Alameda county is noted for its peculiarly rich and fertile soil, which seems especially adapted to the cultivation of all kinds of fruit and vegetables, the size and weight of which are *truly marvelous.* This valley is the currants' home, which are raised in immense quantities. Here is located a large drying establishment, by the Alden process. One man, a Mr. Meek, has a 2,200 acre farm here, on 800 of which, are 250,000 currant bushes, 1,200 almond trees, 4,200 cherry trees, 8,000 prune and plum, 1,500 pear, 2,500 apple, 1,500 peach, and 2,000 apricot trees, besides six acres in blackberries, and many orange trees.

Another *poor* fellow has 120 acres, on which are 2,000 plum trees, 2,000 cherry, 1,500 pear, 1,000 apple, 400 peach and apricot trees, and 25 acres in currants and berries. Still another individual has 100,000 currant bushes, and the fourth, raised in 1877, 200 tons of pumpkins, 800 tons of beets, and 20 tons of carrots. Currants grow as large as filberts; cherries, three inches in circumference; plums, pears, apples, peaches, and apricots, extraordinarily large, while carrots grow three feet long, and weigh 35 pounds; cabbages, 75 pounds; onions, five pounds; water-melons, 85 pounds; pumpkins—well, no scales *can* weigh *them*—pears, 3½ pounds; strawberries, two ounces; and beets—not *hoodlums*—200 pounds. These beets beat in weight those raised in any other country—so far as heard from—and, one of our aged and revered philosophers once said "they beat the devil."—Possibly, we are not sufficiently educated to make any *positive* statement of the kind that could be taken as *reliable.* All kinds of grain yields are enormous.

From Lorenzo, it is 2.7 miles to

San Leandro—This place contains a population of about 2,000. It was once the county seat of Alameda county, but that honor was taken away to enrich Oakland, yet the citizens seem to stand the loss, and do not stop in their efforts to improve and beautify the town, or in raising the finest and largest vegetables in the State.

The next station is 4.2 miles distant, called

Melrose—This is a small station at the junction of the Alameda railroad. Close beside the mountain, on the right, is located the Mills Seminary, for young ladies. The building cost about $100,000. The grounds occupy 65 acres, and are most beautiful, in trees, lawns, etc.

ALAMEDA — Alameda county, is four

miles to the left of this station. Population 1,600. It is situated on the eastern shore of San Francisco Bay, four miles from Alameda Point, and ten miles from San Francisco, with which it is connected by ferry boats, by the San Francisco & Alameda railroad, and with the "Central" at Melrose. It has good schools and churches and elegant private residences. The town abounds in beautiful groves of oaks. The Encinal and other parks are very beautiful. It is a favorite resort for bathers and for picnic parties from San Francisco. The *Encinal* and *Argus*, both weekly papers, are published here.

Passing on to the westward, the traveler will note a race track on the left, where some of the best blooded stock in the State can often be seen exercising.

Near by is a large smelting furnace, for manipulating gold, silver, and other ores.

We are now passing through what—only a few years since—was an open country, *now* the suburbs of a big city, that is known as

East Oakland—once called Brooklyn, 2.3 miles from Melrose. An incorporated part of Oakland—a very thrifty place—is separated from the old city by an arm of San Antonio Creek, but connected by bridges. East Oakland and Oakland are situated on the eastern shore of San Francisco Bay, on ground which slopes gradually back from the bay for several miles to the foot-hills, or base of the Contra Costa Mountains, in their rear. Upon this sloping ground are built many elegant "out of town" residences of the merchants of San Francisco, which command a beautiful view of their city, the Bay, the Golden Gate, and the surrounding country.

A short distance to the northeast, in a canyon of the mountains, are situated the "PIEDMONT WHITE SULPHUR SPRINGS."

These springs are strongly impregnated with sulphur, and it is claimed that they possess medicinal qualities. But *why* Californians should be *sick*, or drink sulphur water, when they have such *good wine*, and *so much of it*, we are unable to understand. The only cotton mill on the Pacific coast is located at this place—the "Oakland Cotton Mills."

Steam cars run regularly between the city and San Francisco, every half-hour, fare 15 cents; by commutation tickets much less. These cars run through the center of the city of Oakland, and *not* on the track used by the Overland train, until the long pier is reached below Oakland Point. When the road was first built, all trains run through the heart of the city, but subsequently a road was built on piles over the shoal water on the edge of the bay, skirting the city front, reaching the old track on the pier, half a mile below Oakland Point, since which time Overland trains, by this route, take that track.

Leaving East Oakland our train runs along over the water, affording a very good view of the bay on the left, and on the right, Oakland, and mountain to the back of it. Two miles further and the train stops at the station for

Oakland—the county seat of Alameda county, has a population of 34,700, and is the second city in size on the Pacific coast. Few cities in the world have ever increased, with as good, healthy, substantial growth, as has Oakland within the last seven or eight years. According to the census of 1870, the population was 11,104, an increase in ten years of nearly twenty-four thousand.

What Brooklyn, New York, is to New York City, so is Oakland to the city of San Francisco. The name of the city is significant of its surroundings, as it is situated in an extensive grove of evergreen oaks, with orchards, parks, gardens and vineyards on every side. Nestling amidst this forest of perpetual green, can be seen, peeping out here and there, the magnificent villa of the nabob, the substantial residence of the wealthy merchant, and the neat and tasteful cottage of the "well-to-do" mechanic, who have been attracted here by its grand scenery, mild climate, and quiet surroundings—being free from dust, noise, or the bustle of a large city devoted to business.

Oakland is lighted with gas; has broad, well-paved streets; is abundantly supplied with water; supports several horse railroads; three daily newspapers—the *Tribune*, *Times*, and the *Democrat*, and is provided with just a score of churches. The Masons, Odd Fellows, and many of the *other fellows* have halls and hold regular meetings.

Public and private schools are ample. The higher educational institutions comprise the University of California, the State University School, the Female College of the Pacific, the Oakland Military School, the Oakland Female Seminary, and the

Convent of "Our Lady of the Sacred Heart." The University of California is at Berkley, four miles distant. It is constructed throughout of brick and iron —they say, earthquake and fire-proof.

Near the University, towards the bay, is located the State Asylum of the Deaf, Dumb and Blind. It is a massive stone building, three stories high, 300 feet above the bay, and commanding a very extensive view. Oakland, besides its attractions as a place of residence, has many manufactories, some very extensive, giving employment to hundreds of people. We have it from the best authority, that there are $4,350,000 invested in the manufacturing business, the annual product of which amounts to an enormous sum. The Central Pacific Railroad Company completed a road called the Berkley Branch, and opened it for business January 9, 1878, which commences at the end of the pier at West Oakland, and runs north and eastward around mountains of the Contra Costa Range, and in full view of San Pablo Bay, to Martinez, 35 miles, thence to Tracy, 48 miles, [see page 178]. By this route all overland trains run during 1879, and by which the trains for Los Angeles, Yuma, and Arizona now run. [S e map, page 120].

The principal hotel at East Oakland is Tubbs'; at Oakland, the Grand Central. At Oakland Point, from which the long pier is built out into the bay, are located the extensive workshops of the Central Pacific. Their yard and ground occupy 120 acres.

Leaving Oakland, our train speeds along through the edge of the city for about half a mile, then gradually curves to the southward, running out to meet the main Oakland pier, which it reaches 1.4 miles from Oakland, 1.6 miles from the Oakland end of the pier, and 2.1 miles from the end of the pier to the westward. Down this long pier rolls our train, directly into the bay—but we must return to Sacramento, for the last time, and take

The New Short Line Route.
See time table

Taking our seat in the cars, in the "Central" depot, the train crosses the bridge—600 feet long—over the Sacramento River, and pass through the town of

Washington—on the west bank, a place of about 1,000 population, then cross the "Tuiles"—a broad belt of overflowed swamp land—on an embankment and trestle bridge, raised above the annual floods, until we reach the highlands or elevated plains. The trestle bridge affords passage for the flood tides.

From Sacramento it is 8.35 miles to

Webster—a small side-track for loading cattle, near by which large numbers of cattle are kept and fed, preparing for market. From Webster it is 4.88 miles to

Davis—the junction of the Marysville Branch, where the road passes on north to Knight's Landing and Willows. It contains a population of about 600, with many fine private residences.

Turning north—from Davis—we pass through large groves of live oak, and highly cultivated fields, and 5.05 miles from Davis, come to a side-track for loading cattle, called MERRIT—but it did not have merit enough for our train to stop, so we passed along 4.34 miles to

WOODLAND—the county seat of Yolo county, situated three miles west of Cache Creek, in the midst of an extensive plain. The town is one of the most thriving in the State. Population about 3,000. Yolo county, in the summer, is one vast wheat field—far, almost as the eye can reach, the waving wheat stretches away on either hand. Huge oaks are seen in every direction, and several large warehouses for wheat appear. The principal hotels are the Crapt, and the Capital. Along the road we will observe a number of vineyards.

Soon after leaving Woodland, on the right, can be seen the race track, where, at certain seasons, some good *time* is made. Again on the left, the track of the Northern railway branches off for Willows. Groves of oak and bands of sheep are numerous.

CURTIS—A side-track, is passed 5.08 miles from Woodland, and 4.05 miles more brings us to

KNIGHT'S LANDING—at the landing of which we stopped on our way "up the Sacramento." (See description, page 171).

Returning to Woodland we take the

Northern Railway.

This road is operated by the "Central," and passes through *one continuous wheat field for the whole distance* from Woodland to Willows, 87 miles. The stations and distances are YOLO, 4.91 miles; BLACKS, 5.88 miles; Dunnigan, 7.52 miles; Harrington, 5.04 miles; Arbuckle, 517 miles; Berlin, 4.6 miles; Macy, .93 miles; Williams, 5.73 miles; Maxwell, 8.85 miles; Delevan, 5.24 miles; Norman, 3.66 miles; Logandale, 2.36 miles, and 5.57 miles more to the "end of the track," at Willows. This road follows the general course of the Sacramento River, at a distance of from ten to twenty miles to the westward.

The Hot Sulphur Springs are situated about ten miles west of Williams, and are quite a resort for invalids. Crude sulphur is found in quantities.

Returning to DAVIS we start again for the south.

Just after crossing Putah Creek we come to the great vineyard belonging to a Mr. Briggs. It contains 500 acres, the greater portion of which is devoted to raisin grapes, the balance is in almonds, figs, apricots and other fruit.

TREMONT is the first station on the hills, 3.79 miles from Davis, but here trains seldom stop.

No. 45 ANNEX.—The Geysers—No. 17 of our large series of views gives a very truthful picture of this wonderful region. Here extremes meet in a most astonishing way, if the diversity of mineral springs can be called extremes, as they are over two hundred in number and possess every variety of characteristics; some are hot and others are icy cold; some contain white sulphur, some black, some red, or yellow; others alum—and boiling alum at that; others iron; others soda; others — oh, well, it's idle to go on particularizing. You have but to name your spring, and it is ready for you. Side by side boil and bubble the hottest of hot springs, and the coldest of cold ones, being frequently but a few inches apart. Indeed, so closely do they lie together that the greatest care must be exercised lest one should step knee-deep into a boiling caldron or an icy bath. Even the rocks become thoroughly heated, and quantities of magnesia, sulphur, alum, epsom salts, and many other chemicals, lie thickly strewn about, making a sort of druggist's paradise. The noises, too, and the smells, are as diversified as the character of the springs; some hiss, some murmur, some roar. Of these springs, one is known as the "Devil's Grist-Mill;" another, the "Calliope;" then the "Steamboat Geyser, the "Witche's Caldron," the "Mountain of Fire," the latter of which contains more than a hundred apertures, and in all of these are shown, each for itself, some interesting and remarkable peculiarity. (See route to the Geysers on pages 198 and 200.)

No. 53 ANNEX. [From pages 210 and 211.] **Yo-Semite and Big Tree**—*Example*, by the new Madera Route, for a trip of less than 5 days. Take Sleeper, and leave San Francisco (say on Monday) at 4 p. m., dine at Lathrop, at 8 p. m., arrive at Madera at 12:10 a. m., Tuesday morning. Rest in sleeper until 5:30, breakfast, take stage and leave at 6 a. m., arrive at Clark's at 3 p. m., and Yo-Semite at 7 p. m. Distances:

From San Francisco to Lathrop.... 94 miles.
Lathrop to Madera.................. 91 "
Madera to Clark's.................. 51 "
Clark's to Yo-Semite............... 24 "

Total........................260.
Time, 27 hours.

Stay in Valley two days.

Returning, leave the Valley at 1 p. m. Thursday; arrive at Clark's at 6 p. m.; leave Clark's Friday, 6 a. m., via Big Trees, and arrive at Madera at 7 p. m., and take sleeper; leave Madera Saturday 4:15 a. m., reach Lathrop for breakfast, and arrive in San Francisco at 12:35 p. m. Special Sleepers run between San Francisco and Madera, giving two full nights' rest, avoiding early and late changes, and many annoyances heretofore experienced by the tourist.

The new wagon road from Clark's into and through the Mariposa Grove of Big Trees—427 in number, the largest being 24 feet in diameter—enables the tourist to spend a portion of a day in the Grove without additional charge and make the same connections.

The old route is to Merced, by same train, stop over at the El Capitan Hotel, and next morning take coaches via either Snelling and Coulterville, or via Mariposa. Taking the Coulterville route, 12 miles, at Marble Springs, is Bowers' Cave; 20 more, Hazel Green. From Hazel Green, elevation 6,699 feet, a fine view of the great San Joaquin Valley can be obtained. Here the McLane wagon road leads off to the Merced Grove of Trees. At Crane Flat, 34 miles from Coulterville, a trail leads off to the Tuolumne Grove of Big Trees, one mile distant. There are 31 trees, the largest being 36 feet in diameter. The first view of Yo-Semite is had at Valley View. 40 miles from Coulterville and 12 miles from Yo-Semite. Distance by this route is about 245 miles.

The Mariposa route is via the town of Mariposa, 46 miles, thence to Clark's 31 miles. At Clark's, a road leads to the Mariposa grove of trees.

STAGE CO. TOURIST TICKET RATES.

Exc. 1. San Francisco via Madera to Yo-Semite and Return..............$59.00
Exc. 3. Lathrop via Madera to Yo-Semite and Return.................... 54.00
Exc. 5. San Francisco via Merced to Yo-Semite and Return............. 55.00
Exc. 7. Lathrop via Merced to Yo-Semite and Return................... 50.00
Exc. 9. Madera to Yo-Semite and Return 45.00
Exc. 11. Merced "   "   "   45.00
Exc. 13. Madern "   "   (Single Trip) 25.00
Exc. 15. Merced "   "   "   25.00

Sam. Miller, Tourist Agent, Palace Hotel, San Francisco.

We give the above "Example" that those whose "time is money," can calculate accordingly. Passengers can leave San Francisco at the same time every day in the week, and make the same time on a round trip, or, can stop over as long as they choose. Tickets good until used.

Dixon—is 4.26 miles from Tremont, and is quite a thrifty town, situated in the midst of a fine agricultural section of Solano county. It has several hotels, and a number of stores. It is 3.35 miles to BATAVIA—and 4.84 miles to

Elmira—which has a pop. of 300, most of whom are agriculturalists. Here the Vaca Valley railroad branches off to the north, through Yolo, the great wheat county, 27 miles to Madison.

Cannon—is a flag station, 3.97 miles from Elmira, and 6.85 miles from

Suisun—near the town of Fairfield, county seat of Solono county, situated on a broad plain, with a population of 1,000, and rapidly increasing. Small schooners come up the slough from the Sacramento River, to near the town.

From Sacramento to this station our train has followed the track of the California Pacific, once called the

VALLEJO ROUTE—It was distant by rail from Sacramento to Vallejo 60 miles, and from Vallejo, via. steamers over San Pablo Bay, 26 miles, making 86 miles to San Francisco. This route is now changed—but let us go and see. From Fairfield, it is 5.38 miles to

BRIDGEPORT—Just before reaching the station, a short tunnel is passed, through a spur of the western range, which is thrown out to the south, as though to bar our progress, or to shut in the beautiful little valley in the center of which is located the station. The grade now begins to increase, as our train is climbing the Suscol Hills, which border San Pablo Bay. These hills are very productive, the soil being adobe. To the tops of the highest and steepest hills the grain fields extend, even where machinery cannot be used in harvesting. In the valley through which we have passed are several thriving towns, but not in sight.

From Bridgeport it is 3.83 miles to

CRESTON—but we will not stop, but roll down through the hills bordering the bay, 8.7 miles to

NAPA JUNCTION—Here we are in Napa Valley, which is on the west; beyond are the Sonoma hills, over which is the Sonoma Valley; to the southward San Pablo Bay. At this junction, we meet the Napa Valley branch, which runs north to Calistoga, 35 miles.

(For a description, see page 196).

Continuing, our route is now south, along the base of the hill we have just crossed, which also runs south to the Straits of Carquinez, the outlet of the Sacramento River, which flows into San Pablo Bay. The FAIR GROUNDS of the counties of Napa and Sonoma are passed, on the right, 3.2 miles, and three miles further appears North Vallejo, 1.1 miles more, South Vallejo, where the boats once laid which conveyed passengers over San Pablo and San Francisco Bays, 26.25 miles to the City of San Francisco. A description of this route across the bay, will be found on page 193. As North and South Vallejo are virtually *one*, we shall speak of them as

Vallejo—The town is situated on the southeastern point of the high-rolling, grass-covered hills bordering Vallejo Bay, which is about four miles long and a half mile wide, with 24 feet of water at low tide. The harbor possesses excellent anchorage, and vessels are securely sheltered from storms. The largest vessels find safe waters; and here are laid up the United States ships when not in use on this coast. The naval force, including the monitors, on this side, all rendezvous here. On MARE ISLAND, just across the bay, are the Government works, dry docks, arsenals, etc., employing 500 men. The finest section dock on the coast is located on the island, just in front of the town; connected by ferry-boats.

The population of Vallejo is 7,000. It has two newspapers, the *Times* and the *Chronicle*, daily and weekly. The Bernard and the Howard are the two principal hotels. It has some fine buildings—churches,and schools. The Orphan Asylum, a fine structure, stands on an elevation to the east of the town. There are some very large warehouses, and a great many vessels are loaded here with grain for foreign ports; it is also the southern terminus of the California Pacific railroad, which connects here with ferry boats that cross the Straits to the Vallejo Junction, on the Overland Route, (see page 187).

Returning to Suisun, it is 5.31 miles to TEAL, 5.18 miles to Goodyears, and 5.76 more to

STEAMER SOLANO.—(See next page.)

Benicia—formerly the capital of the State, at the head of ship navigation, and contains about 2,000 inhabitants. It is a charming, quiet, rambling old town, with little of the noise and bustle of the busy seaport.

The United States arsenals and barracks are located near the town, and are worthy a visit. Benicia is celebrated for her excellent schools. The only law school in the State is located here, and also a young ladies' seminary.

The Straits of Carquinez—pronounced kar-kee-nez—are about one and a half miles in width, through which the Sacramento and San Joaquin Rivers reach San Pablo Bay, five miles west. In front of the city a long ferry slip has been built by the R. R. Co., and another on the west side of the Straits, at Port Costa. These slips are of piles 18 inches in diameter, of an average length of 95 feet, braced and bolted in the strongest manner possible; between these slips plies the

Solano, *the largest steam ferry-boat in the world*. To avoid the heavy grades by the way of Livermore Pass, and the detour necessary to reach Sacramento, via San Pablo and Stockton, at the same time to shorten the route, has for many years been a desideratum with the Railroad company. They finally settled on this route which has the advantage of being 40 miles *shorter* than the Livermore route, and 61 miles *less* than by Martenez and Tracy.

When this route was decided upon the next thing was to cross the Straits and build a boat that could take on board a large number of freight cars, or an entire passenger train.

The "Solano" is the same length as the City of Tokio, and has the greatest breadth of beam of any vessel afloat. Her dimensions are: Length over all, 424 feet; length of bottom—she has no keel—406 feet; height of sides in centre, 18 feet, 5 inches; height of sides at each end from bottom of boat, 15 feet, 10 inches; moulded beam, 64 feet; extreme width over guards 116 feet; width of guards at centre of boat, 25 feet, 6 inches; reverse shear of deck, 2½ feet. She has two vertical steam engines of 60-inch bore, and 11-inch stroke. The engines have a nominal horse power each, but are capable of being worked up to 2,000 horse power each. The wheels are 30 feet in diameter, and the face of the baskets, 17 feet. There are 24 baskets in each wheel, 30 inches deep. She has eight steel boilers, each being of the following dimensions: Length over all, 28 feet; diameter of shell, 7 feet; 143 tubes, 16 feet long by four inches diameter each; heating surface 1,227 feet; grate surface 224 feet; entire heating surface, 9,816 feet; entire grate surface, 1,792 feet. The boilers are made in pairs, with one steam smoke-stack to each pair, 5 feet and 6 inches in diameter. She has 4 iron fresh-water tanks, each 20 feet long, and 6 feet in diameter; registers 483,541, 31-100 tons. She is a double ender, and at each end has four balance rudders, each 11½ feet long and 5½ feet in depth. They are constructed with coupling rods, and each has one king pin in the centre for the purpose of holding it in place. The rudders are worked by an hydraulic steering gear operated, by an independent steam pump, and responds almost instantaneously to the touch. The engines are placed fore and aft, and operate entirely independent, each operating one wheel. This arrangement of the engines and paddles makes the boat more easily handled entering or leaving the slips, or turning quickly when required, as one wheel can be made to go ahead and the other to reverse at the same time. One wheel is placed eight feet forward, and the other eight feet abaft the center of the boat. It has four tracks running from end to end, with the capacity of 48 freight, or 24 passenger cars. In its construction, 1,500,000 feet of lumber were used. Many of the timbers are over 100 feet long; four, the Keelson's are 117 feet long, each measuring 4,032 feet.

Leaving the station, our road bed has, in many places, been blasted through high rocky, narrow spurs. Soon after passing one of these cuts, we catch the first glimpse of San Pablo Bay ahead. Next to the right, MARE ISLAND, and further to the right still, VALLEJO. (See page 185.)

Valona—is 2.55 miles from Port Costa, from which it is half a mile to

Vallejo Junction—opposite the city of Vallejo, between

which ply a line of ferry boats connecting with the California Pacific R. R. for Napa, Calistoga and the Geyser springs.

Soon after our train passed VALONA, a side track, it rolled into a long tunnel cut through one of the largest of the many narrow rocky spurs which slope down from the mountains on the left to the waters of the straits on the right, like so many huge mountain fingers thrust out tantalizingly to bar our progress. However, the annoyance is but momentary, for as we emerge from the dark tunnel to the glorious sunlight, a vision of beauty, one of the most diversified, suddenly appears, as though by magic, before our wondering eyes. What a glorious view! Words can never do justice to the picture. Across the narrow straits to the right, is the harbor and city of Vallejo, with the Suscal Hills rising in the back ground close to the eastward. MARE ISLAND is one mile to the west, across the inlet; to the north, away beyond all, is beauteous Napa Valley, at the head of which, forty miles away, is Mount St. Helena; still further, and more to the left are to be seen the mountains in which are situated the great Geyser Springs of California. Turning now more to the westward, our eye falls upon the Sonoma Hills, Sonoma Valley, Petaluma, Santa Rosa, and Russian River Valleys, the richest and most productive in the world; beyond, and bordering these are the great Redwood Forests of California; still farther rise the long blue outline of the Coast Range. This range, which bounds our vision to the west, extends south to the Golden Gate. The most elevated peak is Mt. Tamalpais, 2,604 feet high. To the southward and left of our train, we behold a beautiful narrow valley, extending for miles, even to the rugged heights of the Contra Costa's. Nestling in the center of all these magnificent surroundings— like a vast diamond—and sparkling from its countless myriads of ripples, is San Pablo Bay, ten miles in diameter, dotted here and there with the keels of commerce, and borderered with the deep evergreen of a semi-tropical country. Travelers write of the beauties of the Bay of Naples, the Lake of Como, etc., but we venture the assertion that for diversity of scenery, extent of vision and magnificent coloring, few views, if any, can compare with the one obtained from this point of San Pablo Bay, and the surrounding country

Running along on the water edge, and crossing numerous creeks and inlets, through another tunnel, passed TORMA, another side track, eleven miles from Martinez, brings our train to

Pinole—a small village with several large warehouses and a long pier extending out into the bay for the accommodation of boats and vessels touching at this place. The country now presents a better agricultural appearance, less rocky, the hills are not so high, are cultivated to the top, and produce abundant crops.

Four miles further we pass SOBRANTE, a side track, cross several beautiful valleys and San Pablo creek and stop at

San Pablo—three miles from Sobrante. The town, of about 500 population, is nearly one mile to the eastward, nestling in beside the mountain foot-hills, embowered in evergreens and surrounded by well-cultivated lands.

We are now opposite the lower end of San Pablo Bay. The neck of land extending three miles out to the westward is the southern boundary of the Bay, the extreme point of which is known as Point Pedro.

Soon after leaving San Pedro station, the passenger will get the first glimpse of San Francisco, the Golden Gate, and their most prominent surroundings, the view improving with each revolution of the wheels.

The side track stations of BARRETT, STEGE, and POINT ISABEL, are each passed in as many miles, and another mile brings us to DELAWARE ST., opposite the town of BERKLEY, which is situated about one mile to the left, beside the same mountains that we have been attempting to "surround" for the last seventy miles.

A short distance before reaching the last station, that building on the high point to the right, is the Powder Works; the large one to the left, 50 yards from the track, is the Cornell Watch Factory. The view obtained at certain points along here of the city of San Francisco, the Golden Gate, the Bay and its Islands, are very fine.

Two miles further we pass the STOCK YARDS, a side track, near which are located extensive yards for stock and several large slaughter-houses, then pass SHELL MOUND PARK,—a "road house"—and two miles further stop at

Oakland, 16th St—(See Oakland page 182.) Its one mile further to

Oakland Point—or West Oakland, and rolls down the pier, two miles in length, toward *Sundown*, to the ferry-boat which conveys passengers over the waters, 3.7-10

miles to the city of San Francisco. (See large illustration of "Birds Eye view of San Francisco, and surrounding country.")

Oakland Wharf—is on the end of this pier. Until the building of a pier at this place, the only harbor of Oakland was to the eastward, at the mouth of San Antonio Creek, the water to the westward being quite shallow for a long distance from shore. The ferry-boats leave and arrive to and from San Francisco, at this wharf every half-hour, and trains, many times composed of 18 or 20 passenger cars, run in connection with the boats to Oakland, Berkley, and other points.

THE PIER—is built of the best materials, and in the most substantial manner, with double track and carriage-way extending the whole length. There are three slips. The one to the north is 600 feet long, and will accommodate the largest ships, the water being 26½ feet in depth at low tide, and 32 at high tide. On each side of the slip are erected large warehouses, one of them 600x52 feet, the other 500x52 feet with tracks running through, for the purpose of loading and discharging.

The next slip south was built to accommodate the "THOROUGHFARE." This steamer was designed expressly for taking freight cars and cattle across the bay. Her capacity is 16 loaded cars and pens for 16 car-loads of cattle—288 head—making 32 car-loads in all. She once made a trip across the bay, loaded, running a distance of three and a half miles in 22 minutes. The boat is 260 feet on deck, 38 feet beam, with flat bottom. The engines are 200 horse power; cylinders, 22x84, and were constructed at the company's shops in Sacramento.

The south slip is the passenger slip, where lands the regular ferry-boat between Oakland and San Francisco. On each side of this slip is a passenger-house—one 30x70 feet, the other 40x50 feet. In these buildings are located the division offices of the Railroad Company. They afford *ample* accommodations for passengers, and the enormous travel, the advance guard of which has only *just commenced* to arrive.

The first ship that loaded at this pier was the "Jennie Eastman," of Bath, England. She commenced loading August 4th, 1870, for Liverpool, with wheat, brought—some from San Joaquin Valley, but the greater portion from the end of the California and Oregon railroad, 230 miles

PALACE HOTEL, SAN FRANCISCO.
A. D. SHARON, Lessee.
See Annex No. 50.

north of San Francisco.

It is hardly understood yet by the people of the world, that the China, Japan, Sandwich Island, and Australian steamships, and ships both large and small, can land at this pier, load and unload from and into the cars of the Pacific railroad; and those cars can be taken through, to and from the Atlantic and Pacific Ocean, without change; that immense quantities of goods are now transported in that way, much of them in BOND, in one-tenth the time heretofore occupied by steamships and sailing vessels. When these facts are fully understood, and the necessary arrangements made, the rush of overland freight traffic will commence, the extent of which, within the next twenty years, *few*, if any, can realize.

From the landing place of the "Thoroughfare," in San Francisco, a rail track leads to the dock of the Pacific mail, and other ocean steamships, and goods are now transferred in that way in bond, but the time is not far distant, when all foreign vessels, with goods for "across the continent," will land at this pier.

The Railroad Company have taken ample precautions against fire on this pier, by providing the two engines that are employed doing the yard work, with force-pump attachments, steam from the locomotive boilers, and supplied with reels of hose and suction-pipe so arranged that water can be used from their tanks or the bay.

Behold!—As we stand at the end of this pier—almost in the middle of San Francisco Bay—and think back only thirty years, we are lost in wonder and astonishment. Here are already two great cities within a few miles of where we stand; the smallest has 34,700, while the largest teems with over 233,066 inhabitants—representatives from every land and clime on the face of the earth. In 1847 not 500 white settlers could be found in as many hundred miles, and not one ship a year visited this bay. Now there are seven large steamships in the China trade, six in the mail service via Panama, thirty-four more regularly engaged on the coast from Sitka, on the north; to South America, Honolulu, Australia, New Zealand, on the south; besides hundred of ships and sailing vessels of every description—all busy—all life. Here, too, at the end of this pier, is the extreme western end of the grand system of American railways which has sprung into existence within the same thirty years. How fast we live! The gentle breeze of to-day was the *whirlwind* of fifty years ago. *Will we—can we—*continue at the same ratio? But why speculate? It is our business to write what is taking place to-day; so we will now step on board the ferry-boat and take a look around while crossing the bay.

GOAT ISLAND, or "*Yerba Buena*," is about one mile distant from the end of the pier, close to the right. It is nearly round, 340 feet altitude, containing 350 acres. It belongs to the Government. Beyond, looking over the broad expanse of water, the mountains of Marin county loom up in the distance, the highest point being Mount Tamalpais, 2,604 feet high. It is in the Coast Range of mountains, at the south point of which is Golden Gate, with Alcatraz Island in the foreground. Directly in front is the city of San Francisco. The highest point to the right is Telegraph Hill—the highest, *far* beyond, a little to the left, is Lone Mountain. In the center, that high building, looming up above all others, is the Palace Hotel; to the left the Bay of San Francisco.

But we are at the ferry; here passengers will find "buses" for all prominent hotels, or street cars that pass them all; fare, five cents.

San Francisco—Ah! here we are at sundown, at the extreme western city of the American Continent. Population, 233,066, and increasing rapidly.

On landing at the ferry-slip in the city, the first thing required is a good hotel. Now, *if* there is any one thing that San Francisco is noted for *more* than another it is for its palatial hotels. The Palace, Baldwin, Lick, Occident, Cosmopolitan and Grand, are all *first-class*, both in fare and price—charges from three to five dollars per day. The Brooklyn, Russ, American Exchange, and International, are *good* hotels, at charges from $2 to $2.50 per day. Then there are a great many cheaper houses, like the "What Cheer," with rooms from 25 to 75 cents per night, with restaurant meals to order.

San Francisco is situated on the north end of the southern peninsula, which, with the northern one, separates the waters of San Francisco Bay from those of the Pacific Ocean. Between these peninsulas is the GOLDEN GATE, a narrow strait, one mile wide, with a depth of 30 feet, connecting the bay with the ocean.

The city presents a broken appearance,

owing to a portion being built on the hills, which attain quite a respectable altitude. From the tops of these hills a very fair view of the city can be obtained.

A large portion of the city is built on land made by filling out into the bay. Where the large warehouses now stand, ships of the heaviest tonnage could ride in safety but a few years ago. To protect this made land, and also to prevent the anchorage from being destroyed, a sea-wall has been built in front of the city.

The climate is unsurpassed by that of any large sea-port town in the United States—uniformity and dryness constituting its chief claim to superiority. There is but little rain during the year—only about half that of the Eastern States. The mean temperature is 54 deg., the variation being but 10 deg. during the year.

San Francisco, in early days, suffered fearfully from fires. The city was almost completely destroyed at six different times during the years of 1849, '50, '51, and 1852. The destruction has been estimated in round numbers to exceed $26,000,000. The result of these fires has been that nearly all the buildings built since 1852 have been built of brick, stone, or iron—particularly in the business portion. The city has many magnificent private residences, and cosy little *home* cottages, ornamented with evergreens, creeping vines, and beautiful flowers. The yards or grounds are laid out very tastefully, with neat graveled walks, mounds, statues, ponds, and sparkling fountains, where the "crystal waters flow."

The *first house* was built in San Francisco in 1835. The place was then called "Yuba Buena"—changed to San Francisco in 1847, *before the discovery of gold*. The city is well built and regularly laid out north of Market street, which divides the city into two sections. South of this the streets have an eastern declination as compared with those running north. The city is situated in latitude 37 deg. 48 sec. north; longitude, 120 deg. 27 min. west.

The principal wharves are on the eastern side of the city, fronting this made land. North Point has some good wharves, but from the business portion the steep grade of the city is a great objection.

The city is amply supplied with schools, both public and private. There is no institution of the city wherein the people take more interest and pride; none, of the credit and honor of which they are more jealous. Some of the finest buildings of the city were built for school purposes, the Denman and Lincoln school houses being the finest of the number.

There are churches of all kinds, creeds, and beliefs, including several Chinese "Joss Houses." The Jewish synagogue is the finest among them, situated on Sutter street.

The NEWSPAPER, and MAGAZINE, are the histories of the present, and the person who does not read them must be ignorant indeed. Californians are a reading people; and he that comes here to find fools brings his brain to a very poor market.

There are in the city 65 newspapers and periodicals, thirteen of which are daily. The dailies are the *Alta Californian*, the *Bulletin, Morning Call, Morning Chronicle, Post, Examiner, Abend Post* (German), *Demokrat* (German), *Courrier de San Francisco* (French), *Mail, Stock Exchange, Stock Report*, and the *California News Notes*, illustrated. The *Golden Era*, and *Spirit of the Times*, are weekly literary and sporting papers. The *News Letter*, and the *Argonaut*, are spicy weeklies. The *Mining* and *Scientific Press*, and the *Pacific Rural Press*, are first-class weekly journals in their specialties. Here, too, is published, the *Journal of Commerce*, the best paper of the kind on the Pacific Coast. The *Coast Review*, is the great insurance authority of the Pacific coast—monthly. Here, also, is Wentworth's *Resources of California*, an invaluable journal. If among all these publications you can find nothing to suit you—*nothing new*—why, then, *surprise* the Bible, by reading it, and you may profit by its teachings.

THE MARKETS of San Francisco are one of the features of the city; those who never saw the fruit and vegetables of California should visit the markets. No other country can produce fruit in such profusion and perfection. The grapes, peaches, pears, etc., on exhibition in the city markets, represent the best productions of all parts of the State.

"FRISCO" BREVITIES—The new City Hall is on Market street. California street is the Wall street of the city. The BRANCH MINT of the United States is located in the new building, northwest corner Mission and 5th streets. THE POST OFFICE AND CUSTOM HOUSE are on Washington street. MERCHANTS' EXCHANGE BUILDING is on California street. The Old Stock Exchange is on Pine street; the New Stock

Exchange is in Leidsdorff street. Horse cars run to all important points in the city; fare, five cents. Mission Bay is two miles south of the City Hall. Market street is the Broadway of San Francisco, though Kearney street disputes the honor.

The Palace Hotel is corner Market and New Montgomery streets; (see description, "ANNEX" No. 50,) the Baldwin Hotel, corner Powell and Market streets. The California Theatre is on Bush street, also the Bush street Theatre. The Baldwin (Theatre) is on the corner of Market and Powell streets. There are three Chinese Theatres, where many of the "tricks that are vain," are performed nightly, which few can understand, yet they are worth one visit.

WATER for the city's use is obtained from Pillarcitos Creek, 20 miles south of the city, in San Mateo county; Lake Honda, five miles south, being used as a reservoir. Yet there are many wells, the water being elevated by wind-mills.

The LIBRARIES are numerous. The Mercantile, on Bush street; the Odd Fellows, on Montgomery street; the Mechanics' Institute, on Post street; the What Cheer, at the "What Cheer House," and the Young Men's Christian Association, are the principal ones, open *free* to tourists upon application. SECRET ORDERS are numerous in San Francisco—too numerous to note here.

SCENERY—The magnificent views of **Cape Horn**, on the Columbia River, **Mt. Shasta**, the **Loop**, **Orange Orchard**, **Woman of the Period**, and many others in this book, were photographed by C. E. Watkins, 227 Montgomery street, who has an enormous collection of views. The views of Mirror Lake, Nevada Falls, and many of those on the line of the Central Pacific, which we have engraved and are to be found in this book, were from photographs taken by Thos. Houseworth & Co., No. 12 Montgomery street, who have views, seemingly, of everything and everybody on the coast.

THE MECHANICS' PAVILION fronts on the corner of Mission and Eighth streets. The Mechanics' Institute own the building and hold their fairs there.

THE DRY DOCK, at Hunter's Point, six miles southeast. is 465 feet long, 125 feet wide and 40 feet deep, cut in solid rock, at a cost of $1,200,000.

PROTRERO SHIP YARDS are located at Protrero, and are re All kinds of small c are built at these ya

CHINA TOWN is s above Kearny; Du mento and Washin son street, between These streets are oc Celestial shopkeepe

THE BARBARY C thieves, cut-throats vile, is situated on Kearny and Dupon precise locality, so *keep away*. Give it value your life.

ANGEL ISLAND, t city, is a mile and three-quarters of a feet. On this island and blue stone, whi in the city for build

GOAT ISLAND, o and a half miles e tains 350 acres; alti

ALCATRAZ ISLAN north, is strongly fo 140 feet above tide, batteries, which co the harbor—a "key islands are all own

POINT LABOSE is FORT POINT is n Hall, five miles at tl Gate. It is the m the coast—on the South Carolina.

TELEGRAPH HIL feet high.

RUSSIAN HILL is CLAY ST. HILL is THE TWIN PEAK rise 1,200 feet. from the summit of of the whole cou around the Golden Pacific Ocean.

FERRY-BOATS run Francisco and Oa Quentin, Berkley, S

Th general office Southern Pacific cor. Townsend and south side of the lines run by them.

The Seal Rocks reached , a beauti

STEAMERS leave r miles; Benicia, 30

NEVADA FALL, YO-SEMITE VALLEY, CALIFORNIA. (See Annex No. 31.)

Cruz, 76; Monterey, 100; Stockton, 110; Sacramento, 125; San Luis Obispo, 209; Eureka, 233; Crescent City, 280; Santa Barbary, 230; San Pedro, 364; San Diego, 456; Portland, 642; Victoria, V. I., 753; Mazatlan, 1,480; Guaymas, 1,710; La Paz, 1,802, Acapulco, 1,808: Sitka, 1,951; Honolulu, 2,090; Panama, 3,230; Yokohama, 4,764; Hiogo, 5,104; Auckland, 5,907; Shanghae, 5,964; Hong Kong, 6,384; Sidney, Australia, 7,183; Melbourne, 7,700 miles.

The PLAZA, WASHINGTON, UNION, COLUMBIA, LOBOS, HAMILTON, and ALAMO Squares, and YERBA BUENA, BEUNA VISTA, and GOLDEN GATE PARKS, are all small, except the last, which contains 1,100 acres, but very little improved. The Oakland and Alameda parks are largely patronized by San Franciscans, who reach them by ferry-boat. But what the city is deficient in parks, is made up by the Woodward Gardens, for an account of which see ANNEX No. 44.

OCEAN STEAMSHIPS—for sailing days and other particulars, see ANNEX No. 27.

For general items of interest, see ANNEX No. 23.

Here we are, on the golden shores of California. We have come with the traveler from the *far* East to the *far* West; from the Atlantic to the Pacific—from where the sun *rises out* of the waters to where it *sets* in the waters, covering an extent of country hundreds of miles in width, and recording a telegram of the most important places and objects of interest—*brief, necessarily, but to the point*—and we feel certain that a pardon would be granted by the reader, if we *now* bade this country farewell, and started on our return trip. But, how can we? It is a glorious country, so let us make a few

Excursions,

say *five*, and then we will start on our trip towards SUNRISE, via the Southern Route.

Route 1.—TO THE SEAL ROCKS, six miles west; procure a carriage. Early in the morning is the best time to start, as the coast breeze commences about eleven o'clock, after which it will not be so pleasant. We will be fashionable—get up early—and drive out to the "Cliff House" for breakfast.

Within the first two miles and a half, we pass a number of cemeteries; some of them contain beautiful monuments and are very tastefully ornamented. The principal ones are the Lone Mountain, Laurel Hill and Odd Fellows. In the Lone Mountain cemetery, on our right, under that tall and most conspicuous monument, which can be seen for many miles away, rests the remains of the lamented Senator Broderick, who fell a victim of the "Code Duello," through jealousy and political strife. Near by are the monuments of Starr King, Baker, and many others, whose lives and services have done honor to the State. On the summit of Lone Mountain, to the left, stands a large cross, which is a noted landmark, and can be seen from *far* out to sea. In a little valley, close to the road, we pass, on the right, surrounded by a high fence, one of the most noted RACE COURSES in the State.

From the city the road leads over a succession of sand-hills; from the summit of some of these we catch an occasional glimpse of the "*Big Drink*" in the distance, the view seeming to *improve* as we gain the summit of each, until the *last* one is reached, when there, almost at our feet, stretching away farther than the eye can penetrate, lies the great Pacific Ocean, in all its mysterious majesty. We will be sure to see numerous ships, small craft and steamers, the latter marked by a long black trail of smoke. They are a portion of the world's great merchant marine, which navigate these mighty waters, going and coming, night and day, laden with the treasure, and the productions and representatives of every nation, land and clime.

Close on our right is the Golden Gate, with the bold dark bluffs of the northern peninsula beyond. The "Gate" is *open*, an invitation to all nations to enter—but beside them are the "Boys in Blue," with ample fortifications, surmounted by the "Bull Dogs" of "Uncle Sam," standing ready to close them at the first signal of danger.

Our descent from the summit of the last hill *seems* rapid, as we are almost lost in admiration of the magnificence spread out before us, until we arrive at the

CLIFF HOUSE—The stranger on the *road*, and at the Cliff House, would think it a *gala day*—something unusual, such grand "turn-outs," and so many. The fact is, this "DRIVE" is to the San Franciscan what the "Central Park" is to the New Yorker—the "style" of the former is *not* to be outdone by the latter. The drive out is always a cool one, and the *first* thing usually done on arriving is to take a drink--water--and *then*, order breakfast--and such nice little private breakfast rooms! Oh, these Californians know *how* to tickle your fancy.

Hark! " *Yoi-Hoi, Yoi-Hoi, Yoi.*" What the deuce is that? *Those hearing us, smile.* We do not ask, but we conclude it must be a big herd of healthy donkeys passing, when two gentlemen enter from the rear, and one of them says: "Colonel, *(there is no lower grade in California)* I will bet you 50 shares in the Ophir or Virginia Conso'idated, that General Grant, that big seal on the top of the rock, will weigh 3,000 pounds." We did not stop to hear more, but rushed out the back door on to a long veranda running the whole length of the house, which is situated on a projecting cliff, 200 feet above, and almost overhanging the waters, when "*Yoi-Hoi, Yoi-Hoi, Yoi*"--and there were *our* donkeys, 500 yards away, laying on, scrambling up, plunging off, fighting, and sporting around three little rocky islands. The largest of these islands is called "Santo Domingo." It is quite steep; few can climb it. A sleek, dark-looking seal, which they call Ben. Butler, has at times attempted it; but away up on the very top--basking in the sun, with an occasional "*Yoi-hoi, boyi*"--lies General Grant, the *biggest whopper* of them all. We knew him at the first sight. He had something in his mouth, and looked *wise*. Often when the din of his fellow seals below become fearful, who are ever quarreling in their efforts to climb up, his "*Yoi-hoi, Boyi*" can be heard above them all--which, in seal language, means, "*Let us have peace.*" Sea fowls in large numbers are hovering on and around these rocks. They, too, are very chattering, but we have no time to learn their language, as here comes a steamer bound for China. (See illustration, page 195.) It steams in close to the islands, and we think we can discern some of our fellow travelers "across the continent" among the passengers. They are on a trip "around the world," and are waving their compliments to the General on the top of the rock.

Just around that projecting point of land to the northwest are FARALLONES ISLANDS, seven in number, thirty miles distant, in the Pacific Ocean, totally barren of everything but seals, sea-lions, and water-fowls. These are *very numerous*. Many of the seals will weigh from 2,000 to 3,000 lbs., and are quite tame (see illustration, page 65), as they have never been disturbed by hunters; the birds -- and they are legion--which inhabit these islands, lay millions of eggs every year, which, until 1871, were gathered and sold in the San Francisco markets. The islands are all rocks; the highest peak is surmounted with a light-house of the first order, 340 feet above the water.

Breakfast is called; being fashionable, we take another--water--and, while eating a hearty meal, learn that these seals are protected by the laws of the State against capture, and something of their habits; then pay our bill, and the ostler *his detainer*, take our seat, and whirl around over a broad winding road, which is blasted out of the rocky bluff on our left to the sandy beach below.

Right here we meet Old Pacific Ocean himself--face to face--near enough to "*shake.*" He is a good fellow when he is himself--*pacific*--but he drinks a great deal, perhaps too much; but certain it is he gets very noisy at times--very turbulent. In driving along the beach, we come to one of the evidences of his fearful wrath. Do you see that ship laying on her side?

One night, after a big carousal, when it was said Old Pacific had been drinking a great deal--more than usual--and was in a *towering passion*, he drove this ship *up almost high and dry* on the beach, where you see her. Not content with that, he chased the escaping occupants far into the sand hills, throwing spars, masts, and rigging after them.

Thank you! We don't want any of that kind of *pacific* in ours.

We will now keep our eye on Old Pacific, and drive along down the beach, by several fine hotels, and then turn into the sand-hills to the left, passing over a high point, where some fine views can be had of the surrounding country, and around to the old Mission Dolores. Here is food for the curious. But we cannot afford to stop here long, as Boreas is getting waked up, and is sliding the sand over the bluffs after us--rather disagreeable. This Mission was founded in 1775, by Spanish missionaries, who, for over 60 years, wielded a mighty influence among the native Californians (Indians). In its most prosperous days, the Mission possessed 76,000 head of stock cattle, 2,920 horses, 820 mules, 79,000 sheep, 2,000 hogs, 456 yoke of work-

SEAL ROCKS—FROM THE CLIFF HOUSE. See page 194.

ing oxen, 180,000 bushels of wheat and barley, besides $75,000 worth of merchandize and hard cash.

The greater portion of all this wealth was confiscated by the Mexican Government, so that when California became a portion of the United States little remained, except these old adobe walls and grounds, together with about 600 volumes of old Spanish books, manuscripts and records.

Returning to the city, we pass many objects of interest well worthy of notice, and through a portion of the city rapidly building up, and in a substantial manner.

Route 2.—At the wharf, beside Oakland Ferry, we will find one of the large steamboats that run in connection with the Vallejo route; let us step on board, and note what can be seen. Leaving the wharf, our course is north, with the Oakland wharf, the route by which we come—far to the right, as also Goat Island. On our left is Alcatraz, with its heavy fortifications, beyond which is the Golden Gate; a little farther to the northward, is the Coast Range, with Mt. Tamalpais as the highest peak; elevation, 2,604 feet.

Looking back, we have a beautiful view of the city; a little further on, Oakland, West Oakland, and Berkley on the right, with the Contra Costa Mountains for a back-ground. Now we pass—on the left—Angel Island, San Quintin, and San Rafael, in the order written. Now comes the "Grandfather," a huge red rock on the left, above the "Old Man and Woman."

Continuing on, we come to the "Two Brothers," on which is located San Pablo light-house; beyond these are the "Two Sisters," making seven rocky islands. Opposite the light-house, on the right, is Point Pedro, which projects out from the mountains on the east, far to the westward, as though to bar our progress.

Rounding this "point," we enter San Pablo Bay, which spreads out to the right and left for many miles. Away to the far right can be seen a portion of the town of Berkley, and further north, San Pablo, through which runs the regular overland trains via Martinez, as noted on pages 186-187-189.

Passing on, we come in front of the Straits of Carquinez, through which flows the Sacramento River—as noted on page 187. We have left a broad expanse of water on our left, over which steamers run to Petaluma—as noted in route No. 3.

Entering through a narrow channel, with Mare Island close on our left, we land at the wharf at Vallejo, take the cars of the California Pacific, and roll along to the Napa Valley Junction, where we were before, while making a trip over the "Vallejo Route." (See page 183.)

NAPA JUNCTION—by this route it is 33.55 miles from San Francisco, and 52.87 from Sacramento. Leaving the Junction we roll up the beautiful valley 3.74 miles to THOMPSON—a signal station, passed by our train, as also many groves of young trees on our right, and beauty on every side. From Thompson it is 4.12 miles to

NAPA CITY—Although this *is* Napa City, county seat of Napa county, on Napa River, and the Napa Valley railroad, the people by no means look sleepy, but as bright as though they had just come *out* of a nap, or *from* a "nip."

This is a lively town, of about 5,000 inhabitants, at the head of tide-water navigation for vessels and steamers of light draught; supporting one daily paper, the *Reporter*, and one weekly, the *Record*.

It is in the midst of a country noted for its mild and genial climate, the great fertility of its soil, and its many well-cultivated vineyards—producing annually over 300,000 gallons of wine and brandy. It is completely hedged in by various spurs of the Coast Range. The valley is about 40 miles in length, by an average width of four miles. This county is much distinguished for its medicinal springs, the most noted of which are the Soda Springs, White Sulphur, and the Calistoga. Near all these springs huge hotels have been erected, which are crowded in summer by residents of this State, as well as tourists from the East, who visit them for health and pleasure.

At the head of this valley—in plain view—is located Mt. St. Helena, an extinct volcano, which rises 3,243 feet above tide. The whole section around-about bears evidences of the volcanic upheaval that once lit up this whole country.

Near Napa City is located one of the two State Insane Asylums—we found the other near Stockton—completed at a cost of over $1,000,000, and capable of accommodating 800 persons. It is of brick, and stands on an eminence about a mile and a half from the city, to the east. Four first-class seminaries and colleges—for the education of girls and boys—besides many public schools, are at Napa, so educational advantages are all right.

Among the good things at Napa, are the Palace and Revere hotels, the first named, a very large house near the depot.

The Soda Springs are situated about six miles to the eastward, on the side of the mountain. The water from these springs has become quite celebrated; a large amount of it being bottled annually, and shipped to all parts of the State.

Stages leave Napa daily for Sonoma, 12 miles west, continuing to Santa Rosa, 12 miles further; also to Monticello, 25 miles; Knoxville, 50 miles northeast.

Leaving Napa, we cross Napa Creek, and roll along through rows of locust trees, planted on each side of the road, and on the right are to be seen a few mammoth cactus pads, close to the track. A run of 5.1 miles and we come to

OAK KNOLL—the country residence of Mr Woodward, of Woodward Gardens, San Francisco. The farm contains 1,000 acres, nearly all under cultivation. Of this farm 120 acres are devoted to fruit and nuts of many varieties.

Crossing Dry Creek, we come to a blackberry ranche of twelve acres, as many people in this valley make a specialty of raising blackberries. Oaks, manzanitas, and pines, now appear in places.

From Oak Knoll, it is 3.52 miles to

YOUTSVILLE—Here, on the left is a large wine cellar, built of brick; near by are large vineyards; further, comes Mason's vineyard of 100 acres, mostly raisin grapes, which are prepared and packed here for market. A run of 3.39 miles and we are at

OAKVILLE—Opposite, on the side of the mountain, can be seen a quicksilver mine, marked by a red formation.

RUTHERFORD—is the next station, 1.95 miles from the last. Fine residences line the foot of the mountain on each side, the whole length of the valley, many completely embowered in shade and fruit trees of several varieties.

Passing on, we find another large wine cellar on the right, and 1.94 miles from Rutherford comes BELLO—a signal station with vineyards and another big wine cellar.

Passing along through this beautiful valley, with huge moss-covered oaks, vineyards and fields on each side, 2.07 miles is

ST. HELENA— This is a town of about 1,200 population, on the western side of the valley, in the midst of vineyards; in fact, there are vineyards and orchards in every direction, some embracing hundreds of acres. Near by is the great vineyard and orange orchard of King.

The town has many neat residences, and one weekly paper, the *Star*. The White Sulphur Springs are situated about two miles west of the town, to which "busses" run regularly. The White Sulphur Hotel is the place to stop.

Leaving St. Helena, we come to a farm of 500 acres, 115 acres of which are in a vineyard. Here, on the left, is another large wine cellar, near the road. This valley, particularly this portion, is called "safe land," meaning thereby that it can always be depended upon for a crop, as the fall of rain is sufficient every year to raise a crop, and irrigation is unnecessary. BARRO, a signal station, is 1.98 miles further, where the valley is about two miles in width, with vineyards extending away up on the side of the hills. From Barro it is 2.1 miles to BALE, another side-track of little interest to the tourist, 1.53 miles from WALNUT GROVE, still another small signal station. Oaks are thick along these bottoms, and present a beautiful appearance. Occasionally we will see the madrone and a few Monterey cypress, with some eucalyptus trees.

Napa Creek, which has been along the road on either one side or the other, the whole length of the valley, has dwindled down to nothing.

Continuing on, up through beauty on every hand, 2.86 miles from the last station, we arrive at the end of the road and

CALISTOGA—which is the most popular of all the summer resorts, near the bay. The springs are just east of the depot, the water of which is hot enough to boil an egg in two minutes, and are said to possess great medicinal qualities, having already won a high local reputation. In the town, every accommodation in the way of hotels, etc., is afforded to the numerous visitors who annually gather here to drink and bathe in the invigorating water, enjoy the unsurpassed hunting and fishing in the vicinity, and above all, to breathe the pure air of the charming little valley, while viewing the beautiful mountain scenery.

The population of the town is about 500; the principal hotels are the Magnolia and Cosmopolitan; the paper which is supposed to furnish "*all the news*" is a weekly, called the *Calistogian*. Calistoga is situated at the head of the valley, 68.15 miles from San Francisco, surrounded on three sides with the mountain spurs of the Coast Range, as well as by vineyards and orchards; wine cellars—well, they are thicker here than quartz mills at Virginia City.

THE PETRIFIED FOREST—is distant about five miles, and consists of about forty acres of ground, covered more or less with petrified trees, some very large, eleven feet in diameter at the stump. These trees are nearly all down, some nearly covered with earth and volcanic matter, while the ground sparkles with silica. They will well repay a visit from the curious.

Stage lines are numerous from Calistoga; first, to the northward, it is 17 miles to Middleton; 20 to Harbern Springs; 20 to Guenoc; 35 to Lower Lake, and 45 to Sulphur Banks, where that *suspicious* mineral can be shoveled up by the cart-load. To the northeast it is 78 miles to Pine Flat; 26 miles to Geysers; 26 miles to Glenbrook; 41 to Kelseyville; 48 to Lake Point, situated on the west shore of Clear Lake, a fine resort at all seasons, but particularly in summer. To the southwest it is five miles to the Petrified Forest, eight to Mark West Springs and 26 to Santa Rosa.

The celebrated Foss, with his stage, leaves Calistoga daily, over a mountain road unsurpassed for grand scenery, en route to

THE GEYSERS—These springs, with their taste smell and noise, are *fearful, wonderful*. We have been told that "California beats the devil." May be, but he cannot be *far* from this place. Here are over 200 mineral springs, the waters of which are hot, cold, sweet, sour, iron, soda, alum, sulphur—well, you *should* be suited with the varieties of sulphur! There is white sulphur and black sulphur, yellow sulphur and red sulphur, and how many more sulphurs, deponent saith not. But *if* there are any other kinds wanted, and they are not to be seen, call for them, *they are there*, together with all kinds of contending elements, *roaring, thundering, hissing, bubbling, spurting* and *steaming*, with a smell that would disgust any Chinese dinner-party. We are unable to describe all these wonderful things, but will do the next best thing. (See large illustration No. 17 and description in ANNEX No. 45, page 184.)

The Geyser Hotel, seen through the foliage in the picture, is the only house which provides accommodations at the springs. Steam baths and other kinds will here be found ample, and board $14 per week. In the region of the springs, are mines of quicksilver, and some silver mines that are being worked to advantage.

Returning to San Francisco, we start on
Route 3.

San Francisco and North Pacific Railroad.

General Offices—San Francisco.

P. DONAHUE.............................*President.*
ARTHUR HUGHES...............*General Manager.*
P. J. MCGLYNN......*Gen. Pass. and Ticket Agent.*

Repairing to the wharf, a short distance north of the Oakland Ferry, we board the steamer Donahue, belonging to this road, and proceed up the bay, as in route No. 2, until Pedro Point is passed, when the course is more to the westward, to the mouth of Petaluma Creek, a very crooked stream, with salt marshes on each side. About six miles from the mouth of the creek, on the right, we come to a double-front cottage, which, when we passed *up* here in January, 1878, stood high and dry, above the marsh. Several days after, on the downward trip, the water covered the whole bottom in one broad sheet, and was apparently on the first floor of the building. When it is understood that the party who settled here did so to demonstrate that he could reclaim the land by an original system of dykes, the joke will be apparent, and to him an aqueous joke.

From the mouth of the creek, it is about ten miles to

DONAHUE—named for the President of the road. It is situated on the east bank of the creek, close in beside the bluffs, or Sonoma Hills, 34 miles from San Francisco. It is simply a landing for the boat where passengers take the cars, which stand under a huge, long building on the end of the wharf.

Leaving the wharf, the Sonoma Hotel is close on the right, almost on the water's edge. Passing along beside the rolling hills, which are cultivated to their summit, one mile brings us to LAKEVILLE, not a very pretentious place, but from which a stage leaves daily for the eastward, over the hills, nine miles to

SONOMA—This town is a quiet, old place, founded in 1820, and contains about 600 inhabitants. Many of the old original adobe buildings are still standing in a good state of preservation. Sonoma has the honor of being the place where the old "Bear Flag" was first raised. It is situated in the Sonoma Valley, one of the richest in the State, and is celebrated for its vineyards and the excellence of its wines. Sonoma is not without its railroad. "It once had the "Prismodial"—single rail—but this has given way to the Sonoma Valley, which runs regular trips in connection with the steamer "Herald" to and from San Francisco and Sonoma daily—distance about 43 miles.

But to return to the railroad, which we left at Lakeville. Rolling hills are on our right, mostly cultivated to their summits, and a few scattering live-oaks; on the left, Petaluma Creek, salt marshes, and in the distance a high ridge of the Coast Range. Seven miles from Lakeville is

PETALUMA—the largest town in Sonoma county, with a population of about 4,500. It contains some fine large business blocks, two good hotels—the American and the Washington. The *Courier* and the *Journal Argus* are two weekly papers, published here. The town is on the west side of the road, situated on rolling hills, by the side of which runs Petaluma Creek, which is navigable for light-draft boats at high tide. Upon and around these rolling hills are some beautiful residences, ornamented with great numbers of trees, among which are the oak, eucalyptus, Monterey and Italian cypress, Norfolk Island pine, and others, presenting, in connection with the mountains and surrounding scenery, a view most charming and delightful.

Stages leave Petaluma daily for Sonoma, thirteen miles east. We understand the Railroad Company have a track graded and nearly ready for the iron, that runs south and west from Petaluma, along the base of the moun ain to San Rafael, where a crossing will be made to San Francisco.

Leaving Petaluma, we cross Pe'aluma Creek and roll along three miles to ELY's, a flag station, amid rolling hills, at the head of Petaluma Valley. One mile further, we come to a beautiful grove of black oaks on a high hill to the right, and we are at PENN'S GROVE. Here we cross the divide and enter the Russian River Valley. GOODWIN'S is half a mile further, a small flag station two and a half miles from PAGE'S, another one of the same importance. Here commences the Cotate Grant, which takes in the hills on each side, four leagues in extent. Two and a half miles from Page's, we are at COTATE RANCHE, a flag station for the ranche near by. At many stations along are to be seen cattle pens and shutes, indicating that raising and fattening cattle for market is one of the industries of the people in this section. Here we find many drooping, moss-covered oaks. Three miles more and we are at the banner town of

SANTA ROSA—population, 4,000. This is the county seat of Sonoma county, situated in the midst of one of the richest valleys in the State. It is fifteen miles from Petaluma, fifteen miles from Healdsburgh, thirty-three miles from Cloverdale, and fifty-seven miles from San Francisco. No city on the Pacific coast has increased faster within the last five years than Santa Rosa, and that increase has been marked by substantial brick business blocks, large manufactories, and beautiful private residences.

The city has water-works, gas-house, railroads, and all the modern improvements. Of newspapers, there are one daily, the *Democrat;* and one weekly, the *Times*. The Grand, and the Occidental, are the two principal hotels.

The streets of the city are broad, set out with eucalypti and other varieties of trees; these, with the surrounding country, afford many very beautiful drives.

Stages leave Santa Rosa daily for Mark-West Springs, ten miles east; Petrified Forest, fifteen miles (see description, page 198); and Calistoga, 20 miles; to the west, Sebastopol is seven miles. Average fare to these places is ten cents per mile. The valley of Santa Rosa, in which the town is situated, Russian River Valley, and Petaluma Valley, really *one,* are 60 miles in length, with an average width of about six miles.

Sonoma county is a very large one, extending to the Pacific Ocean on the west. In the western portion are located immense forests of redwood timber, which we shall, note hereafter. V'neyards are numerous as well as orchards where immense quantities of oranges, lemons, plums limes, apples, English walnuts, almonds, apricots, and other f uits and nuts are raised for market There are nearly 1000 acres in vineyard—5,000,000 vines—which produce annually full 2,500,000 gallons of wine, and 35,000 gallons of brandy. Pomegranite trees do quite well, and never fail a crop. All the lands in this county are classed as "safe lands", owing to their proximity to the ocean the rain-fall is abundant for all purposes.

Darying, is an important industry, yielding, from official reports, 2,750,000 pounds of butter, 400,000 pounds of cheese, and milk—not enough figures; the amount of milk marketed is not recorded.

In mines, Sonoma is well represented; gold, copper, and quicksilver are the principal metals. In grain, the product figures up over 600,000 bushels of wheat; 30,000 bushels of barley; 250,000 of oats; 8,500 bushels of corn, and many other kinds of grain in proportion. As for mineral springs, why, Sonoma county is the home of all kinds of springs, chief of which are the Geysers, Skaggs, Mark West, and the White Sulphur.

Four miles from Santa Rosa comes FULTON—a small town at the junction of the Fulton and Gurneyville Branch. The stations on this branch, are: Meacham's, two miles; Laguna, four miles further; Forestville, another two miles; then two miles to GREENVALLEY; three miles more to KORBEL'S, and another three to GURNEYVILLE—in the midst of a forest of redwood. Along this branch road, and at Gurneyville, are located six sawmills, which cut 150,000 feet of lumber daily. One of the largest of these redwood trees measured 344 feet high, and 18 feet in diameter. The town of Gurneyville is situated on Russian River, on what is known as "Big Bottom," in the finest redwood forest in the State.

Leaving Fulton and crossing Mark West Creek, the first station on the main line is two miles distant, called MARK WEST—but our train will not stop, unless signaled. The valley along here is ten miles in width. Three miles further comes WINDSOR—a small place four miles from Grant, a flag station which is two miles from HERALDSBURG—reached just after crossing the Russian River, on the north bank of which it is situated. The town contains a population of about 2,000. Fifteen miles from Santa Rosa, and 72 from San Francisco, is Russian Valley, in which the town is located, noted for its great yield of wheat, and the extraordinary quickness of its soil, producing potatoes, peas, and many other vegetables w.thin 65 days from the time the seed is planted. The *Russian River Flag*, and the *Enterprise*, are weekly papers, published here. The tourist will find excellent hunting and fishing near by, with ample hotel accommodations.

From Healdsburg the valley gradually narrows, and four miles further brings us to LITTON'S SPRINGS, a signal station, near where are located the springs of the same name. Four miles further is GEYSERVILLE — Here stages leave for Skagg's Springs—a popular resort—situated at the side of the mountain, at the head of Dry Creek Valley, about eight miles west. TRUETT'S is six miles further, another flag station, four miles from the end of the road, at

CLOVERDALE—This town contains a population of about 500, with two hotels, the United States, and the Cloverdale, and one weekly newspaper, the *News*. The town nestles in at the mountain base, at the head of the Russian River Valley, and is 55 miles from Donahue, and 90 miles from San Francisco, being a point from which several stage lines radiate, for the northern and surrounding country.

Stages — four and six horse — leave Cloverdale daily, for the Geysers, 16 miles distant, over one of the finest mountain roads in the State. It is built on a uniform grade of four feet to the hundred. The owners of this line—Van Arnam & Kennedy—are old "knights of the whip," drive themselves, and often make the trip in one and a half hours. The fare for the round trip is $4.50. For description of the Geysers, see ANNEX No. 45, page 184, and the large illustration, No. 17.

Stages run north to Ukiah, the county seat of Mendocino county,—31 miles, where connections are made with all adjoining towns; also, to the northeast, to Hopeland, on the Russian River, 16 miles; Highland Springs, 22 miles; Kelseyville, 25 miles; Lakeport, 36 miles; Upper Lake, 42 miles; and Bartlet Springs—a great medical resort —63 miles. To the northwest, they run to Boonville, 31 miles; North Fork, 50 miles; Navaro Ridge, 63 miles; Little River, 70 miles; Salmon Creek, 72 miles; and Mendicono City, 75 miles; average fare to all, ten cents per mile.

Returning to San Francisco we take *Route 4.*

North Pacific Coast Railroad.

General Offices—San Francisco.
JNO. W. DOHERTY..*President and Gen. Manager.*
DAVID NYE.........................*Superintendent.*
F. B. LATHAM....*Gen'l Passenger and Ticket Ag't*

This road is a three-foot narrow gauge,

built and equipped in the best manner, traversing a section of the country very attractive to the tourist. It runs in a northwesterly direction from San Quentin and Saucilito, on the west side of the bay, twelve miles distant. The road has two southern termini, which unite at Junction, 17 miles from San Francisco. The bay is crossed by ferry from Davis St., for Saucilito, and from San Quentin Ferry—Market St. wharf—for San Quentin. We will take the latter route, which for nearly ten miles will be the same as No. 2; then, the route will be more to the westward. When near the point of Angel Island, on the left, the little town of Saucilito can be seen nestling close in beside the mountain. Between Saucilito and Angel Island runs Raccoon Straits. Mt. Tamalpais now looms up away to the left. Nearing the shore, also on the left, is

SAN QUENTIN—a noted place of summer and winter resort. The resident tourists number from 600 to 1,000, their term of residence varying from six months to a life-time. The quarters for their accommodation are furnished by the State, free of charge. The Lieutenant-Governor exercises personal supervision over the guests, assisted by many subordinates and a company of soldiers. The guests come here, not of their own will, but through their folly, and we believe they would quit the place, *if they could.* By law it is known as the State Prison. The buildings are of brick, large, and readily distinguished, on the point to the left of the landing. Changing for the cars, we glide along on the edge of the bay, with oak and shrub covering the rolling hills on the left, one and a-half miles, and arrive at

SAN RAFAEL—the county seat of Marin county; population, about 3,000. It was settled in 1817 by the Jesuit missionaries. It is situated in a beautiful little valley, on low rolling hills in view of the bay and San Francisco, and of late has become a thriving suburban town.

The town contains several good hotels, and two weekly papers, the *Herald* and the *Journal.* Along the streets, and around the private residences, are many shade trees, among which are the blue gum, oak, Monterey cypress, spruce and pine, which present a beautiful appearance. Proceeding through the town two miles, we reach the

JUNCTION—Here connects the branch track from Saucilito; let us digress long enough to come up on that route. Leaving Davis St. Ferry, in San Francisco, the course is almost due west for six miles to

SAUCILITO—a small town situated close in beside the mountains of the Coast Range, containing a population of about 300. On the trip across the bay, a beautiful view can be had of the northwestern portion of San Francisco, Alcatraz, the Golden Gate, and the forts located there. At Saucilito we take the cars and soon come to the shops belonging to the Railroad Company; three miles further, LYFORD's, another mile, the SUMMIT; two more, across an arm of the bay, is CORTE MADERA; two miles further

TAMALPAIS—Here saddle horses can be procured for a ride up to the summit of the mountain, 2,604 feet, from which the finest view can be had of the Pacific Ocean, San Francisco, and San Pablo Bays, and the surrounding country, that can be obtained at any point. The distance is about eight miles. From Tamalpais station it is two miles to the Junction. From Saucilito the route has been one of beauty. In almost every nook of the mountain-side are residences surrounded with all that money and good taste can provide to make them beautiful and attractive homes.

Leaving the Junction, after 1.5 miles comes the side-track of FAIRFAX, surrounded by rolling hills, covered with an eternal verdure of green. Curving to the right, look! away up there to the left—see our road! Can we get there? Up, up we go, through a tunnel, and roll around the head of the little valley, and then to the left we can look away down and see the road up which we passed only a few moments ago. Keeping around on the southern slope of the hills, with an awful chasm on the left, beyond are high mountains upon the sides of which can be seen an occasional huge redwood tree.

Curving around again to the right, up another little valley, our road again appears *far* up on the opposite side, and again the head of the valley is reached; the curve to the left is again made, and down, far below, is the road bed. There are two "Cape Horns," only not as high as Cape Horn on the Central Pacific. The scenery is very beautiful.

Climbing up, see, on the right, the wagon road to Mt. Tamalpais, *under* which is the tunnel through which we pass; altitude, 565 feet; length, 1,250 feet. Beyond the tunnel, the grade descends, curving around on the side of the hills, down into a little valley through which runs the San Geron mo Creek. Here we find the madrone tree, (see ANNEX, NO. 8, page 164) and many oaks with drooping, moss-covered boughs.

Six and a half miles from Fairfax, we come to NICASIO, a small station with an altitude of 370 feet. The mountains, on the left, are covered with a dense growth of trees, of many varieties, among which are redwood, pine, Douglas spruce, madrone, and buckeye shrubs. Passing LAGUNITAS, a small station, the road enters a narrow canyon, down which we run, with the redwood towering far above; pass the old powder mill and extensive pic-nic grounds on the right, which are visited in the summer by thousands from San Francisco— on, past a big dam, and we arrive at

TAYLORSVILLE—This is a small station, named for a Mr. Taylor, who established here the *first* paper mill on the Pacific Coast, known as the "Pioneer Paper Mill." The canyon is narrow, with some tall redwoods along the creek, and on the side of the mountain to the left. Opposite, the country is rolling with few trees—something of a dairy country.

TOCOLOMA—comes next, three miles from Taylor's. Here a stage line runs to the town of Olema, two miles to the south, over the ridge, and also to Bolinas, fourteen miles distant. Passing on by milk ranches, crossing bridges, through deep cuts, over high embankments, curving around the side of the mountain on the left, the train comes out into a little valley, and 4.5 miles from the last station, and 38.5 miles from San Francisco, stops at

OLEMA STATION—This is an eating station, the only one on the road. Trains stop twenty minutes. Stages for Bolinas, south thirteen miles, leave *every* day, *except week days*.

Leaving, the route is more to the northward, with Bolinas Bay over the hills to the left. The timber to the right has entirely disappeared, and there is but little on the left, with very little cultivated land. We are now approaching a section which is almost entirely devoted to dairying. Soon we come to Tomales Bay, a portion of which is crossed on a long pile bridge, where are extensive beds of planted oysters, the boundaries of which are marked by poles. Ducks are very abundant, and white pelicans can often be seen as well as wild geese.

This bay is about twenty miles in length, with an average width of one mile. Our train runs along on the edge of this bay, around rocky points, through spurs of the bluffs, and across little inlets for about sixteen miles, where the road turns sharp to the right, up an arm of the bay. In this distance we find the following stations: Wharf Point, three miles from Olema MILLERTON, two miles further, and MARSHALLS, nine more; then comes

HAMLET—Here the regular passenger trains meet. All these side-track stations along the bay are for the accommodation of the dairymen living near, who ship large quantities of milk and butter to San Francisco daily.

Tomales Point is on the opposite side of the bay, which is here only about three and a half miles from the ocean.

Turning to the right, our road follows up a narrow little valley around rocky points, with high grass-covered hills on each side— makes one great rainbow curve, away around the head of the valley, and comes to a stop at

TOMALES—This station is 55 miles from San Francisco. Here the Railroad Company have large warehouses for storing grain, from which large quantities are shipped annually. Tomales consists of a few dozen buildings, devoted to merchandizing, with a surrounding country well cultivated. Mt. St. Helena can be seen on the right, and, in a clear day, *far* beyond the snow-capped Sierras. Leaving the station, the road passes through the fourth tunnel, crossing a small creek on a high trestle bridge, and then a small inlet from the ocean, where we leave Marin county, enter Sonoma, and come to

VALLEY FORD STATION—Here a stage leaves daily for Petaluma, eighteen miles east. Years ago the section we are now entering was the southern border of the great redwood forests. Here the lumberman began his labors, and as years passed, step by step he penetrated this great lumber region, leaving in his track stumps, fire, smoke, and finally the clearing, broad, rich fields and well-cultivated farms, from the productions of which he subsists while persistently following up his receding prey—the redwoods.

The waters from Bodega Bay sit back to near the station, on the left. Three miles further, we come to BODEGA ROADS, and one mile more to FREESTONE, over a heavy grade. Here we come to another great horse-shoe curve, around the head of a small valley. First, the road-bed is *far above*, then *far below*, with a deep gorge on the left, in which grow madrone, redwood, and oak trees. Now we come to a **trestle bridge, 300 feet long and 137 feet high**, over a frightful gorge; and then to the Summit Tunnel, 610 feet long, beyond which is

HOWARD—The principal business at this station is burning coke. Passing on, we enter "Dutch Bill Canyon," called so in early days after Mr. Howard, who there wrestled with the big redwoods that it then contained. Redwoods now appear on each side, as also saw-mills. STREETEN MILL is passed on the left, then another tunnel— there are five tunnels in all, on the road, aggregating 3,850 feet—then a long wood shute, and

TYRONE MILLS—Here are extens've saw-mills on the left, with side-tracks running to them, with a capacity of 40,000 feet of lumber a day.

Leaving this mill, on a down grade, through towering redwoods, 300 feet high, we roll down past another large mill, on the right, to the Russian River, just after passing an unimportant side-track of that name.

The river at this place comes down through a perfect forest of towering redwoods, and is about 300 feet wide, with an average depth of two feet. The train runs along on the southern bank, past a beautiful little cottage on the right, away up on a high spur of the mountains, that projects out into the river, and which has been left, as it were isolated by the cutting made by the Railroad Company in building the road. It is *one* of several country residences belonging to the President of the road. Beyond this point a short distance, is the MOSCOW MILLS STATION, opposite which comes in from the north, Austin Creek, abounding in redwoods. A short distance further, Russian River is crossed on a bridge 400 feet long, and the train stops at the end of the road at

DUNCAN'S MILLS—Here are located extensive saw-mills, in the midst of great forests of redwoods. The station is 80 miles from San Francisco, and consists of one large hotel, the Julian—a good station building, some shops of the Railroad Company, several stores and a dozen or more residences, some of which are very good. Game of various kinds is abundant, such as deer, bears, etc., and some *w ld*

hogs. Fish—well, *this* is the fisherman's paradise. From Duncan's Mills it is six miles to the Ocean, reached by boats on Russian River, which is near the station, also by a good wagon road. Stages leave Duncan's Mills daily, except Mondays, for the following places: Fort Ross, 16 miles; Henry's, 16 miles; Timber Cove, 20 miles; Salt Point, 25 miles; Fisk's Mills, 30 miles; Stewart's Point, 34 miles; Gualala, 44 miles; Fish Rock, 50 mi'es; Point Arena, 60 miles; Manchester, 66 miles; CUFFEY'S COVE, 80 miles; Navarra Ridge, 86 miles; and Mendocino City, 96 miles; average fare ten cents per mile.

Along the line of this road are located several large saw-mills, which produce for market, 200,000 feet of redwood lumber daily.

In conclusion; the ramble about Duncan's Mills will be found by the tourist, a very pleasant one, in fact, the scenery along the whole line is very interesting. The rapid changes and the great variety are charming, instructive, and when once made will ever live in pleasant memory.

Returning to San Francisco, we start on
Route 5.

Southern Pacific Railroad
General Offices, San Francisco.

CHAS. CROCKER, ... *President.*
GEO. E. GRAY .... *Chief Engineer.*
A. C. BASSETT, ... *General Superintendent.*
H. R. JUDAH, ... *Gen. Pas. and Ticket Agent.*

This company own the road from Goshen, in the San Joaquin Valley, and, including the Goshen Division, to Los Angeles and Yuma, in Arizona, but it is leased to and operated by the "Central" Company. This leaves the Southern only the line from San Francisco to Soledad, 142 miles, and the Trespinos division of 18 miles, Monterey 16, making 176 miles, over which we propose journeying.

Leaving the depot, which is situated opposite the general office, corner Townsend and Fourth, the route is south, through the city for over four miles, most of the distance built up with business blocks, manufactories, large wool warehouses, shops and private residences.

The company's machine shops—extensive works—are situated about two miles from the depot; another mile is Valencia street, where is a horse-car line to the more central part of the city. Then we move another mile, through some deep cuts and high hills on the right, and are at

BERNAL—a small station 4.6 miles from our starting point. Some gardens and vegetable fields now appear, and a short distance from the station is the Industrial School, on an elevation to the right.

SAN MIGUEL—is two miles further, among the sand-hills, where are some well-cultivated gardens. To the right is Lake Mercede and the city water works.

Continuing along through the hills, which in places are close on each side—with the San Bruno Mountains in the distance on the right—down a little valley, then through deep cuts, past COLMA, a side-track, and Baden Bay, all in quick succession, we come to the signal station of BADEN, or as often called "Twelve Mile Farm." At this place Mr. Chas. Lux, of Lux and Miller, the largest cattle dealers on the Pacific Coast, resides; and on Mr. Lux's "Twelve Mile Farm" can be found at all times, some of the best cattle in the State.

Two miles further, and we are over the hills and down on the edge of San Francisco Bay, which is on the left, and at

SAN BRUNO—This station consists of a good hotel, and *four* targets, as it is a great resort for shooting at target. The targets are on the edge of the bay to the left; distances, 200, 500, 800, and 1,000 yards each. Here the "sports" gather to try their hand. The San Bruno Hotel is on the right of the road, where all the targets are at *shorter range*, and the shots always certain to hit the red.

MILLBRAE is the next station, 17 miles from San Francisco. To the right of the road, half-a-mile distant, is the residence of D. O. Mills, President of the Bank of California. It will be recognized by the two tall towers. A little beyond the station is **Millbrae dairy, with large yards and buildings.** On the left, in the bay, are great beds of planted oysters. Soon after leaving Millbrae, we pass Burlingame, designed and laid out by the late Mr. Ralston in long streets and avenues, extending for two miles along the road, and from the base of the mountains, on the right, to the bay on the left, about another two miles. Beside these streets and avenues, are double rows of planted trees, most of which are eucalyptus and Monterey cypress. There are some beautiful residences here and there along the base of the mountains on the right.

Two miles from Millbrae, we pass OAK GROVE, a small station named for the grove of oaks near by.

One peculiarity of this country *is;* no matter how much ground is shaded with oaks, it makes no difference with the crops, all kinds of which seem to grow equally well in the shade and in the sun.

SAN MATEO—(pronounced Ma-t-o). Here are some of the finest private residences and grounds in the State. This town contains a population of about 1,500. Oaks and orchards are EVERYWHERE. Stages leave San Mateo daily on the arrival of the train from San Francisco for Half-Moon Bay, 14 miles west; Purissima, 23 miles; Pescadero, 30 miles. At the latter place connections are made tri-weekly for Pigeon Point, seven miles; Davenport's Landing, 38 miles, and Santa Cruz, 40 miles; average fare ten cents per mile.

Leaving the station, we pass—on the right—a beautiful park, and the Young Ladies' Seminary; also a race track. To the left the bay lies close, and the land is of little value, until reclaimed, but on the right is beauty, spread out with a lavish hand. Live oaks are scattered around in all directions, with buckeye in the ravines coming down from the mountains on the the right. Windmills are numerous the whole length of the valley.

BELMONT—which is 25 miles from San Francisco, comes next. At this station the guests of the late Mr. Ralston were wont to alight to visit his residence. This place is located a half-mile to the west, up a little valley, just out of sight from the railroad. It originally contained about 100 acres, which, upon the death of Mr. Ralston, came into possession of Senator Sharon, who presented 40 acres of the land, including an elegant cottage, to the widow, Mrs. Ralston. Leaving Belmont, the Phelps estate, is on the right, and double rows of eucalyptus on the left, for two miles. The country between the hills and the bay is flat, and under a high state of cultivation.

REDWOOD CITY—comes next, 3.5 miles from Belmont. It is the county seat of San Mateo county, and a thriving place. It was named from the great redwood forest on the west, a large quantity of which finds its way to market in the shape of lumber, wood and bark, from this station. The city is supplied with water from an artesian well. The county buildings, schools, churches and hotels, are all said to be *first-class*, as well as the weekly paper, the *Times and Gazette*. Stages leave daily for Scareville, seven miles; La Honda, 16 miles, and Pescadero, 30 miles.

Passing FAIR OAKS, a small station in the midst of beautiful residences, surrounded with parks, gardens, orchards and moss-drooping oaks, we come to

MENLO PARK—near which reside a score or more of millionaires, including Ex-Gov. Stanford, Milton S. Latham, J. C. Flood, Albert Grand, Faxon Atherton, Maj. Rathbone, M. D. Sweney, Col. Eyre, and many others. Menlo Park Hotel is situated on the right, and is embowered in trees, vines, and flowers. On the left, leaving the station, is "Thurlow Lodge," a palatial residence, situated in the center of princely grounds, with the most costly surroundings, consisting of deer park, trees, gardens, orchards and shrubbery. A little further, on the right, comes the 500-acre farm of Ex-Gov. Stanford, President of the Central Pacific railroad. Here is the home of "Occident," and some of the finest blooded stock on the Pacific coast.

MAYFIELD—a town of 1,000 inhabitants, 34.9 miles from San Francisco, is situated in the widest part of Santa Clara Valley, embowered in "blue gum" oaks, and other trees. It is 4.2 miles from MOUNTAIN VIEW—a small station, so named from the extended view which it affords of the Coast Range on the west, the Contra Costa, on the east, as well as the whole surrounding country.

The great oaks add an indescribable beauty to this country, and grow in great profusion, particularly on the Murphy Grant, through which we are now passing. This grant originally covered some thousands of acres, in this, the richest portion of the Santa Clara Valley. MURPHY'S STATION—for the accommodation of the grant—is located near its center.

We are now opposite the head of San Francisco Bay—on the east—and the little town of Alviso, which is noted for its strawberries and fruit, as well as being a point from which immense quantities of produce are shipped on the boats that land at its ample wharf.

We pass on through a section, where every foot of land is in a high state of cultivation, for two miles, and come to LAWRENCE—a small place 3 5 miles from the beautiful

SANTA CLARA—This is a beautiful and quiet old town of about 4,000 inhabitants, originally founded by the Jesuits, in 1774. It is situated near the center of Santa Clara Valley, one of the loveliest in the world, possessing a soil of surpassing richness. It is celebrated for the salubrity of its climate, and the excellence and variety of its fruits; is thickly settled, and as a wheat-growing valley it has no superior. In point of improvements, good farm-houses, orchards, vineyards, etc., it has few, if any, equals.

Churches and schools are numerous; Santa Clara and San Jose—three miles apart—are both noted for their educational institutions, where some of the finest in the State are located. The convent of Notre Dame, the San Jose Institute, the State Normal School, and the new building of the University of the Pacific, Methodist, Female Seminary, and the Catholic Collegiate Institute, stand as monuments to attest a people's integrity and worth.

There are two weekly papers published at Santa Clara—the *Index* and *News*. Stages leave daily for Los Gatos, seven miles; Lexington, ten miles; and the Congress Springs, thirteen miles; fare, ten cents per mile. These springs are resorted to by those suffering with pulmonary complaints.

South Pacific Coast R. R., narrow gauge, now completed from Alameda, opposite San Francisco, to Santa Cruz, 80 miles, passes through Alvarado, a manufacturing town on the east side of the bay, about 10 miles west of Niles, to Santa Clara, thence southwest, through a long tunnel, under the Coast Range of mountains, 37 miles to

SANTA CRUZ, situated on an arm of Monterey Bay, and is often called the "Newport" of California, being a noted summer resort for sea bathers, who find good accommodations in the shape of hotels, bathing houses, etc. It is the county seat of Santa Cruz county, population, 3,000; connected by rail with the Southern Pacific at Pajaro 21 miles and, with Fulton eight miles; and by stage, with all adjoining towns up and down the coast, and by steamer to San Francisco.

Returning to Santa Clara, we can, if we choose, step into the horse-cars, or take a carriage for San Jose, and ride over the most beautiful avenues in the State, it is bordered on each side with two rows of poplar and willow trees, planted by the early Jesuit missionaries nearly 100 years ago.

Behind these trees are elegant cottages,

beautiful orchards, nurseries, and gardens, containing almost every variety of vegetables, fruits, and flowers.

By steam cars it is 2.6 miles from Santa Clara to

SAN JOSE CITY—(Pronounced San O-za); population, 18,000. This is the county seat of Santa Clara county, and is the largest town in Santa Clara Valley, in population being the fourth in the State. It was first settled by the Spanish missionaries, in 1777. The city is lighted with gas; the streets are macadamized, and ornamented with rows of shade trees on each side. Artesian wells, and the "California Wind Mill," together with a small mountain stream, abundantly supply the city with good water. The *Alameda*, or grove, was planted in 1799. It is by far the prettiest grove of planted timber in the State, and by many people it is claimed that San Jose is the prettiest city in the State. It is certainly one of the best improved, and there are none more beautiful. Its orchards, vineyards and shade trees; its fine private and public buildings, and the delightful climate of the valley, render it a favorite place of summer resort.

San Jose has numerous church edifices—ample public and private schools, hotels, and newspapers. The *Mercury* and *Independent*, both daily and weekly; the *Patriot*, daily; and *Argus*, weekly, are published here. The Auzerais, St. James, Exchange and Lick, are the principal hotels. The city is connected by railroad with Solidad, 72 miles, south, and San Francisco by two lines—the one we came on, through the thickly settled and well-cultivated Santa Clara and San Mateo counties; distance, 50 miles, and by Central Pacific via Niles and Oakland.

The new road to Mt. Hamilton—20 miles distant—leaves San Jose, and can be seen winding up the side of the mountain, on the east. It was for the erection of a college on the summit of Mt. Hamilton—altitude, 4,400 feet—that the millionaire, James Lick, left $150,000 in his will. The building has been completed, and reflects much credit on the doner.

Stages leave San Jose daily for the noted

NEW ALMADEN QUICKSILVER MINES—These mines are very extensive, and should be visited by the curious. They were discovered by an officer in the Mexican service during the year 1845, who, seeing the Indians with their faces painted with vermilion, bribed one of them, who told him where it was to be found. The following year, several English and Mexicans formed a company for working the mines, large sums of money were expended, and many difficulties had to be overcome; but finally, by the introduction of important improvements, the mines have proved to be very valuable. The different mines furnish employment for, and support from 1,000 to 1,500 persons. Nearly all the miners are Mexicans.

It is supposed that these mines were known and worked by the native Indians of California, long before the country was known by white men. They worked them to procure the vermilion paint which the ore contained, for the purpose of painting and adorning their villainous persons, and to "swop" with the neighboring tribes. Near the mines are the springs, where is put up the New Almaden Vichy Water, so noted for its medicinal qualities. The Guadalupe Quicksilver mines are ten miles distant.

Both San Jose and Santa Clara are embowered in trees, among which are the oak, eucalyptus, poplar, spruce cedar, Monterey and Italian cypress, orange, pepper, sycamore, and many others.

Leaving San Jose, the State Normal School building is on the left in the center of a block, surrounded by beautiful grounds. Several miles further on is the Hebrew Cemetery. Here the road to Mt. Hamilton can be plainly seen; it is 22 miles long and 30 feet wide, with a uniform grade of five feet to the hundred.

Away to the right, on the side of the mountain, marked by a red appearance, is a quicksilver mine, but the water prevents work. Still further and below, is the New Almaden mine, marked by columns of steam that are always ascending.

Coyote Creek is now on our left, in a broad, low bottom. The small stations of EDEN VALE, COYOTE and PERRYS, are soon passed, and 13.8 miles from San Jose, we are at

MADRONE—The country passed over is well settled, and many fine residences are scattered along the valley, which is about one mile in width, with low rolling hills on the west.

Leaving Madrone, on the right a large sharp cone rises up out of the valley 1,000 feet in height. We call it Johnson's Peak, named for the enterprising newsman of this road.

TENNANTS—is four miles further, beyond which is the most magnificent moss view that one could conceive. Sycamore and moss-drooping oaks are very plentiful, reminding one of the appearance of a New England apple orchard after a storm of snow and rain, where all the limbs and boughs are borne down with icicles and snow.

GILROY—is seven and a half miles from Tennant and 80.3 miles from San Francisco; a regular eating station, where trains stop twenty minutes for meals, which are *very good;* price, 50 cents. Gilroy contains a population of about 2,000, most of whom are engaged in agricultural and pastoral pursuits. Tobacco is raised in large quantities, and dairying is made a specialty by many of the people. The principal hotels are the Southern Pacific and the Williams.

Stages leave Gilroy for San Fillipe, 10 miles; Los Banos, 48 miles; and Firebaughs, 80 miles east; fare ten cents per mile. Stages run daily to the Gilroy Hot Springs, a very attractive resort, 15 miles east. From Gilroy it is 2.2 miles to

CARNADERO—a small station where passenger trains meet, and from which a track branches to the left and continues up the Santa Clara Valley, 11.8 miles to

HOLLESTER—a thrifty town of 2,000 inhabitants, most of whom are agriculturists. From Hollester it is 6.2 miles to TRESPINOS—the end of the track.

From this point large quantities of freight are shipped for the New Idria Quicksilver, Picacho and other mines in the country, to the south and east. Stages run tri-weekly to San Bruno, 25 miles; New Idria, 65 miles; Picacho, 75 miles; fare about ten cents per mile.

The original route of the Southern Pacific railroad was from this point, via San Benito Pass to Goshen, in the San Joaquin Valley. From Goshen the road is built a distance of 40 miles this way, to Huron. Whether the link between the two divisions will be completed and *when*, we will *never tell*, till we know. The distance across to Huron is, to San Benito Pass, 60 miles; to Huron, 100 miles.

Returning to Carnadero, we soon come to the great Bloomfield Ranche, which takes in many thousand acres, crossing the valley and over the mountains, on each side. It is the home of Mr. Miller, of Lux & Miller, the great cattle men. At Baden, twelve miles from San Francisco, we pass Mr. Lux's place, the "Twelve Mile Farm." On this ranche are kept and fattened great numbers of cattle, for the market of San Francisco.

Continuing up the valley, which is here narrowed to one mile in width, with low grass covered hills on each side, we come to the residence of Senator Sargent, on the right, and a short distance further,

SARGENT STATION—in the midst of a dairy country. Stages leave here for San Juan, south, six miles distant, up a little valley to the left, distinctly seen a few miles further on our way.

Soon after leaving the station, we turn more to the westward, and the little valley is completely crowded out by the bluffs, and we run along on the bank of Pajaro River, up a narrow canyon, and cross the line between Santa Clara and Santa Cruz county, at the point where Pescadero Creek comes in on the right. Continuing up, between high bluffs, we cross a bridge over the Pathro River and are in San Benito county, then dive through a tunnel 950 feet long, and come out into the beautiful Pajaro Valley, which is nine miles long and four wide, a portion of the Aroma Grant, once a very extensive one. The Santa Cruz Mountains are high, on the right, and covered with a dense growth of redwoods. Passing Vega, a signal station, we come to

PAJARO—(pronounced Pah-a-ro) thirteen miles from Sargent's, and 99.4 from San Francisco.

WATSONVILLE—is one mile to the right from this station, and contains a population of 4,000, and is a thrifty town, situated three miles from Watson's landing, on Monterey Bay, where steamers and other vessels land regularly. It contains two weekly papers, the *Pajaronian* and the *Transcript*. The Lewis House is the principal hotel.

From Pajaro, the Santa Cruz, narrow-gauge railroad connects with the Southern Pacific. This road is 21.15 miles long and runs through Watsonville, Aptos, and Soquel, to Santa Cruz. (See map, page 120.) The lumber business is, next to the agricultural, the most important interest in this section of the country. From Pajaro, our course will be east of south, to the end of the road.

Rolling down this beautiful valley, we come to Elkhorn Slough, over which our road is built on piles for a long distance. To the right, down this slough, is Moss Landing, nine miles distant, between which and a pier, close on our right, a small steamboat plies regularly, for the transportation of freight and passengers for the regular coast steamboats that stop at this point.

We are now running along, over and beside a salt marsh, inhabited by cranes, pelicans, ducks and mud-hens, with peat bogs and stagnant pools for immediate surroundings, while to the left, a half-mile away, is high rolling prairie, covered with cattle and sheep, beyond, the long range of the Gabilan Mountains, while to the *far* right, a glimpse can be had of the Ocean.

From Pajaro, 10.3 miles, brings us to CASTROVILLE—one-half-mile to the west of the railroad; population about 800. The town is situated at the northern end of Salinas Valley, in Monterey county, one of the most productive in the State. It is recorded in the Agricultural Bureau in Washington, that the largest yield of wheat ever known was grown in this valley, in 1852, being 102 bushels to the acre. That year whole fields averaged 100 bushels to the acre; an ordinary crop is from 40 to 50 bushels. In 1873 Monterey county produced 800,000 bushels of wheat, 400,000 bushels of barley, 70,000 bushels of oats, and other productions in proportion. Sheep and cattle in large numbers are raised. The wool-clip for 1866 amounted to 1,500,000 lbs; butter, 360,000 lbs.; cheese 120,000 lbs.; average value of land, $8 per acre. The lands in this valley are mostly 'safe lands,' will produce without irrigation.

In the spring of 1880, a branch road was completed from Castroville to Monterey, 16 miles. It is of standard gauge, and takes the place of the old narrow gauge from Salinas.

MONTEREY.—This place is situated on the southern extreme of the bay of Monterey, the most capacious on the Pacific coast, 136 miles from San Francisco by rail, and about 100 by steamer. Immediately to the westward of the city is Point Pinos, jutting out to the northward four miles, to meet Point Santa Cruz, another long promontory extending from the north, between which and the main land—land-locked as it were—is the broad bay of Monterey. This bay was first discovered by Cabarillo in 1542. In 1770 the site was occupied by the Jesuits, under the leadership of Padre Junipero, who,

June 3, of that year, held the first mass. The bell which called the faithful together was hung from a tree, the location of which is now marked by a cross, erected on the centennial day of its celebration, bearing the legend, "JUNE 3D, 1770." On the hill, near this cross, are the ruins of an old fort, near a Mexican fort of a later date; and higher up the hill is where the American fort of 1846 was built, when the Americans seized the country.

Monterey is a quiet, sleepy old town, where every person seemed satisfied with himself, apparently believing the world is completed; living on in the dreamy self-satisfied consciousness that the spirit of progress is at an end—a present tangible heaven of eternal sunshine. It is a glorious place to spend a few weeks; having done so, the pleasurable memories of the sojourn will ever remain a ray of soft sunshine, while plodding through the cares, trials and perplexities of active business life. Monterey—as one might well suppose—is a favorite resort in the summer for the better classes of citizens of the State, as well as for tourists, who find ample accommodations.

Returning to Castroville, to the east, beside the mountains, can be seen, at certain points after leaving Castroville, the little villages of Natividad Sodaville, and the Alisal race-track.

SALINAS—is 7.9 miles south of Castroville, situated to the right of the road, and on the east bank of Salinas River, with a thriving population of 3,000, and many fine stores, hotels and private residences. The Abbot is the principal hotel, and the *Index* and *Democrat* are two weekly papers.

Stages leave daily for New Republic, east, three miles; Natividad, northeast, six miles; fare, ten cents per mile.

Starting once more for the south, we find this to be the widest portion of Salinas Valley, which is about 90 miles in length, with an average width of eight miles. The valley is situated between the Gabilan mountains, to the eastward, and Santa Lucian Range on the west, about 20 miles from the Pacific Ocean, from the winds of which it is protected by the mountain named.

CHUALAR—is 10.9 miles from Salinas, and consists of several stores, hotels, saloons and a dozen or more resi-

SUMMIT SIERRA NEVADAS, DONNER L

W SHEDS AND TUNNELS. (See Annex No. 42.)

dences. Here are cattle pens and shutes, indicating that we are in a country where cattle are shipped to market; the same might be said of GONZALES a station six miles further, only there are a few more people, "scratching" the soil, which is greatly abused by this shiftless method of farming. Proceeding on 8.4 miles further, we reach the end of the track at

SOLEDAD—This is a small place of 100 or more inhabitants, with a few stores, hotels, saloons, stage-stables, freight warehouses, and some private residences. It is a point from which a large amount of freight is shipped to the southward, and from which a regular daily line of stages run to the following places: Lowe's 28 miles; Solon, 40 miles; Paso Robles, Hot Sgrings, 80 miles; San Louis Obispo, 10 miles; Arroyo Grande, 125 miles; Gaudalupe, 140 miles; Santa Barbara, 220 miles; San Buena Ventura, 250 miles, and Newhall, 300 miles; average fare, eight cents per mile.

To the westward of Solidad, seven miles, away up a cosy nook of the Coast Range, is situated the PARAISO SPRINGS, which it is claimed, possess medicinal qualities. We could hear of no analysis of these waters, which boil up in close proximity to each other. Some are very hot—others very cold, but soda, iron and white sulphur are the principal ingredients.

A hotel will be found at the Springs, where bathing in the waters, hunting, fishing and inhaling the pure mountain air can be enjoyed.

In conclusion, this is one of those trips where a great diversity of scenery, numerous objects of interest and the wealth and beauties, and the varied productions of the State can be seen and contemplated.

Returning to San Francisco we start

TOWARDS SUNRISE.

Ho! for Yo-Semite, the "Big Trees," over the "Loup," across the Mojave Desert, down through Solidad and to Los Angeles; then, over the San Barnardino, down *under the sea*, over the great Colorado desert, into Arizona, through New Mexico and on to the "Father Land"—around the circle.

Leaving San Francisco, the route is via Oakland, Martinez, and Tracy, to Lathrop, over the track of the Overland line, as described, commencing on page 178. Just before reaching Lathrop our

No. 31 ANNEX. **Nevada Falls.**—In order to form a proper idea of the superb picture, No. 13, of the large series, it will be necessary to premise that the Yo-Semite Valley is an immense gorge, in the western slope of the Sierra Nevada Mountains, about seven miles in length, from east to west, and from one-eight of a mile to two miles in width, from north to south. The walls surrounding this mighty chasm are nearly perpendicular, and from 2,000 to 6,000 feet high. The various streams that find their way into the valley flow over this tremendous wall on entering. At the eastern end of the valley proper, it divides into two canyons, projecting still eastward, but diverging as they mount the Sierras. It is through the south of one of these canons, that the main branch of the Mercede River flows, and on entering the valley, it makes two leaps. The lower one, or Vernal Fall, of 250 feet in height, 100 feet wide, and from three to four feet deep, where it leaps the square-edged barrier. Continuing up the canyon for a mile, above the Vernall Fall, amidst the wildest scenery imaginable, and we reach the Great Nevada Fall, the subject of our picture. The canyon narrows, in a wedge-like form, to quite a point, and just at the right of this vortex is the fall. It is 900 feet high, 75 feet wide at the brink, and 130 feet below. Regarded as to its height, volume, purity of water, and general surroundings, it is one of the grandest objects in the world. The spectator facing the east will observe on his left the "Cap of Liberty" lifting its rounded summit of smooth and weather-polished granite, 2,000 above the Fall, 5,000 feet above the valley below, or 9,000 feet above the sea. To paint in words, in the space allowed us, the beauties of the Fall, the bolder scenery, the foliage, mosses and ferns, always moist from the spray, and brilliant green in summer, the roar and rush of the fast-flowing river, the majestic grandeur of the rocky frame-work, which towers above and around it, is simply an impossibility; we shall not try, but refer the reader's imagination and judgment with these statistics to the beautiful picture, which we have engraved from a faithful photograph.

No. 40 ANNEX. **Mirror Lake. Yo-Semite Valley**—In the large illustration, No. 12, is presented one of the most wonderful, as well as charmingly picturesque scenes to be witnessed in this most romantic valley. As will be seen, it represents one of the most bold and striking views of a charming little sheet of crystal water of almost a couple of acres in extent, in which numerous schools of speckled trout may be seen gaily disporting themselves.

The waters are as still as death, as though awed by the wondrious grandeur of its surroundings. Close to the southeast stands the majestic "South Dome," 4,590 feet in altitude above the lake. On the north and west lie immense rocks that have become detached from the top of the mountain, 3,000 feet above; among those grow a large variety of trees and shrubs, many of which stand on and overhang the margin of the lake, and are reflected on its bosom, as in the picture.

14

train leaves the track of the Overland and turns to the right, leaving the station building between ours and the Overland track.

At LATHROP our train stops 10 minutes for supper, and then turns to the right, up the Great San Joaquin Valley The general direction of our road, for the next 350 miles, is to the southeast.

SAN JOAQUIN (pronounced San Waw-Keen).—This valley embraces portions of nine counties, and is larger than many kingdoms of the old world, and *far richer*, extending to Visalia, county seat of Tulare county. The amount of grain and stock raised in this valley, and the hundreds of smaller ones tributary to it, is almost incredible, for a country so recently settled. The valley is about 200 miles in length, and averages about 30 miles in width; comprising near 6,000,000 acres of the richest agricultural lands in the State, besides near a million acres of tuiles and salt marsh lands, which, when reclaimed, prove to be the most fertile lands in the world.

Morano—is 5.5 miles from Lathrop, important only, as many other stations on this road are, as a shipping point for grain, with side-track and great storage warehouses.

Ripon—comes next, 4.7 miles further, near which the Stanislaus River is crossed, and three miles more comes

Salida—another small station, with accommodations for shipping and storing grain. From Salida it is 6.8 miles to

Modesto—the county seat of Stanislaus county. It was laid out in 1870, and now contains a population of over 1,500, while the county contains about 11,000 Agriculture is the chief occupation of the people

Leaving Modesto, we cross the Toulumna River, and in 4.6 miles reach CERES, a small, unimportant side-track, 8.5 miles from TURLOCKS, another small station, 10.1 miles from CRESSEY, reached just after crossing the Mercede River. Continuing on 6.7 miles, ATWATER is reached. At this station, as well as those we have passed are large buildings for storing grain, as grain-raising—wheat—is the only occupation of the settlers. From Atwater it is 7.5 miles to one of the most important places so far on the road,

Merced—the county seat of Merced county; population about 3,000; has many fine buildings, including a $75,000 court house and a large first-class hotel, the El Capitan, Col. Bross, proprietor. There are two weekly papers published in Mercede, the *Argus* and the *Express*.

From Merced it is ten miles east, to the foot-hills and thirty to the western edge of the valley, at the base of the Contra Costa Range.

The county of Merced is the richest in the valley; it had 360,700 acres of land under cultivation in 1876, which yielded a little over 4,500,000 bushels of wheat, besides large quantities of barley, rye, corn, peas, beans, potatoes, hay, tobacco, cotton and many other kinds of crops The county contains a population of 65,000, most of whom are tilling the soil In this county was raised the finest cotton in the State. The value of these lands ranges from $2.50 to $10 per acre.

The game is plentiful in the river bottoms and along the foot-hills. Irrigating canals convey water over a great portion of the land. Some of these canals are quite extensive; one, the San Joaquin & King's River Canal is 100 miles long, 68 feet wide and six feet deep.

For several years the greater portion of the travel for Yo-Semite Valley and the big tree groves took stages at Merced passing over the route, via Coulterville or Mariposa, but a new route (see map, page 120) has been laid out from Madera, 33 miles further south, which, it is claimed, makes the distance by stage much shorter, and over a better road However, we shall give both routes in ANNEX No 53, and tourists can decide which they will take. Should they go in on one and out on the other route, little of the scenery will be overlooked. (See page 184.)

MARIPOSA—county seat of Mariposa county, is 45 miles east, reached by stage from Merced. This town contains about 1,000 inhabitants Once it was noted for its rich placer mines, but now quartz mining is the principal occupation of the people. In Bear Valley are the mills and mines (or a portion of them) belonging to the "Las Mariposa Grant," or the Fremont estate, as it is usually called. The Benton mills are on the Mercede River, about two miles from the town, reached by a good dug road, down a very steep mountain In Mount Ophir and Princeton, mining towns near by, are large quartz mills, belonging to the estate and extensive mines.

Leaving Merced. it is 9.9 miles to

Athlone—is a small station near the crossing of Mariposa River, beyond which the Conchilla River is crossed, and MINTURN is reached 6.4 miles from Plainsburg, in the extreme western edge of Fresno county. For a long distance the foot-hills of the Sierras on the left have appeared to be close and very rugged. The peaks of Mt. Lyell and Ritter loom up on the left, full 60 miles away, and a little further southward Mts. Goddard, King, Gardner, Brewer, Silliman, Tyndall and others can be distinctly seen with their summits covered with snow.

Berenda—is reached 9.5 miles from Minturn soon after which we cross the Fresno River, and many broad, sandy, dry creeks, and, 7.5 miles more, come to

Madera—This is a busy town of about 800 population. Here we find a large "V" flume, 53 miles long, for floating lumber down from the saw-mills in the mountains at the end of the flume. It was completed in 1876, and does an immense business.

From Madera, a new road has been completed into the Yo-Semite Valley, via Fresno Flats, through Fresno and Mariposa big tree groves. For map of route, see page 120, and for description of route, ANNEX No. 53, page 184.

No visitor to this coast ever thinks of leaving it without viewing the wonderful.

YO-SEMITE VALLEY AND THE BIG TREES—The grandest scenery on the American Continent, if not in the world, is to be seen in the valley of the Yo-Semite, (pronounced Yo-Sem-i-te; by the Indians, Yo-Ham-i-te). This valley was discovered by white men in March, 1851, first by Major Savage. It is about eight miles long, and from one-half to a mile in width. The Merced River enters the head of the valley by a series of waterfalls, which—combined with the perpendicular granite walls which rise on either side from 2,000 to 6,000 feet above the green valley and sparkling waters beneath—presents a scene of beauty and magnificence unsurpassed. except, *possibly*, in childhood's fairy dreams.

Here is *majesty—enchanting—awe-inspiring—indescribable!*—the lofty cloud-capped waterfalls and mirrored lakes; the towering, perpendicular granite cliffs and fearful chasms, strike the beholder with a wondering admiration impossible to describe.

We have often desired to take our readers with us, in a pen and pencil description of this most remarkable valley, and the "Big Trees," but in view of our limited space, the magnitude of the undertaking, together with our conscious inability to do justice to the subject, we have contented ourselves by giving a number of beautiful illustrations, which **include** the great Yo-Semite Falls, Nevada Falls, Mirror Lake, and a map of the routes and the surrounding country, showing the relative position of the valley, trees, and adjoining towns to the railroad.

The most notable falls in Yo-Semite Valley are: the Ribbon, 3,300 feet fall; the Upper Yo-Semite, 2,634 feet; the Bridal Veil, 950; the Nevada, 700; the Lower Yo-Semite, 600; the Vernal, 350 feet. The South Dome is 6,000 feet high; the Three Brothers, 4,000; Cap of Liberty, 4,240; Three Graces, 3,750; North Dome, 3,725; Glaciers Point, 3,705; El-Capitain, 3 300; Sentinel Rocks, 3,270; Cathedral Rocks, 2,690; Washington Tower, 2,200; and the Royal Arches, 1,800 feet high.

The Fresno Grove of Big trees has not heretofore been accessible to the tourist, and will therefore form a new and attractive feature to this modern route. Like the Mariposa Grove, it is divided into two groves, usually called the Upper and Lower, about one mile apart, and covering a mile square each—together they contain from 800 to 900 trees of the *Sequoia Gigantea* of all sizes. One in the Upper Grove measures 88 feet in circumference 6 feet from the ground. In the Lower Grove there is one that is 95 feet in circumference 3 feet from the ground.

From Madera, we find a grazing country; large herds of sheep abound.

The old Fresno placer mines are to the eastward, along the foot-hills, but little is being done with them, by the whites; the Chinese are working them over, as they are many of the abandoned placers throughout the State.

Borden—on Cottonwood Creek, is the next station. This place is 2.8 miles from Madera, with about 100 population. Here irrigating ditches appear on each side, and much of the land is under cultivation. Nine miles further comes SYCAMORE, a side-track of little account, just at the crossing of the San Joaquin River, which is here a small stream, with very little water. The country is now quite flat, with many little round mounds from ten to thirty feet in diameter, and from two to five feet in height. They present a very peculiar appearance, somewhat resembling a prairie dog town, only much larger, and without the hole in the top.

From Sycamore it is 9.8 miles to

Fresno—the county seat of Fresno county, a county the most diversified in the State; where land can be purchased for from $3.00 to $10 per acre. The town of Fresno has a population of about 800, has a $60,000 court-house, some good business blocks, two weekly newspapers—the *Expositor* and *Review*—and is a thrifty, growing place. The soil about the town is largely clay, producing well when irrigated, but never a seed without.

A most beautiful view is here to be obtained of the mountains on the left; the principal peaks rise from 12,000 to 14,000 feet above this valley, covered with snow the year round. Planted timber appears at places, and some of the private residences are surrounded by trees, mostly eucalyptus.

A stage line runs to Centreville, seventeen miles east. Several schemes are in hand for building large irrigating canals, taking the water from the San Joaquin and King's rivers, which, when completed, will be of great benefit to this people.

To the southwest, three and a half miles, is located the California Colony of about 125 families. The colony's land is regularly laid out for a town, with 40 acres of ground for each family. The canal that supplies water to the colony is crossed about five miles after leaving Fresno. Little of the land along here lying near the road is cultivated, but when the irrigating canals spoken of are completed, they will all be found occupied and yielding large crops.

From Fresno it is 9.6 miles to

Fowler—a small station where trains seldom stop, there being only half-a-dozen buildings—so we roll on 10.5 miles further to

Kingsburg—where there are several stores, and about one dozen buildings.

All along this valley numerous windmills are in operation, for irrigating and domestic purposes, that raise abundance of good water from a depth of from fifteen to forty feet. Soon after leaving Kingsburg, the road is built on an embankment which extends to King's River, which is crossed on a long trestle bridge.

This river rises in the high Sierras, to the northeast, and after reaching this valley, has a broad, sandy bottom, is very crooked, its course being marked, far above and below, with trees and willows which grow thickly along its bank. King's River, where the railroad crosses it, is the boundary line, beyond which lies the county of Tulare.

Sheep ranches, fenced fields—some very large—are now noticeable extending to the right and left—well we don't know how far, as the valley hereabout is full 40 miles in width, and sheep and fences, and fences and sheep, extend as far as the eye can distinguish the appearance of the land, the soil of which is clay and sand, in places somewhat alkaline. Cattle are also raised to some extent in the foot-hills, and pens and shutes for shipping are to be seen at many of the stations on the road.

The next station is CROSS CREEK, 8.1 miles from Kingsburg, and 58 miles from

Goshen. Here we come to the Southern Pacific railroad—Goshen Division—the track of which could be seen on the right, just before reaching the station. This division is only completed 40 miles, and is designed, eventually, to connect with the line extending towards it from Gilroy, which is now completed to Trespinos 100 miles south of San Francisco, referred to on page 207.

The stations on the Goshen division are: HANFORD, 12.9 miles from Goshen; LEMOOR, eight miles further; HEINLEN, 1.6 miles, and 17.5 more to

HURON—whole distance, 40 miles from Goshen, distance from Huron to Trespinos, 100 miles. This division runs through what is known as the "Mussel Slough" country, a section where the land is very rich, adjoining Tulare Lake, on the north, where the yield of all kinds of crops is marvelous. Reports say some of these lands have yielded as high as $250 per acre in a single year; that *five* crops of Alfalfa a year is common, and vegetables—well, we will *never tell you*—the yield is IMMENSE! two hundred pound pumpkins, eight feet in circumference; potatoes twelve pounds in weight, and cornstalks 20 feet high, are *some* of the figures. The price of land ranges from $20 to $100 per acre.

At Goshen, a track branches off to the left, on which cars are run seven miles to

VISALIA — the county seat of Tulare county. It contains about 1,600 inhabitants, and is situated in the midst of the most fertile land in the State, and on the Kaweah River. The country round about presents to the eye a beautiful appearance. Large oaks cover the plain in every direction, and orchards, gardens, vineyards, and well-cultivated fields are to be seen on every hand. Visalia is the center of the rich section once known as the "Four Creek Country."

The town boasts of a $75,000 court-house, some good stores, gas-works, several big saw-mills, six hotels, three weekly newspapers—the *Delta*, *Times*, and *Iron Age*—one bank, a flouring mill, a normal school, and a number of public schools, and churches of various denominations. Stages run from Visalia to Glenville, 65 miles.

From Goshen, Visalia is entirely obscured from view by the tall oaks that abound in this section of the country on every side. These oaks are old and ragged, many are fast decaying, and when gone, the country will be nearly bare, as there are few young trees growing to take their places.

At Goshen, is the end of the Visalia division of the "Central," and the commencement of the Tulare Division of the Southern Pacific—operated under a lease by the "Central" company. Although *this* is the nominal end of divisions, all changes, usual at such stations, are made 10.5 miles further at

Tulare—This is a new town, as it were, built up under the stimulating influences of a railroad point where are located extensive shops, round house, warehouses, and station buildings, incidental to its being the end of divisions. The town contains about 500 population, and is situated in the midst of a broad plain about 20 miles east of Tulare Lake, and is a thriving town. It is a point from which large amounts of freight are shipped on wagons, to the adjoining country, and where wool in great quantities, is brought for shipment to San Francisco.

The company's shops and grounds at this place—as is the case in some other localities—are surrounded with rows of beautiful trees, chief of which is the "blue-gum." These trees, from a distance, give the place more the appearance of grounds surrounding some palatial residence, than where several hundred men are employed manipulating iron. These grounds are also covered with green sward, which is watered when necessary, by long hose connected with the works.

Soon after leaving Tulare, we cross Deep and Tulare creeks, both narrow streams with steep banks, rich soil, and lined with trees; the land is covered with a thick growth of short grass. Passing the neigborhood of these creeks, the country seems to suddenly change, and at

Tipton—10.4 miles from Tulare, presents a barren appearance. To the right, left and front, sheep abound, but not a tree or shrub. Five miles beyond Tipton, are groves of eucalyptus trees, immense numbers of which are on both sides of the road. The lands here, that are irrigated at all, are supplied with windmills. Twelve miles from Tipton comes

Alila—just after crossing Deer Creek.

TULARE LAKE, is about seven miles west of this station, and is a body of water covering an area of about 7,000 square miles, is nearly round, or 30 miles long by 25 miles in width, in which fish in great varieties abound, as do ducks, geese, and other water fowl.

OWENS LAKE—another large sheet of water, but not as large as Tulare by about one-fourth—is 78 miles from Alila, in a northeasterly direction.

Passing on over White River, 8.3 miles, we come to DELANO, a place of a half-dozen buildings, just in the edge of Kern county. The country along here is treeless and not very inviting. From Delano it is 11.8 miles to Poso, and 11.8 miles more to

Lerdo—To the southwest, about 40 miles, are located the Buena Vista Oil Works, in a section of country where great quantities of oil are found in holes and ditches in the ground, where it is now waiting for enterprise to sink wells, build tanks for saving and marketing, when it will yield immense returns. This oil region is about eight miles by three in area.

Passing on about nine miles, we come to Kern River, which we cross on a long trestle bridge. This river is one of the largest flowing from the Sierras, and even in a dry season, carries a large amount of water.

Sumner—is the next station reached, 12.4 miles from Lerdo. This is a very busy place of about 250 population, it being the distributing point for a large amount of freight. To the westward, one and a half miles, and connected by "buses" hourly, is

BAKERSFIELD—This town is the county seat of Kern county, and contains a population of about 800. It is situated at the junction of the two branches of Kern River, has a $35,000 court house, a bank, several hotels, a flouring-mill and two weekly

BIRDS-EYE VIEW OF THE LOOP, TEHACHAPI PASS.

newspapers. Kern Lake is 14 miles south of west from Bakersfield, and is about seven miles long by four wide. Six miles further is Buena Vista Lake, some larger. Around these lakes, and Tulare Lake, the land is exceedingly rich.

Kern Valley, in which Bakersfield and these lakes are situated, is one of the richest in the State, being composed almost wholly of sedimentary deposits. Vegetables grow to fabulous proportions, the soil being of the same nature as that in the "Mussel Slough Country" before named.

The irrigating canals are extensive. One is over 40 miles in length, with a width of from 100 to 275 feet, eight feet deep, cost $100,000. Besides the canals there are many farms that are irrigated by wells and wind-mills. There is one ranche, nine miles from Bakersfield, that contains 7,000 acres, on which are two flowing artesian wells, of seven inches bore, one 260 and the other 300 feet deep. From these wells the water rises twelve feet above the surface, and discharges over 80,000 gallons per day.

On this ranche are over 150 miles of

No. 43 Annex. Mount Shasta—as shown in No. 15, of our large views, is a prominent feature in the landscape of the Sacramento Valley, at the head of which it is located. The view is looking to the northeast. In the foreground is the broad Valley of the Sacramento, then come towering forest trees, massive rocks, and a variety of foliage, upon which alternate patches of shade and sunlight are thrown with striking effect. Above all, towering high in mid-air, Mount Shasta springs, in a series of graceful curves, far up into an almost unclouded heaven, its sides and summits enfolded in the eternal snows. The contrast between the verdure-clad valley and the cold, wintry peaks of old Shasta, king of mountains, is a chief interest in the picture, reminding the spectator of some of the most striking effects of Alpine scenery. Mt. Shasta is 14,440 feet high. (See page 169.)

No. 46 Annex. The large view. No. 18, of San Francisco and the Golden Gate, is a real *multum in parvo*—a complete bird's eye view of the city of San Francisco and its surroundings, covering a scope of country about twelve miles in diameter—showing the Golden Gate, portions of San Francisco Bay, the Pacific Ocean in the distance, and the Pier of the Central Pacific railroad in the foreground, from whence passengers are transferred across the bay to "Frisco." This beautiful picture has been prepared and engraved expressly for this book. It shows what the Goddess of "American Progress"—as represented by view No. 1—has accomplished within the past few years, and is a very appropriate illustration with which to close our series of large views from Ocean to Ocean.

Crofutt's Grip-Sack Guide tells all about Colorado. Sold on the trains.

CROSSING THE LOOP OVER TUNNEL NO. 9, TEHACHAPI PASS.

canals and irrigating ditches, 32 miles of hog-tight board fence; 4,000 acres are under cultivation, 3,000 of which are in alfalfa, from which four and six crops a year are cut. Nearer Bakersfield, the same party, Mr. H. P. Livermore, has another large ranche, with 500 acres in alfalfa, and 3,000 in wheat and barley. On these ranches are 8,000 sheep, 4,000 stock cattle, 300 cows, 350 horses, 100 oxen, 70 mules, and 1,500 hogs. The same party makes all his own reapers, mowers, harvesters, plows, harrows, threshing machines and cultivators—everything in use on the place, except steam engines. He has one plow, the "Great Western," which is said to be the largest in the world. It weighs something over a ton and is hauled by 80 oxen, cutting a furrow five feet wide and three feet deep, and moving eight miles a day. Another plow called "Sampson," is used for ditching, and requires 40 mules to work it.

Another party in the county has 40,000 sheep, 2,000 acres in alfalfa, and raises 60,000 bushels of grain. Another *poor* fellow raised, in 1877, 84,000 lbs. of pumpkins and sweet potatoes; some of the former weighed 210 lbs., and of the latter some weighed 15¼ pounds. While attending to these *little* vegetables, he would occasionally

No. 37 ANNEX. Falls of the Willamette River.—The scene of the large illustration, No. 9, represents the Falls of the Willamette River, at Oregon City, Oregon, where the hills approach the river on each side, forcing the river through a deep canyon, and over a fall of from 30 to 40 feet. The cliffs on either side of the river rise abruptly hundreds of feet in height, and are covered at the top and less precipitous places, with a growth of evergreens. Locks are built on the Oregon City side of the river, large enough to admit the passage of boats 200 feet long and 40 feet in width. Water power is also supplied from the same source of 4,000 horse powers, which is used for running woolen mills and other manufactories at Oregon City.

look after a small band of *sixteen* thousand sheep.

But enough of this. We could fill our book with these and many other astonishing figures. "Well," you will say, "these California farmers should be contented and happy men." One would think so, but they are not. They are the most inveterate grumblers of any class of people in the world. All Californians will, in the intervals between grumbles, express the opinion that there is no place under the blue canopy of heaven so good for a *white man* to live in as California. Ah, well! are they correct? *Personally*, were it necessary, our *affirm* could be forthcoming.

Returning to Summer, twelve miles, brings us to a small place called PAMPA, and 1 7 3 miles more to

Caliente—Since leaving Summer, the grade has increased; the valley has been gradually narrowing by the closing in of the mountain ranges on each side, leaving only a narrow strip of land. Nearing this station, it still more contracts, until a deep canyon is reached, in the mouth of which is located Caliente, surrounded by towering cliffs. There are several stores, one hotel and a large station and freight warehouse at this place. A large amount of freight is re-shipped at this point, on wagons, for the surrounding country. Stages leave this station daily for Havilah, 25 miles; Kernville, 45 miles; fare about 14 cents per mile. These stages carry passengers, mails and express. Tourists should now note the elevations; Caliente is 1,290 feet above sea level; within the next 25 miles the train will rise to the summit of Tehachapi Pass, to an altitude of 3,964 feet, an average of over 106 feet to the mile. Within this distance we shall find some of the grandest scenery on the whole line; will pass through *seventeen* tunnels, with an aggregate length of 7,683.9 feet, and then "OVER THE LOOP," one of the greatest engineering feats in the world; feat where a railroad is like a good Roman Catholic—made to *cross itself*. But here, the difference is in favor of the railroad, as these Californians will always be a *lee-tle* ahead; it does its crossing on a *run, up grade, toward heav n*. [Any design to indicate the route of the good Catholic is disclaimed.] See illustrations on pages 214 and 215.

Away up the canyon, the grade of the road can be seen at a number of places where it winds around the points of projecting mountain spurs, from which points we will soon be able to look down upon Caliente.

Leaving the station, our route will be found illustrated on page 214. Caliente is at the foot of the mountains, at the extreme further end of the dotted line, which indicates the course of the road, and shows its windings, the Loop and the surrounding country, on a flat surface. As we ascend the narrow canyon, the road gradually commences to climb the side of the cliffs on the right, leaving the bed of the canyon far below, on the left. Up, up, around rocky points and the head of small ravines, over high embankments, through deep cuts, and tunnels "One" and "Two," a distance of 5 3 miles from Caliente, we arrive at

Bealville—This is a small station named in honor of General Beal, late minister to Austria, who owned 200,000 acres of land in this county.

Oaks, cedar and spruce trees are to be seen in the gorges and on the mountain side, where a sufficient soil is left between the rocks and an occasional shrub of the manzanita, along the road. Continuing our climb, the ravines are deeper at every turn; tunnels No. three, four and five are passed through, each revealing in its turn, new wonders and rapid changes. No. five tunnel is the longest on the "Pass," after passing which and No six tunnel, the canyons on the left become a fearful gorge.

Just after emerging from the sixth tunnel, by looking *away* down the canyon, Caliente can be seen, and at the rounding of nearly every mountain spur for some miles further. Continuing our climb, winding around long rocky points and the head of deep ravines, twisting and turning to gain altitude, the scenery is wondrous in its rapid changes. The old Los Angeles and San Francisco wagon road can be seen in places, where it, too, winds around the side of the mountain, and in others, along the little ravines and larger canyons.

The opposite mountains now loom up in huge proportions, rocky, peaked and ragged, a full thousand feet above our heads, and double that amount above the bottom of the canyon below. Soon after passing tunnels seven and eight; again we look down from dizzy heights into *fearful, fearful* chasms. Up a long curve to the right, and we are at a point where the mountains, from ten to twenty miles to the south and westward can be seen, the peaks of many covered with snow.

Keene—is reached 8.3 miles from

Ilealville. This station is not an important one, and trains do not always stop, but pass on, across two bridges in quick succession, many deep gravel-cuts, and then, after curving to the right, we approach the "Loop" and tunnel No. nine. Passing through this tunnel, we start on the grand curve around the "Loop," and soon find ourselves *over* the tunnel and in the position of the train as illustrated on page 215.

This "Loop" is 340 miles from San Francisco, is 3,795 feet in length, with an elevation of 2,956 feet at the lower and 3,034 feet at the upper track, making a difference between tracks, of 78 feet.

Leaving the "Loop," our train continues to climb and curve, first to the left, then to the right, and after passing through two more tunnels, Nos. 10 and 11, comes to

Gerard—a station 5.4 miles from Keene, more in name than fact. Here the old Tehachapi Pass stage road appears. The mountains are not as high above us, but are rough, broken, and ragged, covered in many places with stunted, scrubby pines and cedars. Rolling on, we pass through, in quick succession, tunnels Nos. 12, 13, 14, 15, 16 and 17, besides a number of short bridges, and come to more open ground; pass Graceville, once an old stage station—on the right, and 6.4 miles from Gerard arrive at

Tehachapi Summit—elevation, 3,964 feet—the highest on the road. This station consists of one store, a hotel, telegraph office, and half-a-dozen buildings. To the southeast about five miles distant, a marble quarry is reported, of good quality. The station is situated on a high grassy plateau, of a few thousand acres, with high mountain ranges to the east and west, and although near 4,000 feet in altitude the climate is so mild and agreeable that some years the crops are very good, and grazing excellent. Many sheep are to be seen in the valley and on the hills which are covered with fine grass.

Leaving the summit, we run along this plateau for a few miles, and then commence a gradual descent towards the Mojava Desert. To the right is a small lake—dry in summer—where salt can be shoveled up by the wagon load. About eight miles from the Summit, the little valley down which we have been rolling, narrows to a few hundred feet with high canyon walls on each side.

Cameron—is the first station from Summit, 9.2 miles distant, of little account. Near this station we find the first of the species of cactus, as illustrated on page 221. In this country they are called the

YUCCA PALM—These trees grow quite large, sometimes attaining a diameter of from two to three feet, and a height of from 40 to 50 feet. They are peculiar to the Mojava Desert, where they grow in immense numbers, presenting the appearance, at a distance, of an orchard of fruit trees. Everything is said to have its uses, and this cactus, or palm—apparently the most worthless of all things that grow, is being utilized in the manufacture of paper, and with very good results. One mill is already in operation at Ravena, and considerable shipments have been made. The supply of "raw material" is certainly abundant, and if the quality of the paper is as good as reported, the Mojava Desert may be able to show cause why it was created.

Leaving Cameron, our train speeds along lively, and 5.2 miles we come to

Nadeau—soon after emerging from the canyon. It is a side-track surrounded by sage-brush, sand hills and cactus. Away to the left are several lakes, dry the greater portion of the year, but having the appearance of water at all times, owing to the water being very salt, and leaving a thick deposit on the bed of the lake when dry.

From Nadeau, it is 5.6 miles to the end of the Tulare division, at

Mojava—(Pronounced Mo-ha-vey.) At times, the "Mojava Zephyr" is anything but a *gentle* zephyr, yet, by using both hands, any person of ordinary strength can keep their hat on. The surroundings of this place are not very beautiful, situated as it is on a desert; but for its size, it is a busy place. It is a regular eating station where trains stop half an hour, and good meals are served, at the Mojava House, close to the depot—*on Main St.* The place consists of several stores, one hotel, large station building and freight warehouse, a 15-stall round-house, a repair and machine shop and about a dozen private residences.

The water used at the station comes in pipes from Cameron station, eleven miles north. A large amount of freight is reshipped from this station on wagons to Darwin, 100 miles, and Independence, 168 miles, in Inyo county—to the northeast—on the east side of the Sierra Nevada Mountains. Returning, these, wagons are loaded with bullion from the mines. The

Cerro Gordo Freighting Co., who do most of this freight hauling, employ 700 head of animals.

Stages leave Mojava every alternate day, carrying passengers, mails and express to Darwin, 100 miles; Cerro Gordo, 125 miles; Lone Pine, 150 miles, and Independence, 168 miles to the northeast; fare, 14 cents per mile.

Mojava is the commencement of the Los Angeles Division. The proposed route of the Southern Pacific railroad—as successor of the rights granted to the old Atlantic and Pacific Co., to the Colorado River, at the Needles—diverges at this point, and runs due east. The lowest point of the Mojava Plains crossed by the railroad survey, is at the sink of the Mojava River, 133 miles east. Its elevation is 960 feet, the highest point being 3,935 feet, at the summit of Granite Pass. The crossing of the Colorado, at the Needles, is 254 miles east from Mojava.

Leaving Mojava, our course is south, over the desert, from which rise great numbers of round buttes; they are of all sizes, from a half-acre at the base, to several acres; from one hundred to five hundred feet in height. Most of these buttes run to a peak, and are grooved or worn out by the elements into small ravines, from summit to base, presenting a peculiar appearance. The cactus, or palms, are very numerous.

Passing GLOSTER, 6.6 miles from Mojava, where there is not even a side-track, and 7.2 miles further, we arrive at

Sand Creek—where trains seldom stop. To the left, ten miles, is Mirage Lake, which looks like water, but is mostly sand and alkali. (For a description of this remarkable phenomena, see page 142.)

Soledad Mountain can now be seen on our right, through which our road finds a way, but *where* and *how*, does not appear. Large numbers of sheep range over these plains at times, and appear to thrive.

From Sand Creek, it is eleven miles to LANCASTER, a side-track, and 10.9 miles further we come to

Alpine—For the last twenty miles the palms have been very numerous, but we shall soon leave them and the desert. Scrub cedar, sand cuts—some very deep—are now in order, while rapidly climbing up to the summit of the Soledad Pass, which we reach four miles from Alpine, crossing it at an elevation of 3,211 feet, and then descend to

Acton—a distance of 9.7 miles from Alpine. This is an unimportant station

No. 32 ANNEX. Pioneer Mail Enterprises— CROSSING THE SIERRAS ON SNOW-SKATES.—The rapid settlement of the fertile valleys lying at the eastern base of the High Sierras of California, created a want for mail facilities in advance of regular methods. It is well known that previous to the winter of 1854, the fearless settlers of this isolated inland world were shut out from communion with the great throbbing heart of civilization on the outside for three or four months of every year, by that almost inaccessible and snow-clad range. Those whose temerity let events bid defiance to this battleground of the storms, and sought to scale its snowy-ramparts, too frequently became snow-blind, or foot-frozen; or, still more frequently, lay down to that sleep which knows no waking—their only mantle the fast-falling snow.

In this emergency one brave heart, at least, was found to dare the perilous task of carrying the United States mail to those enterprising pioneers. It was Mr. John A. Thompson, a Norwegian. Early education and habit had made him an adept in the use of the snow skate. Without hesitation he made a contract with T. J. Matteson, of Murphy's Camp, Calaveras county, to continue postal service in winter, as well as in summer, over the route, via the Calaveras grove of big trees (the only grove then known), to Carson City, for $200 per month, without regard to the depth of snow.

Our illustration introduces our hero in *propria persona*. It will be seen at a glance that the *snow-skate* is totally unlike the Indian or Canadian *snow-shoe*—the latter being adapted mainly to a light, loose snow and level country; and the former to compact masses and mountainous districts. The "shoe," moreover, is of slow and laborious use; whereas, the "skate" is of exceedingly rapid and exhilarating adaptability—especially on down grades, when its speed is frequently equal to the ordinary locomotive. The motion is a slide -not a step. The pole in the mail carrier's hand acts as a brake on down grades, and as a propeller up hill.

In Sierra county, California, where snow often falls to the depth of ten or twelve feet, the snow-skate is a great favorite, becoming a source of pleasant recreation on moon-lit evenings—visits of from ten to fifteen miles being made after tea, and returning the same evening. Here, too, snow-skating forms one of the most popular of pastimes—racing. A belt, studded and set with silver, becomes the prize of the successful racer. Sometimes young ladies will challenge gentlemen to a race for a pocket handkerchief, or a pair of gloves—which, of course, is always accepted. The accidents which sometimes occur throw no damper on the sport. See page 63.

The Sierra Nevada Mountains—are about 500 miles long, and from 60 to 100 miles in width, their general direction northwest and southeast. The height of the principal peaks are —Mt. Whitney, 15,088 feet; Williams, 14,500; Shasta, 14,444; Tyndall, 14,386; Rawenh, 14,000; Gardner, 14,000; King, 14,000; Brewer, 13,886; Dana, 13,227; Lyell, 13,117; Castle Peak, 13,000; Cathedral Peak, 11,000; Lassen's, 10,578 feet.

near the head of the infamous Soledad, Canyon, known as the "Robbers' Roost." This canyon is a deep gorge, with rugged, towering mountain cliffs rising on each side, in places from 500 to 2,000 feet above the bed of the canyon, the fronts of which look as though they had been slashed by the hand of the great Architect, from summit to base, into narrow, deep ravines, and then left, presenting as wild, gloomy and dismal gorges as the most vivid imagination can conceive. These, with the dense growth of pines, cedar and shrubs, make the mountains almost impenetrable, and all that the most wary villain could desire.

The canyon is about 25 miles in length, inhabited mostly by Mexicans. It was the headquarters and *home* of the noted Vasques, and his robber band, who was hung at San Jose, March 19, 1875. Later, a band of a dozen or more raided Caliente, binding and gaging all who came in their way, and after loading their riding animals with all they could carry, returned to *this* their rendezvous. By a shrewd plan, five of the number were captured, and lodged in jail at Bakersfield, from which they were taken by the citizens and hung without much expense to the county. But with all their devilment, the trains and railroad property have always been secure. At the next station, the brother of this noted chief resides, against whom, as we understand, there stand no accusations.

Passing on down, the canyon widens, and cottonwood, sycamore and a few oaks and willows line the little creek, which ripples over the sands. Mining, to some extent, is carried on by the Mexicans living here, but in a primitive way, using arastras, with water, horse, hand, and, in three cases, steam power.

Ravena—is the next station, 3.7 miles from Acton. Here are located a village of several dozen log, sod and stone houses, belonging to the Mexicans, and the paper mill, before alluded to, as utilizing the yucca palm for making paper.

We were told at this place that "moss agates and grizzly bears abound," but just *why* the two should be coupled together, we are not informed.

About one mile below Ravena, on the left, away up on the side of the mountain, 600 feet above our train, is a huge rock, called George Washington, from the fact that it bears a *striking likeness* to the "father of his country," who, it seems has left his impress all over his country. Continuing down, the canyon narrows; the bluffy walls on each side assume more formidable features, and in fact is the most formidable portion of the canyon, the rugged spurs shooting out as though they would bar our farther progress.

Two of these spurs did bar the progress of our way, until tunnels were completed through them, which aggregate 596 feet in length.

Timber can be seen on the tops of the mountains, and in the largest of the deep ravines, but inaccessible, from the unusual ruggedness of its surroundings. Lime-rock abounds and game, both large and small, is very numerous, including the grizzly bear. When we passed this way in January, 1878, Mr. Lang, of Lang's Station, close ahead, had killed one of these bears that weighed 900 pounds, and Lang called it a *small* one.

Lang—is a small station, 8.5 miles below Ravena, and about half-a-mile west of where the "last spike" was driven, Sept. 5th, 1876, that united the line, building from Los Angeles and San Francisco. The bottom, below the station, widens, sand hills and sand beds appear, as well as sheep, on the adjoining hills, which are now lower, with grassy sides; and 13.1 miles from Lang, and our train stops at

Newhall— a small station named for a Mr. Newhall, who owns 50,000 acres of land in the vicinity, on which range thousands of cattle and sheep. The Southern Hotel with accommodations for 150 guests, a beautiful park, and a planted grove of trees are among the late improvements.

Stages leave this station daily for Ventura, 50 miles; Santa Barbara, 80 miles; San Louis Obispo, 190 miles; Paso-Robles Hot Springs, 220, and Soledad, 300 miles, at the end of the Southern Pacific railroad, in Salinas Valley, as noted in excursion No. 5. These stages carry passengers, mails and express.

We now confront the San Fernando Mountains on the south, which rise up before us, towering to the skies, in one great black solid mass, apparently presenting an impenetrable barrier to our further progress. Such *was* the case until the engineers of this road, failing to find any way *over* them, resolved to pierce *through them*, which was done, resulting in a tunnel 6,967 feet long, built in a straight line and timbered all the way. These mountains, as stated, are high, rising up out of the valley from

2,500 to 3,000 feet, but narrow—a huge "hog-back" ridge. Leaving Newhall, it is 1.6 miles to

Andrews—To the west of this station, about four miles, are located several oil wells, in a region said to be very rich in oil. Two refineries have been established at this station, which furnish for shipment about one car-load per day. Live oaks and some white oaks are numerous along the road and on the sides of the low-hills, for the last fifteen miles, making the country look more cheerful than it otherwise would.

Leaving Andrews, we soon commence to ascend, passing through deep cuts to the

SAN FERNANDO TUNNEL—This tunnel, as before stated, is 6,967 feet in length, timbered all the way, and is reached from the north up a grade of 116 feet per mile; grade in tunnel, 37 feet per mile; grade beyond tunnel—south—for five miles, 106 feet per mile; elevation of tunnel, 1,469 feet. The view, from the rear end of the car, while passing through the tunnel, is quite an interesting one.

The light, on entering the great bore, is large and bright, the smooth rails glisten like burnished silver in the sun's rays. Gradually the light lessens in brilliancy; the rails become two long ribbons of silver, sparkling through the impenetrable darkness; gradually these lessen, the light fades—and fades, and fades—the entrance is apparently not larger than a pin's head, and then all light is gone and darkness reigns supreme—and still we are not through. It is the history of many a life: *the bright hopes of youth expire with age.*

As we emerge from the tunnel, the valley of San Fernando dawns a bright vision of beauty upon us. Here we enter, as it were, a new world of verdure and fruitfulness—a land literally "flowing with milk and honey." From the tunnel we have descended rapidly, 5.2 miles to

San Fernando—named for the famous old mission of San Fernando, located about two miles to the right, embowered in lovely groves of orange, lemon and olive trees. It is in the middle of the valley of the same name, surrounded by mountain ranges. The San Fernando Mountains are on the east and north, the Coast Range on the West, and the Sierra Santa Monica on the west and south. The greater portion of the western and central part of the valley is under a high state of cultivation, but the eastern, along where our road is built, is covered with sage-brush, cactus, grease-wood, small cedars and mesquite shrubs.

The station is of little account—only a few buildings, a store, hotel, cattle pen and shutes make up the place. Leaving the station, we pass groves of planted trees; those on the right, of the eucalypti species.

Sepulveda—is the next station, 12.5 miles south of San Fernando. It is situated on the east bank of Los Angeles River, where passenger trains meet and pass. Continuing along down the valley—which now begins to present an improved appearance—8.6 miles we come to East Los Angeles.

Los Angeles Junction—is situated about one mile east of the city, from which street-cars run regularly; fare, 10 cents or four tickets for 25 cents. The principal hotels, the Pico and St. Charles, charge from $2 to $3 per day; the United States and Lafayette from $1.50 to $2.00, all of which send buses to the depot, on arrival of trains.

Los Angeles!—Ah, here we are at the "City of the Angels!" Los Angeles is the county seat of Los Angeles county, situated on the Los Angeles River, 24 miles north from the port of San Pedro; but the principal shipping point is at Wilmington, about two miles above San Pedro, at the head of the bay, with which it is connected by railroad 22 miles distant. It is also connected with Santa Monica by rail, 18 miles to the westward, where steamers land from up and down the coast. The city contains a population of about 16,000—has many fine business blocks, three banks, several large, fine hotels, chief of which is the Pico. The churches and schools are all that could be desired, both in numbers and quality. There are four daily, seven weekly, and a number of miscellaneous publications. The dailies are: the *Star, Express Herald,* and the *Republican.*

Water for irrigation in the city is supplied by Los Angeles River, and by windmills. The manufactories are not very numerous, the shops of the Railroad Company being the principal ones. The town is a railroad center, commanding an extensive trade at present, and in the future it fears no rival. It is already connected with Santa Monica, on the west, 18 miles; Wilmington, on the south, 22 miles; Santa Ana, on the southeast, 33 miles; Yuma, on the east, 248 miles, and San Francisco, north, 470 miles. Los Angeles is an old town, having been settled in 1771. It is located at the southern base of the Sierra Santa Monica range on a gradual slope, and is completely embowered in foliage. The vineyards, in and around the city, are very numerous; they are to be seen on all sides, equaled only by the number of orange, lemon, and fruit orchards. It is really a city of gardens and groves. Then, as one rides to the westward, or the southward, magnificent plantations stretch away as far as the eye can reach. Here is the wealth of the tropics; here can be seen the orange, lemon, lime, pomegranate, fig, and all kinds of tropical and semi-tropical fruits, attaining to the greatest perfection; here will be seen the huge palm-tree, the banana, the beautiful Italian and Monterey cypress, the live oak, pepper, and the eucalyptus, as well as the orange and lemon trees in the grounds and parks, gardens and lawns, of almost every citizen's residence. One orchard—situated in the heart of the city, the "Wolfkill"—contains 100 acres. In this orchard are 2,600 orange trees, 1,000 lime, and 1,800 lemon trees; besides, there are adjoining 100 acres in vineyard. But why particularize? Look where you will, and you will see vineyards and orchards laden with luscious fruits, and will be ready to exclaim: "Why, oh, why was 'mother Eve' driven out?"

Leaving Los Angeles, we will take the cars on the

LOS ANGELES AND INDEPENDENCE RAILROAD — under the management of the "Central" Company, of which Mr. E. E. Hewett is Ass't Superintendent, and speed away to the westward. The first few miles is through the edge of the city, and then past a succession of vineyards,

YUCCA PALM OF MOJAVA DESERT. See page 217.

orange and fruit orchards, nurseries and groves of planted trees. Then come broad fields and pretty little farm-houses; then through a succession of deep sand cuts, and the broad ocean appears, and then

Santa Monica—called by some the "Long Branch of the Pacific Coast." It is certainly a beautiful location, and if it does not attain the same popularity as its namesake, on the Jersey shore, it will not be for lack of natural advantages. Its location is one of surpassing loveliness—in front the Pacific Ocean; in the background the noble range of the Sierra Madre. Far out to the seaward looms up mistily the island of Catalina. The facilities for bathing could hardly be better. The beach is fine, the sand hard and smooth, and the slope gradual, with no terrors of undertow to appal timid swimmers. The place is protected from cold winds by a prominent head-land, and the climate is very equable.

The following table shows the mean temperature of January and July in California and other States and countries, taken from reliable sources:

Place.	Jan'y	July.	Difference.	Latitude
	Deg'e	Deg'e	Degre	Deg. min.
San Francisco..	49		8	37 48
Monterey	52	5	6	36 36
Santa Barbara..	54	7	17	34 24
Los Angeles ...	52	75	23	34 04
Santa Monica...	52	69	17	34 02
San Diego	51	72	21	32 41
Sacramento	45	73	28	38 34
Humboldt Bay..	40	58	18	40 44
Sonoma	45	66	21	38 18
Vallejo	48	67	19	38 05
Fort Yuma	56	92	36	32 43
Cincinnati	30	74	44	39 06
New York	31	77	42	40 37
New Orleans	55	82	27	29 57
Naples	46	76	30	40 52
Honolulu	71	78	7	21 16
Mexico	52	65	13	19 26
London	37	62	25	51 29
Bordeaux	41	73	32	44 50
Mentone	40	73	33	43 41
Marseilles	43	75	32	43 17
Genoa	46	77	31	44 24

It will be seen by referring to the above table that Southern California possesses a climate unexcelled in equability by any portion of the world, and of the happiest medium between the extremes of heat and cold. Santa Monica has these advantages of temperature in a special degree, the air being modified by the ocean to a point most agreeable and invigorating, both to the pleasure-seeker and the invalid.

The bathing house, situated on the beach, about fifty feet above the water, is the finest on the coast. It is a large building supplied with baths of all kinds, where the bathers have within reach, faucets by which a supply of either fresh or salt water, hot or cold, can be instantly obtained by the effort of turning them on. Here, too, are steam, swimming, and plunge baths, besides the ordinary ocean baths, accommodations for which ample provision is made.

Santa Monica was first laid out as a town in 1875, and in two years attained a population of 800. It has some good stores, and quite a number of good hotels, chief of which are the Santa Monica Hotel, and Ocean House; the latter has accommodations for about 50 guests, and the former for 125. These houses are so situated as to command a most extensive view. Their charges are from $12 to $18 per week. **Santa** Monica *had* its newspaper once —the *Outlook*—but we hear it has moved, and is now a *look-out* at Anaheim.

Point Dumas, a prominent head-land to the northwest, is 13 miles distant. Point Vincent, to the southwest, is 20 miles distant. Santa Rosa Island, west, is 91 miles distant; Santa Barbara Island, south of west, is 25 miles distant; San Nicholas Island, 37 miles in the same direction, and Santa Catalina Island, south, is about 40 miles distant. These islands are a great protection to Santa Monica from the wrath of old Pacific, when he becomes excited.

The wharf, which was built from the end of the railroad to deep water, affording a landing for coast steamers, was destroyed in 1878, but we presume will be rebuilt.

In the range of the mountains on the north, game of many varieties can be found, and in the lagoons south of the town, ducks, geese, snipe, curlews, and other varieties of game are abundant.

The drives are very fine, being along the beach for many miles, and then, on the high plateau 500 feet above, extending for many miles, affording a most extended view; or, up to the natural springs on the side of the mountain, which furnish the town with water, bubbling up like a fountain, and is caught in a large basin or pond, for city use.

A popular excursion is up Santa Monica Canyon to Manville Glen—a wild, rugged mountain-place covered with old forest trees, down which ripples one of the neatest little brooks imaginable. The point of the mountain above has become a very popular camping ground, where camps are made, and parties spend months in rambling over the mountains and enjoying the ocean baths, etc.

There are some beautiful country residences about Santa Monica, among which is one of Senator Jones, of Nevada.

Returning to Los Angeles, we take the WILMINGTON DIVISION—and start directly south through a succession of vineyards, gardens, orange and fruit orchards, to FLORENCE, six miles from Los Angeles. At this station the track of the San Diego Division branches off to the left. But we continue south, through broad, well-cultivated fields, where the good effects of irrigation are shown, by large crops of vegetables, which abound in the section we are now traversing. Gradually the rich soil gives place to alkaline and salt flats, and sloughs, with occasionally a few bands of sheep on the more elevated lands.

About two miles before reaching Wilmington, we pass, on the right, embowered in trees, the old headquarters of the Military Department, of Southern California and Arizona, abandoned in 1870. When the Government had no further use for the property it was sold, and is now used by the Protestants, and called Wilson's College.

Wilmington contains a population of about 500, most of whom are engaged in the shipping interests. At the long wharf are great warehouses, beside which, vessels drawing twelve feet of water, can lay and load and unload from and into the cars of the railroad, which run the whole length of the wharf. Vessels drawing 15 feet of water can cross the bar, two miles below, but are unable to reach the wharf, and are unloaded two miles below.

Inside the bar is a ship channel, perfectly sheltered, several miles in length, with a width of from 400 to 500 feet, and a depth, at low tide, of from 20 to 25 feet, shoaling at its head to 12 feet.

The Government has expended over half a million of dollars to improve the harbor at this place; the breakwater is 6,700 feet long, and when completed, it will be of incalculable advantage to the people of this section of country.

Wilmington is a point where immense quantities of ties and redwood lumber are landed from the Humboldt Bay country, 200 miles north of San Francisco, on the coast, and also where are landed large quantities of coal from the Liverpool vessels that come here to load with grain. The coal is brought for ballast, more than for profit.

Rattlesnake Island is in front of the harbor—sand principally San Pedro Point is two miles south, and Point Fermin, around that point to the west, reached by wagon-road around the beach or over the bluffs, six miles distant. Deadman's Island is a small, isolated rocky peak, where commences the breakwater improvement below Point Pedro.

Fermin Point is on the most prominent headland on the west, surmounted with a light of the first order, [which is kept by two ladies.] Near this point, *in stone*, is the subject of our illustration, below, called San Pedro's Wife or the "WOMAN OF THE PERIOD."

The distance from Wilmington by steamer to San Francisco is 387 miles; to San Diego, 95 miles; to Santa Catalina Island, 20 miles. This Island is owned by the Lick estate, is 35 miles long and ten wide, on which are some gold mines, and great numbers of sheep and goats.

SAN PEDRO'S WIFE OR, THE WOMAN OF THE PERIOD.

The Island San Clemente is 80 miles further, a long, narrow strip of land, on which there is no water, where range thousands of sheep and goats, which seem to thrive better than on Santa Catalina Island, where water is abundant.

Returning again to Los Angeles, we start over the

San Diego Division.

Leaving Los Angeles, the course is the same as over the Wilmington route to Florence, six miles south, where our route turns to the left.

Leaving Florence, we cross the Los Angeles River, along which are some broad, rich bottom lands, passing large groves of eucalyptus trees, and 5.5 further come to

DOWNEY—This is a thrifty town of agriculturalists, about 500 in number, with some good buildings. The Central Hotel is the principal hotel. The country is flat, and vineyards and orange orchards are to be seen at different places, over which the waters of San Gabriel River are conducted in numerous canals and ditches

Leaving the station, we soon cross San Gabriel River, note the existence of many sycamore trees, some oaks and many "Gum-trees," and four miles are at

NORWALK—This is a new station, in the center of a broad fertile valley, with only the smaller portion under cultivation Continuing on, over a grassy plain, where are a few trees, and a few alkali beds, we pass COSTA, 6.3 miles from Norwalk, and roll along through an improving country The La Puente Hills are on our left, beyond which rise the San Gabriel Mountains. From Costa it is 3.6 miles to

ANAHEIM—Here we are at a live town of 1,500 population, which, from the car windows, presents a beautiful appearance, with its long rows of trees and beautiful fields. A run through the town will reveal the fact that it contains many fine buildings, some of which are devoted to merchandising, besides good churches, fine schools, two good hotels—the Planters and the Anaheim, and one newspaper—the *Gazette*. The town is embowered in foliage; tall poplar trees, cypress, eucalyptus, orange, pepper, castor bean, palm and many other trees are among the number seen everywhere.

Here we find extensive irrigating canals and a complete net-work of ditches, conducting the water through the streets and over the grounds in all directions, A great number of the private residences are painted white, (not a very common thing in California,) and look very cheerful. Leaving Anaheim, we cross a sandy bottom, and then Santa Anna River, over a long bridge, pass ORANGE, a small hamlet on the left—where is a grove of planted trees—and 4.9 miles from Anahiem, and two miles further come to

SANTA ANA—This town is 33.3 miles southeast of Los Angeles, and about half a mile west of the depot, where is now the end of the road, and where a town is being laid off, called East Santa Ana. Santa Ana is situated about one and half miles south of Santa Anna River, and like Anaheim, is embowered in trees and surrounded by vineyards, orchards and the best of land, under a high state of cultivation. There are some large stores in the town and good brick buildings, several fine churches, good schools, three hotels—chief of which is the Santa Ana Hotel—one daily and two weekly newspapers: the *News* and the *Times* are weekly, and the *Free Lance* is a small, *live* daily.

Newport Landing is eight miles west of Santa Ana, where most of the steamers call, on their way up and down the coast. A good wagon road leads from Santa Ana to the Landing, and also extends eastward to San Bernardino, 40 miles distant. The road was built by the counties of Los Angeles and San Bernardino.

The new Black Star coal mines are situated about twelve miles northeast, and are said to be extensive and the coal of good quality. To the east is the high range of the Sierra De Santa Anna Mountains, on the eastern slope of which are located the Temescal Tin mines.

Some of the lands surrounding Santa Ana and to the south and west for many miles, called "safe lands," will raise a good crop without irrigating, but the greater portion requires the water—to supply which a company is now engaged building a canal to take the waters of the Santa Anna River away to the eastward. The canal will be 18 miles long, and will furnish ample water for 20,000 acres of land.

Stages leave Santa Ana daily for San Juan Capistrano, southeast 24 miles; fare $2.50; San Louis Rey, 65 miles; fare, $5.00; also to San Diego, 100 miles, and all intermediate points.

SAN DIEGO—As this is reached from Santa Ana, the nearest point by rail and stage, it seems to be the proper place for a short description of the town. San Diego

MOUNT SHASTA, SACRAMENTO VA

, CALIFORNIA. (See Annex No. 43.)

was first settled by the Jesuit missionaries, in 1769, and is the oldest town in the State. It is a port of entry, and the county seat of San Diego county. It is situated on San Diego Bay, which, for its size, is the most sheltered, most secure and finest harbor in the world. The bay is 12 miles long and two miles wide, with never less than 30 feet of water at low tide, and a good, sandy bottom. By act of Congress, it is the western terminus of the Texas & Pacific railroad, but *when* that road will be built, if ever, is a problem, the solution of which, all the citizens of San Diego, about 5,000 in number, are exceedingly anxious to have demonstrated, and there is little question but what they would all elect to have it built without delay. The city is connected by steamer with San Francisco, 456 miles north, and by stage to all inland towns. It is 14 miles north of the dividing line between Upper and Lower California, and is destined to make a city of great importance. Tropical fruit of every variety is produced in the county, and the climate is one of the finest in the world, the thermometer never falling below 40 deg. in the winter, or rising above 80 deg. in the summer. The country is well timbered and well watered, producing large crops of all kinds of grain, fruit and vegetables. Gold, silver and tin ores have recently been discovered, which promise at this time to be very extensive and profitable. Several quartz mills have been erected. Two weekly papers are published at San Diego—the *World* and *Union*.

SAN JUAN CAPISTRANO, is a quiet, sleepy, conservative old town, twenty-four miles from Santa Ana, situated in the center of a beautiful little valley, hemmed in on three sides, in a variegated frame-work of emerald hills, with the broad Pacific Ocean on the west, gleaming like a mirror at mid-day, and glowing like a floor of burnished gold at sunset: Here is located the old mission, which gave its name to the town. It was founded in 1776, and is situated on an eminence, commanding a view of the surrounding country, with extensive orchards of orange, lemon, olive and other trees, planted nearly 100 years ago, which continue to bear abundantly. To the south of the town is the Rancho Boca de la Playa, of 7,000 acres; Rancho Neguil, of 12,000 acres, and the Rancho Mission Viejo, on the east, of 46,000 acres. These ranchos include a great deal of good agricultural land, but *now* the greater portion is used for pasturage.

GOSPEL SWAMP—This singularly productive region is situated a few miles north of west from Santa Ana, the soil of which is very similar to that about the "Mussel Slough" and Lake Tulare, heretofore noted. The soil is wholly composed of the richest sedimentary deposit, the decomposition of vegetable matter that has been going on since the creation of the world. In this section, all kinds of vegetables attain immense proportions, *so large* that we dare not give the figures. This is the pumpkin's *home*. Pumpkins weighing 320 to 340 lbs. are not uncommon in this region. A single vine produced in 1877, 1,400 lbs. of pumpkins without any further care than putting the seed in the ground—and it was a poor year for pumpkins at that. Corn is the principal crop, in gathering which they find much difficulty, owing to the height of the stalks. If some enterprising Yankee would invent a portable elevator with a graduated seat and revolving buckets for holding the ears of corn, he could find in this section an extensive *field* in which to operate.

Returning once more to Los Angeles, and for the *last time*, we take our old seat, and start for SUNRISE, at Yuma. See Time Table.

Leaving Los Angeles, our course is south about one mile—on the track we have been over several times—then to the left, and finally *due east*, crossing the Los Angeles River, just beyond which is the 80-acre vineyard of Mr. Sabichi, and follow up a little valley. On the right are low, rolling grass-covered hills, around which are many little cottages nestling cosily beneath a wreath of foliage, consisting of orange and other fruit trees. We are now on an ascending grade, and shall continue to be, for the next 80 miles.

To the left, about four miles, is located PASADENA—(Key of the Valley)—quite commonly known as the "Indiana Colony," a new and beautiful settlement northeast from Los Angeles about seven miles, and three miles from the old mission of San Gabriel. Five years ago this position was occupied only by the one adobe house of a Spaniard, Garfias, who once owned the ranche. A company of eastern men, largely from Indiana, purchased the tract, with an abundant water privilege arising in the Arroyo Seco Canyon, and nearly every one of the sub-divided tracts of 7½, 15 or 30 acres each was taken within a year by actual settlers, and these, almost without

15

exception, eastern families of the highest class and of comfortable means. Young orange orchards, just commencing to bear, now form the principal feature of the town; its abundant mountain water is distributed to hydrants, bath-rooms and fountains in and about each house; the dry-bed of the Arroyo, on its western edge, furnishes abundant wood; the Sierra Madra or San Fernando range bounds and guards its northern side, and its site overlooks the whole San Gabriel Valley.

The *Lake Vineyard Ass'n* has more recently opened up a fine tract, bordering Pasadena on the east, and the two settlements, now blending into one, have some seventy houses, many of them very handsome, a Presbyterian and a Methodist church, two school houses, stores, shops and a daily mail. Not alone those who have their pretty homes and orange groves there think it the most desirable of all California's delightful spots, but unprejudiced travelers, who have seen the whole, acknowledge that here, indeed, as its Spanish name asserts, is the "*key of the valley*" and that valley the far-famed and Eden like San Gabriel.

In visiting the orange groves and old Mission Church of this locality it will more than pay to turn aside the two or three miles necessary in order to see Pasadena and Lake Vineyard.

To the right, before reaching the next station, several huge palm trees can be seen, like those shown on the foreground of our illustration, on page 10. They are the *fan* palm, great numbers of which are to be seen on our route hereafter.

Passing up through the little valley, 9.2 miles from Los Angeles, we come to

San Gabriel—The station is on a broad plateau gently sloping from the mountains on the left. Far to the right, away down on the San Gabriel River, embowered in all kinds of fruit trees, and surrounded by vineyards, is the old, Old San Gabriel Mission, founded Sept. 8, 1771. All the old missions in California—twenty-one in number—were founded by members of the Order of San Francisco, who were sent out by the college of San Fernando, in the City of Mexico, who were of the order of Franciscan Friars. The orange orchard at the Mission was the *first* planted, as the Mission was the *first* founded in California by the old Padres. Some of the trees are very large, and continue to bear the best of fruit. The "Wolfkill" orchard in Los Angeles is the next in age, and the second in size. To the north of this station, two miles distant, is situated the

LARGEST ORANGE ORCHARD IN CALIFORNIA—It is owned by L. J. Rose, Esq., and contains 500 acres. In this orchard are orange trees of all sizes, loaded with fruit the year round. Besides oranges, great numbers of lemon, lime, almond, English walnut, and many other varieties of fruits and nuts are raised here to the greatest perfection. Pomegranates, 5,000 in

No. 23 ANNEX. California—was first discovered in 1542, by a Portuguese, Juan R. Cabrillo, while in the Spanish service. It was held by the Spanish then by the Mexican Government, until 1848, when by treaty it became a portion of the United States. It was admitted as a State in 1850. It covers an area of 160,000 square miles, divided about equally into mining, agricultural, timber, and grazing lands. All kinds of grain, fruit, and vegetables grow in profusion. The grape culture has occupied the attention of many of her people, who find that they can produce wine surpassed by none in this country, and few in the old. Large quantities are used throughout the United States, with a yearly increased shipment to European markets. Her manufactures are of a high order, and attract favorable notice at home and abroad. The spirit of enterprise manifested by her citizens has deserved and won success. Under the liberal, farseeing policy of the younger class of capitalists and merchants, who appeared about the time of the inauguration of the great railroad, a new order of things arose. Men began to regard this land as their future home.

From this time, money expanded, trade, agriculture, mining and manufactures began to assume their proper stations, and a brighter era opened to the people of the Pacific slope.

The Coast Range—is the range of mountains nearest the Pacific Ocean, extending the whole length of the State, broken at intervals with numerous small rivers, and narrow, fertile valleys. The principal peaks are—Mt. Ballery, 6,357 feet high ; Pierce, 6,000; Hamilton, 4,450; Diablo, 3,876; Bauch, 3,790; Chonal, 3,530; St. Helena, 3,700; Tamalpais, 2,604 feet. Mount St. Bernardino, away to the southward, in the range of that name, is 8,370 feet in height.

The Rainy Season—on the Pacific coast is between the first of November and the first of May, the rain falling principally in the night, while the days are mostly clear and pleasant. At Christmas, the whole country is covered with green grass ; in January with a carpet of flowers ; and in April and May with ripening fields of grain. During 15 years of observation the average has been 220 clear, 85 cloudy, and 60 rainy days each year. The nights are cool the year round, requiring a coverlid.

number, are growing here, planted by Gen. Stoneman.

The town of San Gabriel is located about one and a half miles north of the station, and is completely embowered in foliage, among which are all the varieties of ornamental trees, fruit trees, vines, and flowers, grown on the Pacific Coast, the citizens seemingly having taken great pains, to procure some of every kind of tree and shrub, with which to beautify their otherwise beautiful town.

We have referred to the old, Old Mission, now we will refer to the *Old* Mission Church, which is located close on our left, just before reaching this station. It is in a dilapidated condition, but the bells are still hanging in plain view from the cars, which were wont to call the faithful to their devotions, long before the "blarsted Yankees" invaded the country.

The Sierra Madre Villa is a finely appointed hotel, situated about three miles from the station, away up on the foot-hills 1,800 feet above the level of the sea. It is in a most beautiful location, overlooking the whole valley of Los Angeles, Santa Monica and Wilmington, with thousands of acres in orange and fruit orchards, and in vineyards, in the foreground, and in the rear the towering mountains. From springs in these mountains the sparkling waters are conducted in pipes, and compelled to do duty in the fountains in front of the Villa, in every room in the house, and for irrigating 3,000 orange, lemon, and other fruit trees adjoining the hotel. This is a lovely place to sojourn—if not *forever*, certainly for a season. At this Villa is the best of accommodation for about 50 guests, at charges from $12 to $15 per week.

Close to the station, on the left, the tourist will find a variety of cactus not heretofore seen on this route. There are over two hundred varieties—so we are told—of these cactus plants. The ones at this station grow about ten feet high, and are of the *pud* species, *i.e.*, they grow, commencing at the ground, in a succession of great pads, from eight inches in width to fifteen inches in length, and from one to three inches in thickness. These pads are covered with sharp thorns, and grow one upon the other, connected by a tough stem, round, and about two inches in diameter. These cacti bear a kind of fruit of a pleasant flavor, which is used principally by the Indians or Spanish-Mexican residents.

From San Gabriel, we continue up the plateau, with the valley of San Gabriel River on the right, 2.5 miles to

Savanna—where are well-cultivated fields, groves and vineyards. Passing on 1.4 miles further is

Monte—This is a thriving town of several hundred families in the most productive portion of San Gabriel Valley. Here corn and hogs are the staples, and hog and hominy the diet. The settlers raise immense fields of corn, and feed great numbers of hogs for market—in fact, *this* is the most *hogish* section yet visited, but we suppose the Monte men would *bristle* up if they were told so.

Passing on, more to the southward, we soon cross San Gabriel River, which here has a broad, sandy bed. Sheep are raised in great numbers in this and the section of country traversed for the next 50 miles.

Puente—is the next station, 6.2 miles from Monte, where trains only stop on signal. It is situated on the east bank of San Jose Creek, beyond which and the west is the La Puente Hills. Most of the bottom land is fenced and cultivated, the settlers being mostly Spanish or Mexicans.

Coursing around to the left, up San Jose Creek, along which will be found many Mexican houses and herds of sheep, ten miles brings our train to

Spadra—elevation 706 feet. This is a small place of a score or more of dwellings, several stores, and one hotel, and is the home of an old Missouri gentleman, familiarly called Uncle Billy Rubottom, whose house is in a grove just opposite the station on the right, a few hundred yards from the depot. He has lived here near 30 years, and keeps "open house" for all his friends, in real old Southern style. He can often be seen at the depot mounted on his mustang, under a sombrero, something smaller than a circus tent, and as happy as a bevy of New England girls would be in a Los Angeles orange orchard.

Passing on up the creek, which is gradually dwindling, beyond which are a succession of buttes, or low, grass-covered hills, 3.5 miles brings us to

Pomona—This is a promising little town of about 600, with some good buildings. Garey avenue—the principal one—is planted on each side, with Monterey cypress and eucalyptus trees, and presents a beautiful appearance. Four artesian wells supply the town with water, and for irrigating purposes, these wells range from 26 to 65 feet in depth, and flow an immense amount of water, which is as pure as crystal. A reservoir holding 3,000,000 gallons is kept full, as a reserve at all times. Here, too, we find many orchards of orange, lemon, fig, and fruit trees. From Pomona it is 9.5 miles to the side-track and signal station of

Cucamonga — elevation, 952 feet. Two and a half miles north is the Cucamonga Ranche, celebrated for its wines. To the South, ten miles, is Rincon Settlement, a rich agricultural region, under a most complete system of irrigation, the water being supplied by the Santa Anna River, which carries a large volume of water at all seasons. A run of 15.2 miles through a section of country where are a few good ranches we come to

Colton—This place was named for the late vice-President of the Southern Pacific, and is a regular eating station for trains from the East and West. The town is not a very large one at present —about 200 persons will be the full number—yet it is quite a busy place, as it is the nearest station to San Bernardino, on the east, and Riverside, on the southwest. The Trans-Continental is the principal hotel, and a very good one. Colton has a newspaper— the *Semi-Tropic*, that makes its bow weekly.

MARBLE—To the west of the station, half-mile distant, a round butte rises from the prairie to the height of 500 feet, about 115 acres in area, in which has been discovered an immense body of what has been pronounced a very fine quality of marble, besides lime and cement in great abundance. A stock company has been formed, a rail track is to be laid to the mine, and the marble will soon be in the market; the demand for which, it is said, is already very great.

THE COJON PASS—(pronounced ko-hoon) through the San Bernardino Mountains, is due north from Colton, and we hear there are plans maturing to build a railroad through this "Pass" to Mojava, a distance of 70 miles. Should this ever be done, the distance from Colton to Mojava and the north will be shorter by 90 miles than the present line via Los Angeles. The grade is said to be easy, and the work of building, light.

Stages leave on arrival of trains, for San Bernardino, four miles east; fare, 50 cents; to Riverside, eight miles southwest, fare, 75 cents.

THE RIVERSIDE COLONY—is located on 8,000 acres of the best agricultural land in the State, most of which is under irrigating ditches, and is in a very thriving condition; in fact, it is the most prosperous, wealthy, and successful colony on the Pacific coast. Land that in 1868 was worth but a few dollars per acre, *now* would sell readily for from $100 to $150 per acre —*verily, this country is the poor man's paradise,*—and there are millions of acres *full as good,* now unoccupied, awaiting his advent.

San Bernardino—four miles east, and 61 miles east of Los Angeles, is the county seat of San Bernardino county, the largest in the State. It was settled by a colony of Mormons in 1847, and the town laid out in the same manner as Salt Lake City, with water running through all the principal streets from a never-failing supply obtained from numerous springs and creeks, in, and coming down from the San Bernardino Mountains on the east, close to the base of which, the town is located. All the Mormons now living here are "Josephites," Brigham, some years since, having called home to Salt Lake all who were devoted to him. The town contains a population of about 6,000, most of whom are engaged in fruit raising and agricultural pursuits. Fruit trees of all kinds, with vineyards, gardens, and groves, are the rule, and, altogether, it is a very beautiful town.

San Bernardino is on the old trail, through the Cajon Pass, to the mining regions of Nevada and Arizona, now of little use. The valley of San Bernardino contains 36,000 acres. Crops of all kinds grow in this valley. Much of the land produces two crops a year—barley for the first, and corn for the second; of the former, fifty bushels to the acre is the average yield, and of the latter, from fifty to sixty bushels. Of alfalfa, from five to six crops a year are grown.

Six miles north of San Bernandino are Waterman's Hot Springs. These springs are said to be almost a sure cure for the rheumatism; they are 700 feet above the valley, and 1,800 above sea level.

Near San Bernardino are the Mountains, —East, the most prominent peak of which is 8,750 feet above sea level.

Returning to Colton, another engine is attached to our train, and we proceed to

☞ No! I'll *never* tell! but ask Butler, freight agent at Colton, to show you his *white owl*—it's a great curiosity.

climb the San Gorgonio Pass; so we bid adieu to the orange groves, the beautiful fruit orchards, the luscious vineyards, and the glorious climate of California, as we shall see no more of those attractions on this trip. "Fare-thee-well, and if forever, still, forever fare-thee-well."

Leaving Colton, we cross Santa Anna River, and 3.4 miles from Colton come to MOUND CITY, a signal station, with an elevation of 1,055 feet. The road now runs up a narrow canyon with low hills on each side.

Sheep are the only things of life now noticeable. Eleven miles further comes

El Casco—another signal station, situated in a ravine extending to the Pass. Up this ravine the average grade is 80 feet to the mile; elevation, 1,874 feet. We are now in a section where large quantities of peaches are raised.

Continuing up the mountain 8.5 miles brings us to the Summit of the Pass, 2,592 feet, at

San Gorgonio—There are some good agricultural lands near, when irrigated, and a scheme is on foot to bring the water from the mountains to the northeast, twelve miles distant, for that purpose.

To the west, seven miles, is the great SAN JACINTO NUEVA RANCHO, containing 47,000 acres. This property is now being sub-divided into 10, and 20 acre farms, and sold on easy terms. We have visited the Rancho and are free to say the greater portion is a soil fully as rich and productive as any in the State, easily irrigated where necessary, from the San Jacinto River which runs through the property, or by wells; abundance of water being obtained within from five to twenty feet of the surface.

From this station it is down grade for 6.2 miles to BANNING, a signal station, and 5.7 miles more brings us to

Cabazon—(pronounced Cabb-a-zone), which means "Big Head," named for a tribe of Indians who live in this country; elevation, 1,779 feet. We are now in the Coahulian Valley. To the right are the San Jacinto Mountains, covered with timber. From Cabazon it is 8.5 miles to WHITE WATER, an unimportant signal station, 1,126 feet altitude, where we enter the cactus and desert country, and from which station it is 7.5 miles to

Seven Palms—elevation 584 feet. This station was named for seven large palm trees, situated about one mile north of the station. They are from 40 to 60 feet in height, with very large, spreading tops. The water at this station is the first and best on the west side of the desert, and in the days when emigrants traveled this route with teams, it was one of the points looked forward to with much pleasure.

From this station to Dos Palmas, a little over 50 miles, the palm trees are abundant.

Indio—is 20.8 miles from Seven Palms, with a *depression* of just twenty feet *below sea-level*. The palm trees along here are many of them 70 feet in height. When we commenced to descend below the sea level, three miles before reaching Indio, we left the sand-belt and entered a region more adapted for agricultural purposes, strange as it may seem. The cactus grows luxuriantly, and the mesquite shrub and palms cover the face of the land. From this point we descend lower and lower at every revolution of the wheels, down, *down under the sea!* Methinks we can see the huge ships sailing over our heads, and many of the leviathans of the deep, with an eye cast wistfully down upon us; then we think of Jonah, and wonder if we will come out as he did; then, along comes the freebooter, Mr. Shark, and appears to be taking our measure with a knowing wink of his left fin—he rises to the surface as though to get a fresh breath and a better start for a grand dive, looking as hungry as a New York landlord, as enterprising as a Chicago drummer, and as "cheeky" as some of the literary thieves who pirate information from our book, without giving credit.

In some points of the depression, where we first enter it, three miles north of Indio, fresh water can be obtained by sinking from twelve to sixteen feet. Here, vegetation is very luxuriant; mesquito, iron-wood, arrow-wood, grease-wood, sage and other woods and shrubs abound. Further to the south, from Walters to Flowing Wells, a distance of over 40 miles, the country is completely barren, in fact, is a "howling wilderness." Through this section, the water obtained by digging is very salt.

The beach surrounding this depression is 40 feet above high water; the lines are the same noticeable around any salt beach, the pebbles laying in rows, away around the different water-lines, as though left but yesterday by the receding waters. Marine and fresh water shells are numerous, indicating a fresh water lake here, subsequent to its being a part of the ocean.

Walters—is 13.3 miles from Indio,

where passenger trains meet and pass. At this point we are 135 feet *below* the level of the sea, and *still going down*. Ten miles further and we are 266 feet *below*; gradually we ascend, and at the next station, 17.4 miles from Walters, are at

Dos Palmas—only 253 feet below. From Dos Palmas, *desolation reigns supreme*, and 10.9 miles brings us to

Fink's Springs—Here we are seven feet lower than at Dos Palmas, being 260 feet below; a little further it will be 263 feet, when we commence to rise.

Five miles south, are twenty-five square miles of mud springs. The first is about 100 yards east of the road, and is cold. Then to the right, from one to six miles, are many springs, both hot and cold. Some are 200 feet in diameter, boiling up as though in a huge caldron, just on a level with the ground. Others are smaller, cone-shaped, rising in some cases 25 feet from the ground, a kind of miniature volcanoes. The mud in these springs is much the same consistency as ordinary mush, bubbling up as in a pot, over a slow fire. The smell, coupled with an occasional rumbling sound, reminds one of a region of which our modern teachers deny the existence.

The railroad track does not cross this depression in the lowest place, as an area west from Dos Palmas is twelve and a half feet lower. This has been called a Volcanic country. There are no signs that would indicate it ever to have been disturbed by volcanic eruptions, except the presence of the mud springs; on the contrary, most of the rocks surrounding this basin for fifty miles are granite, which is unusual in a volcanic section of country. What few rocks there are here, that are not granite, show no appearance of volcanic matter. Spurs of San Bernardino Mountains have been on our left, up to this point, after which they dwindle to small, isolated sand hills, here and there.

Flowing Well—is the next station, 17.7 miles from Fink's Springs. We have risen, so that we are now only 45 feet *below* sea level. At this station the Railroad Co. sank an artesian well 160 feet deep, and got an abundance of water, through a six-inch pipe, but it was too salt for use.

Six miles further we pass

Tortuga—a signal station, 183 feet altitude, and 6 miles further come to

Mammoth Tank—so named from a natural water tank in the granite rocks on the left, five miles distant, which holds 10,000 gallons, filled by rains, and nearly always has water in it.

It is said there are several hundred varieties of cactus on this desert, and we are ready to admit the statement without hunting further proof than what can be seen from the car window. They are *here*, of all sizes, shape and form. Eleven miles further, we come to another signal station called

Mesquite—so named because there is no mesquite near or in the immediate vicinity. Next comes—13.8 miles—

Cactus—elevation, 396 feet, named for a variety of cactus called "occtilla," which grows in great numbers, near.

To the east, from this station, can be seen Chimney Peak—a conglomerate rock—a huge cone, 160 feet in diameter, which rises from the summit of some low hills, 700 feet in height, beyond which, 40 miles away, can be seen the Castle Dome Mountains. They are on the east side of the Colorado River, from the summit of which rises Castle Dome, a granite column, 500 feet above the mountain range, which presents the appearance of a monster, square, flat-roofed building, but which in reality, is a long, narrow column, when viewed from a point to the southward of the Dome.

Mesquite, sage, and grease-wood shrubs are now to be seen on all sides. Directly ahead is a tall, round butte, called Pilot Knob, on the east side of which are located some lead mines. This butte is just seven miles north of the Mexican boundary line. Passing on 13.6 miles, we come to a signal station, called

Pilot Knob—From here, our course changes a little more to the eastward, and we soon come in view of the Colorado River, with a wide, sandy bottom covered with willows and mesquite. From Pilot Knob it is 9.4 miles to Yuma, about five of which brings to us the first view of the river, and the next four to the west end of the bridge. To the left, before crossing the bridge, is Fort Yuma, a Government post, occupied by about one dozen "boys in blue." It is on a high butte, overlooking the surrounding country. To the *right*, on the opposite side of the river, on a high bluff, is located the Quartermaster's Department. Crossing the bridge, which has a draw for river boats, and through a deep cut, we are in Arizona, and at

Yuma City—This is unlike any city we have heretofore visited. It contains a population of about 1,500, one-fifth of whom are Americans, the balance Spanish, Mexicans, and natives—Indians. The buildings are all one story high, made of sod, adobe, or sun-dried brick, the walls being from two to four feet thick, with flat roofs. The roofs are made by a layer of poles, covered with willows, sometimes a covering of cloth, or rawhide beneath them, and then covered with dirt to a thickness of from one to two feet. On all sides of these houses verandas project from ten to twenty feet, built of poles, like the roof, some with dirt, others with only the brush. These verandas are built for protection against the powerful rays of the sun. In summer the heat is intense; often the mercury marks 126, and once, some years ago, we learn from a reliable authority, it was 130 degrees in the shade. As might be supposed, snow and frost are unknown in Yuma. In summer, the American, Spanish and Mexican residents wear as little clothing as possible, while the native Indians' covering, will not exceed the size of a small pocket handkerchief, adjusted in the mother Eve fashion, with sometimes a long trailing strip of red material dangling from the rear belt, *a la* monkey.

In the hot weather, which is intense for about eight months in the year, the people sleep on the roofs of the houses, covered by the drapery furnished by nature—darkness.

Yuma, with all its varieties of citizens, is a very orderly city. The great majority of the people are Roman Catholics, that denomination having the only church building in the city. There are a few stores, with quite extensive stocks of goods. The hotels are not very extensive, such only in name; the Palace and Colorado are the two principal ones. Yuma has one weekly newspaper—the *Sentinel*.

Most of the Spanish and Mexican houses are surrounded with high fences, made of poles, set in the ground close together, to a depth of three or more feet, and secured together about four feet from the ground, with narrow strips of rawhide interwoven, when soft, around and between the poles, so when the hide dries the fence is very strong. Many of these fences present a very ragged appearance, as the poles range in height from four to twelve feet above the ground. The more enterprising of thet people saw these poles off to a uniform height, when they present a much more artistic and finished appearance.

The Railroad Company have large warehouses here built of lumber, for the accommodation of both the railroad and steamer business. The boats on the Colorado River are all owned by the Railroad Company, and are run in connection with the trains.

Just above the railroad bridge, on the west bank of the Colorado River, is situated Fort Yuma. It is located on the top of a bold, round butte about one-fourth of a mile in diameter, rising about 200 feet above the river bottom, and projecting into the Colorado River to meet a promontory of about the same height on the east side. Between these bold points flows the Colorado River, about 300 yards in width. The Colorado River reaches this point from the northward, and the Gila (pronounced Hee-le) from the east, forming a junction close above the points named. It is proposed by those managing the interests (so we hear) of the Texas & Pacific railroad, to build a bridge across the Colorado River at these bluffs, some work of grading having been done in the fall of 1877, just previous to the locating of the present railroad bridge, a few hundred yards below.

From the high butte above named, a view can be had of Yuma, the valleys of the Colorado and the Gila rivers, the mesas, and the surrounding country for many miles.

COLORADO RIVER STEAMERS.

Passenger and freight steamers leave Yuma for Aubry, during the summer season, weekly, commencing the first Saturday in May and continuing until the last of October, from that time until January following, they will leave every alternate Saturday. Steamers for Camp Mohava leave every fifth Wednesday, commencing about the middle of January. These steamers run to El Dorado Canyon, from May 1st to the last of October (stage of water permitting).

Distance from Yuma, per river steamer, to Castle Dome, 35 miles, fare, $5.00; Ehrenberg, 125 miles, fare, $15.00; Aubry, 220 miles, fare, $28.00; Camp Mohava, 300 miles, fare, $35.00; Hardyville, 312 miles, fare, $35.00; El Dorado Canyon, 365 miles, fare, $45.00.

The Colorado river is the largest in Arizona. Its principal tributaries are the Grand River, which rises in the Middle Park of Colorado, and the Green River, which rises in the eastern portion of Idaho. From the junction of the Grand and Green rivers, the stream is called the C l orado, and with its windings has a length of 3,000 miles to where it enters the Gulf of California. It is navigable at all times about 500 miles, and in a season of high water about 150 miles further to Callville. The time is not far distant when a trip to the Grand Cañon of the Colorado will be one of the most attractive and popular in America—if not in the world. Along this cañon for nearly 300 miles the channel of the river has been cut through the mountain walls that rise up on each side from 1,000 to 3,500 feet, forming the longest, highest and grandest cañon the eye of man ever beheld.

Stages leave Yuma daily, carrying passengers, mail and express for Castle Dome, 30 miles; Horse Tanks, 58 miles; Tyson's Wells, 93 miles; with branch line to Ehrenberg, 28 miles further; to Wickenburg, 128 miles, and Prescott, 193 miles; average fare, 16 cents per mile.

Leaving Yuma, our course is due east, with the Gila River on the left, or north side. The river bottom is from two to five miles in width and covered with white sage, greasewood, mesquite shrubs, willows, small cottonwoods and some ironwood. The soil is a mixture of loam, sand and clay, with alkali beds in places. Very little of the land is cultivated, yet there are a few Mexican or Spanish settlers, who "tickle the ground" a little within the first ten miles after leaving Yuma. Their irrigating ditches are crossed in a number of places, and we are told the vegetables and early wheat raised are very good. On the north side of the river, five miles away, a Spanish settler has a large ranche, which is quite productive.

About ten miles east of Yuma, the bluffs on each side close in on the river, and our road is built through a succession of rocky points or spurs which extend to the river bank. To the right or south side our view is wholly obstructed; but to the northward, beyond the river, the country is very much broken with cañons and ravines coming down from the high rocky bluffs which overtop each other in the distance, some of which must reach an altitude of 1,500 ft. above the valley.

A few miles through rock cuttings and our train will reach the river bank and afford us a view of Los Flores, a small mining camp on the north side of the river, the "drifts" showing plainly. A two-stamp mill is the extent of the machinery used.

Gila City—is 15.7 miles east of Yuma, inhabited principally by Papago Indians, with a small sprinkling of whites, most of whom are engaged in "dry washing" for gold in the cañons and ravines south of the station. The gold is fine and not very abundant.

Leaving the station, within a few miles we will see the *first* of a kind of cactus peculiar to Arizona. It is certainly the "Boss" cactus of the world. (See Annex No. 55 and page 235.)

Leaving Gila City, the country is more open, the river bottom is several miles broad, and covered with small cottonwoods, willows, and underbrush; much of this land would produce crops with irrigation, but the river could not be depended upon to supply the water at the time it would be required.

By looking away to the southward, the first glimpse is obtained of a peculiar sharp needle-pointed rocky butte, which in general formation is found in our travels only on the Gila Desert, where they are very numerous. These buttes are of volcanic formation, completely isolated, many of

No. 38 Annex. Cape Horn—is a bold promontory, situated on the north side of the Columbia River, in Washington Territory, about midway between the Cascade Mountains and the Dalles. This promontory is of basaltic formation —like most others on the Columbia—and rises near 250 ft. perpendicular from the water's edge, and extends about one mile in length, the lower part projecting several hundred feet out into the river. Cape Horn derives its name from the danger in passing it. Our large illustration, No. 10, represents a small party of pleasure and curiosity seekers on a pleasant afternoon, when the winds had lulled, who have successfully rounded the cape.

No. 39 Annex. Wood Hauling in Nevada—No. 11, of the large views, is a beautiful engraving, representing a ten-mule team loaded with wood. The three wagons are coupled together like a train of cars—called "trail wagons" on which are loaded twenty-four cords of wood, At the point represented in the picture, the team is about on the dividing line between Gold Hill, down the canyon to the rear of the wagons, one-fourth mile—and Virginia City, directly ahead, about the same distance around the point of the mountain. This plan of coupling wagons is quite common on the Pacific Coast for all kinds of heavy hauling. The picture was engraved by Mr. Bross, of New York, from a photograph.

INDIANS WATCHING THE "FIRE WAGONS."—SEE ANNEX NO. 49.

which rise abruptly from the plain to an altitude of 2,000 feet. In color, they vary from dark brown to black, and in general appearance resembling iron slag. Some of these buttes take the form of narrow "hog-back" ranges, very sharp, and very steep, extending several miles. The view between the buttes or ridges are on a level with the plain and extend as far as the eye can reach; where they overlap each other the appearance is like one continuous range.

From Gila City, it is 14 miles to **Adonde**—a side track station, with one building, several tents and a big water tank.

The railroad company have to haul all the water they use, on the first 150 miles of their road east of Yuma, in water-cars, from either Adonde or the the Colorado River at Yuma. The water from the Colorado is preferred over that from Adonde, as the latter is strongly impregnated with alkali.

Leaving Adonde we leave the Gila River far to the left, and will soon realize that we are fairly out upon a vast expanse of desert, inhabited solely by rattlesnakes, lizards and owls, with an occasional woodpecker. Sage knolls, ironwood, mesquite, greasewood, clay, and sand—the latter very heavy—is now the rule, with an occasional bunch of white calette grass. The surrounding peaks are now prominent in all directions, on both sides of the river; many on the north side are castellated and of a peculiar sombre appearance.

Passing several buttes close on the left,—between our train and the river —the largest of which is known as Antelope Peak, and along over a sandy waste, we approach Mohawk Summit, 26 miles from Adonde, but there is no station, no signs of life. This summit is simply a low pass in one of those long, rocky, narrow ridges which here runs north and south, across our path. Just before reaching the summit our road is bridged over a dry sandy depression, which apparently, was once the bed of a broad stream of water. Along the banks are many trees, among which we notice the Paloverde, with its smooth, bright yellow bark, otherwise much resembling the madrone tree heretofore described. Ironwood is also to be seen as well as the "boss" cactus, in great numbers. For description, see Annex, No. 55, and illustration opposite page.

To the east of the summit, the evidences to prove that this country was once lighted by volcanic fires, are abundant. The whole surface of the country is covered or underlaid with lava. It crops out in every ravine, and at every cutting. Where the lava is exposed to the air, it is soft, and readily broken in pieces in the hands. By the action of the wind and rain much of the surface lava has become reduced to dust which covers the ground; disagreeable at all times, but when wafted by a Gila zephyr is terribly annoying.

Texas Hill—is 7.6 miles east of the summit—a side track, and section house now comprises the station. Continuing eastward, the general appearance of the country is unchanged, except as to its volcanic evidences which are more noticeable.

Stanwix—is an unimportant side-track 22.7 miles east of Texas Hill. The Gila River is here about 10 miles to the northward, the bottom lands of which, as we ascend the river are improving, and with irrigation, raise good crops of wheat and vegetables.

Sentinel—is another side track 4.6 miles from the last, but it is a lone Sentinel, opposite the place on the river where the Oatman family were murdered by the Tonto Indians in 1851. A run of 13.9 miles brings us to

Painted Rock—so named for the noted land mark on the north side of the river. Called by the natives "Pedras Pintados." (See Annex, No. 48.)

Gila Bend—is 13.9 miles from Painted Rock—and derives its name from its location near the great bend of the Gila River, and from an old stage station of the same name, a few miles to the northward. The appearance of the country bordering the line of Railroad—since crossing the San Barnardino Mountains—up to this station, in an agricultural point of view —particularly, to an east-of-the Missouri River farmer—is not very encouraging. Yet, *with irrigation*, there are millions of acres of productive lands. At this "bend" of the Gila River, we strike the edge of one of the richest and finest bodies of land in Arizona—but it must be irrigated—and the Gila affords abundance of water for that purpose. With a proper system of canals and wind-mills, oranges, lemons, vineyards, nuts, and all kinds of tropical and semi-tropical cereals, can be raised in abundance;—and, within our knowledge—we know of no section of the trans-Missouri country where a more promising opportunity for the investment of capital in a safe, legitimate, and growing business, than is here indicated. Wood is a scarce article in many parts of Arizona—but is plentiful about Gila Bend and along the river bottoms,—which, in a country devoid of coal, is an item of no small consequence.

Continuing eastward, the side track of ESTERELLA is 18.8 miles, and 18.2 miles more to

Maricopa—where the first through train from San Francisco arrived May 12th, 1879. This town of Maricopa, located as it is in the center of great mineral wealth, the distributing point for a vast region of country—north and south of it—is destined at an early day, to become one of prominence. It now contains several large mercantile houses, hotels, restaurants, etc. The Railroad Co. have a good depot, and a large freight building for the accommodation of the great amount of merchandise arriving here for distant points—mostly to the northward,—Phoenix, Vulture, Wickenburg, Prescott. etc. Ores and bullion are also received here as return freight, for shipment to San Francisco and the east

Between Gila City and Maricopa there are few buildings, except those used by the Railroad Company. The "section houses" are all alike, built of lumber with double, or sun roofs. The upper roof is supported by upright timbers and is elevated about two feet above the lower roof, over which it extends, on all sides, about four feet. The space between the roofs allows the air to circulate freely, and to a great extent protects the occupants of the buildings against the powerful heat of the sun, which often, in the

CACTI GIGANTI.—THE "BOSS" CACTUS OF ARIZONA.—See Annex No. 55.

summer, marks 115 to 130 degrees on these plains.

The old stage station of Maricopa Wells is situated about ten miles to the northward, on the Gila River, and not far distant is the Gila Indian Reservation, where live the Pima and Maricopa Indians, numbering 4,328. This reservation contains 70,000 acres of as rich and productive lands as there is in the Territory, much of which is cultivated by the Indians, who are self sustaining.

For interesting historical matters regarding Arizona, see Annex, No. 64.

Distances: Maricopa to Yuma, 156 miles; Tucson, 91 miles; San Francisco, 887 miles; El Paso, 399 miles; Phoenix, 35 miles, Vulture, 90 miles; Wickenberg, 90 miles; Prescott, 152 miles. Stages leave daily with passengers, mails and express for Phoenix, Prescott and intermediate places—fare, about seventeen cents per mile. The general direction of our road from Maricopa changes from the east to the south-east for the next 140 miles, when it again turns to the eastward.

From Maricopa it is 14.9 miles to a side track called SWEET WATER, and 11.1 miles further to

Casa Grande—this like all railway stations, when they are at the "end of the track," was a very busy place.—Temporary wooden buildings, canvass tents, and shanties of all kinds, and for all purposes, were scattered in all directions; immense quantities of railroad material of every description covers many acres of land; ponderous "prairie schooners" were loading merchandise for distant points while others were unloading ores and bullion; stage coaches with passengers, mails and express were leaving and arriving loaded to their utmost; and people of every nationality, color, dress and occupation, were to be seen on every side intent on some kind of business. Such was Casa Grande January 1st, 1880. But when the road was extended it settled down as a shipping point for the mining region to the northward — and only such buildings remain as are necessary for that business.

This station is named for the old ruin of Casa Grande, situated about 14 miles to the northward. (See Annex, No. 47.)

The general features of the country along the road for the last fifty miles, in an agricultural point of view, is much improved; sage, grease wood, and mesquite trees, together with grasses of various kinds, cover the face of the land; while herds of cattle, sheep and horses are not uncommon.

Stages leave Casa Grande daily for Florence, 25; and Silver King, 57 miles; fare, about seventeen cents per mile.

Toltec—is the next station "down on the bills," 9.6 miles from Casa Grande, and 9.1 miles from

Picacho—a small station from which large quantities of coke, and merchandise is shipped on wagons for the mines, to the north- and eastward. We are now following up the lower portion of the Santa Cruz Valley, along which there is no running water; but, judging from the rank growth of sage, mesquite, and greasewood, which cover the land, it would not be a very difficult task to sink wells and find water sufficient for irrigating purposes. After a few miles run from Picacho station, we arrive opposite "Picacho Peak," a noted land-mark, and rocky butte on the right. It was here, at the base of this "peak" in May, 1862, where the first and only battle was fought in Arizona between the Confederate and Union forces. In the summer of 1861, the Union troops were withdrawn from this Territory, and on the 27th of February, following, Cap. Hunter of the Confederate forces arrived at Tucson, from Texas, and took possession; soon after the news reached San Francisco that the Confederates had control of Tucson, Genl. Carlton, of the Federals — California column — started for this Territory, and was met by the Confederates at this "peak" as above stated. The battle resulted in a victory for Genl. Carlton and the abandonment of the country by the Confederates.

Red Rock—a side track—is 13.9 miles from Picacho, and 15.5 miles from RILLITO, another small station, on a little Creek of that name, 17.1 miles from

Tucson — pronounced Tu-son. — Had we visited this place 322 years ago, we would have been classed with the "Old Pioneers," instead of a "tenderfoot" of 1882.

Records show that Tucson is the second oldest town in the United States; Santa Fe, New Mexico, being the first. The first settlements were made by the Spaniards in 1560, and a presidio or fortification was constructed to protect their settlement at San Xavier; and from the appearance of many of the old adobe buildings, and the aged look of some of the citizens, we are not disposed to dispute the records, or doubt the fact that a few, at least, of the earlier settlers are still living.

Tucson is the county seat of Pima county, situated on a *mesa* or table land, gradually sloping to the westward—overlooking the Santa Cruz Valley—in lat. 32 deg. 20 min. north and long. 110 deg. 55 min. west of Greenwich. Elevation 2,239 feet. It is 978 miles from San Francisco; 220 miles from Deming; 308 miles from El Paso, Tex.; 75 miles north of the Mexican boundary; and 370 miles from Guaymas, Mexico.

Sorin, in his sketch of Tucson says: "The Santa Cruz River is one of those erratic streams, common in this Western Country, which run for a distance on the surface, then beneath the ground, again on top, and so on. In its strange course it so happens, that the river comes to the surface about two miles south of Tucson and runs past the mesa on which the town is built, and thus makes some three thousand or more acres of land capable of irrigation and consequently of cultivation. In this rich bottom years ago the old mission church of Tucson was built by the Jesuits, and to protect the cultivators of the adjoining fields a presidio or military camp was established; and for self-protection incoming settlers congregated about the garrison and thus the town grew upon its present site."

The City of Tucson was incorporated February 7th, 1877, and the Southern Pacific Railroad was completed to it, March 10th, 1880. Its present population is estimated between 8,000 and 9,000; composed of Spanish, Mexican, Indian, American, and English speaking people. The streets are regularly laid out, are narrow with the usual Mexican Plaza. In the older portion of the city the buildings are constructed of adobe, one story, in the old Spanish-Mexican style (where one goes out of doors to get into each room) with an occasional one of wood, sandwiched in here and there, and occupied by the most enterprising business men,—or more recent arrivals—those who come with the Railroad.

The business portion of Tucson, is about half a mile west of the depot, between which, and the depot are some fine private residences of wood, one large hotel—Porter's—commodious depot and freight buildings, and many other modern structures in course of erection. The Railroad Co. have a round-house and quite extensive machine and repair shops located here.

The city supports three daily newspapers, the *Citizen*, *Star*, and the *Journal*, besides several weeklies.

Gas, Water, and Street Railroad Companies have been chartered and the present prospects are, that the citizens of Tucson will soon be able to enjoy all those luxuries. There are quite a number of hotels, principal of which are Porter's at the depot, and the Palace, at the old town. There are two banks; three flouring mills; two breweries; two ice manufactories; one foundry and machine shop; six churches and church organizations; four schools—public and private; eight wholesale dry goods houses; sixty-six dry goods and grocery stores and the usual number of shops of all kinds found in a city of the size. As a law-and-order-city, Tucson has few equals. The carrying of weapons and drunkenness is *severely* punished by fine and imprisonment.

The United States Depository for the District of Arizona and the United States Custom House, and the Deputy Collector of Internal Revenue, as well as the Surveyor Gen'l Office of Arizona, is located here.

There are about 3000 acres of land in the vicinity of Tucson susceptible of irrigation; but it is all taken up and title can only be had by purchase from private individuals. The valley of Santa Cruz, in which most of the land referred to is located, is very rich, and with irrigation, capable of producing two crops annually—corn in the spring and wheat in the fall.

Game is not abundant in the vicinity of Tucson, but bear, deer, antelope and wild turkeys can be found in the foot-hills and mountains.

The road south from Tucson, along up the Santa Cruz Valley, has been for near 300 years the great highway between Mexico and Arizona, leading directly to the harbor of Guaymas. We understand a plan is now maturing by capitalists, to parallel this old road with iron rails and the time is not far distant, in the nature of things, when this route will be traversed by the "Iron Horse."

Resorts,—in and around the city:—
SILVER LAKE, is southwest of the city, half a mile distant; is caused by a dam in the Santa Cruz River, and extends over several acres; a race-track is adjacent. Boats, bath-houses, swimming baths, groves, pavilions, hotels, etc., are provided for the accommodation of visitors.

LEVIN'S PARK—situated on the west side and near the heart of the city, in a grove of cottonwoods, seven acres in extent, in which are located a theatre, music pavilion, billiards, bowling, bar, baths, brewery, restaurant, shooting gallery, etc., and is patronized, at times, by all classes.

SAN XAVIER DEL BAC—is an old mission—nine miles south of the city, in Santa Cruz Valley, over 100 years old, erected by the Jesuits, for the purpose of saving the souls of the Papago Indians. Travelers visiting Tucson usually take a run down to this old mission—where, strange as it may seem—the Mexicans are wont to congregate at certain seasons of the year, to witness bull-fights that take place in the vicinity.

AQUA CALIENTA—Mineral warm springs—are situated 14 miles east of the city at the foot of the Mountains, and are said to possess medical qualities. The water is 88 degrees Fahrenheit, and contains soda, magnesia, iron and sulphur. Cottages and ample hotel accommodations are provided for the public.

CAMP LOWELL—Military headquarters for the Arizona—is seven miles east from the city, and is much visited by the citizens of Tucson.

The mountain system as viewed from Tucson is quite extensive. To the east, and north-east, is the jagged mountain range of Santa Catarina, rising from the plain, about twelve miles from the city, to the height of near 2,000 feet. Turning to the south, the Santa Ritas, boldly appear in a succession of peaks, the highest, Mt. Wrightson, over 10,000 feet above the plain, from twenty-five to fifty miles distant; while more to westward, can be seen the Atacoso Mountains, at the base of which is located the old town of Tubac, and the old mission of Tumacacori. Returning to the immediate vicinity of the city, the Sierra Del Tucson—close the view to the westward, rising from just across the valley, completing one of the most beautiful and interesting landscapes of mountain and plain; which with the wonderous hues of Arizona's gorgeous sunsets, completes a picture that none but the hand of the Great Maker can produce.

Stages leave Tucson as follows: Arivaca, 65 miles, and Oro Blanco, 77 miles—three times a week—Monday, Wednesday and Saturday. Tubac, 60 miles, and Calabasas, 67 miles, twice a week—Tues. and Sat. Silver Hill, 46 miles, and Silver Bell, 55 miles, twice a week—Mon. and Thurs. Old Hat District, 45 miles, three times a week, Mon., Wed. and Fri. Fort Lowell, 9 miles, and San Xavier, 7 miles—daily. Magdalena, 130 miles, Hermosillo, 275 miles, and Guaymas, 370 miles, twice a week—Tues. and Sat. Altar, 150 miles, and Guaymas via Altar, 420 miles, twice a week—Mon. and Wed. Fare, from six to twenty cents per mile, varying with competition.

The "life of trade" at Tucson, is derived from the mining industry. It is the great outfitting point for nearly every mining district in the territory, also, for many of the mines and camps in Sonora. There are 29 mining districts within a radius of 100 miles from Tucson; the greater number of which purchase all their supplies in that city. Some of the mines are exceedingly rich in gold, silver, lead and copper, and the rapid increase of precious metals is most wonderful. Wells, Fargo and Co's report of the yield for 1880, was $4,472,471; for 1881, $8,198,766, an increase, in one year, of

THE SANTA RITAS—NEAR MT. WRIGHTSON.

$3,726,295. Arizona is not only rich in precious metals, with a mild and healthful climate, but is sufficiently dry and warm to convince the most skeptical in the authenticity of certain old bible versions which shall be nameless in this connection. Suffice it to say, below we give the minimum and maximum of Rainfall and Temperature, as recorded at the following Government Forts and Camps in Arizona for a term of years:

NAME.	RAINFALL.	TEMPERAT.
Fort Yuma, (Yuma City)	3.84 inch.	35 to 112 deg.
Fort Prescott, (Prescott)	27.09 "	10 to 91 "
Fort Bowie	14.60 "	21 to 103 "
Camp Lowell, (Tucson)	10.83 "	19 to 113 "
Camp Grant	22.54 "	16 to 109 "
Camp Apache	13.21 "	6 to 104 "
Camp McDowell	14.09 "	18 to 114 "
Camp Mojava	13.40 "	27 to 118 "
Camp Verde	14.20 "	5 to 113 "
Average	14.07 inch.	17 to 100 deg.

But the hardy miner and prospector does not seem to give the weather a passing thought. We meet him everywhere, going right along with his pockets full of "prospects," selling his claims; buying his "grub;" punching his "burro," and taking a "smile" regardless of the weather or anything else.

For Arizona items of interest, see Annex No. 64.

Leaving Tucson, our course is south-east, over a broad plain covered with sage, mesquite, and greasewood, 14.6 miles to PAPAGO, a small sidetrack station, from which we run up Rillito Creek 13.5 miles to

Pantano—a small station of half a dozen buildings, and one store, besides good depot and freight buildings. This is the nearest shipping point on the railroad for several important mining districts, towns, and camps. Chief of which are: Total Wreck, 4 miles; Harshaw, 50 miles; Patagonia, 60 miles; and Washington, 64 miles. Daily stages run to all these places; fare, from 10 to 15 cents per mile.

Since leaving Tucson, we have been climbing the world, and at Pantano are 1,297 feet higher, or 3,536 feet elevation.

Mescal—is the next station, 9.3 miles from Pantano, and 8.6 miles from

Benson—At present this is a lively place. It is situated in San Pedro Valley; elevation, 3,578 feet; and is the shipping point for the celebrated Tombstone Mining District and many thrifty mining towns to the southward; several large stores and forwarding houses are located here; a hotel, several small shops, a large depot, and extensive freight warehouses together with an immense amount of railroad construction materials; as this is the initial point from which the Atchison, Topeka & Santa Fe Railroad Co. are building a railroad southward, some people say, to Guaymas, on the Gulf of California; but, we could procure no definite information. Certain it is, the road is completed to Contention, 18 miles, and still going forward; yet, at the time of our visit, Jan. 14th, 1882, no passenger trains were running; stages were leaving Benson daily for Contention. 18 miles, and Tombstone, 30 miles; fare, $2.00 and $3.00 respectively. Freight for the Mexican state of Sonora is forwarded from Benson, in immense quantities, the passenger travel is also an important item.

San Pedro Valley is one of the richest stock raising portions of Arizona, grass being abundant, and water sufficient for that purpose. The lands are mostly owned by the Spanish-Mexican settlers, who are "like the dog in the manger," opposed to new comers, cultivating only small patches of ground and raising only what they need for their own subsistance.

The Tombstone Mining District, has attracted more attention than any other in the territory. The principal mines of this district, lie about eight miles east of the San Pedro River, in a low cluster of hills, called the Tombstone Mountains.

Sorin says: The region of country embraced in the Tombstone District, has long been known to contain mineral. The first discovery of silver in this locality was at the "Old Bronco Mine," six miles southwest of Tombstone town. The exact date of the first location is not known, but the old Bronco mine has been worked in years gone by, and produced some good ore. There is a dark history connected with this mine, and it is said no less than sixteen men have been killed or murdered there. The discovery of the new mines was made in February, 1878, and the extraordinary richness was soon noised abroad, and prospectors from all parts of the country flocked in and many hundred claims were recorded. There are four towns in the Tombstone District, Tombstone, Richmond, Charleston, and Contention. Tombstone, the principal town, is near the Tough Nut group of mines, and is already a thriving city of several thousand people. Richmond, about one and a half miles south of Tombstone, has a number of business houses. Charleston, on the San Pedro River, where the Tombstone and Corbin mills are located, is quite a thriving village of from 500 to 600 population. Contention City, is also on the San Pedro, nine miles below, at the Contention Mill, is an important place, connected with Benson by railroad, and is growing rapidly. The principal ore producing mines in the district are: The Tough Nut group; the Lucky Cuss mine and group; Contention, Grand Central, Empire, Sunset, Emerald, and many others that prospect rich. Leaving Benson our direction changes to the northeast, and we commence to climb the Dragoon Mountains; passing O-CHOA, a side-track in 9.7 miles, from which it is 9.4 miles more to

Dragoon Summit—altitude, 4,614 feet. This point is a natural pass, apparently designed by nature for a railroad, between the Dragoon Mountains, on the south, and the Limestone Mountains on the north; the grade is easy and the work of grading was light. Reports, locate recent discoveries of rich minerals in the mountains near this station.

Cachise—is ten miles east of the summit, named for a noted Indian chief, who for twelve years was the head devil of the Apache Indians, and made his headquarters in the mountains near. He believed that he and his tribe had suffered great wrongs, and most fearfully did he revenge them. He has been dead but a few years, and the remnant of his tribe are now eating at "Uncle Sam's" table on the San Carlos reservation.

Descending into Sulphur Spring Valley, 10.8 miles from Cachise, we reach

Willcox—a thriving town of about 250 population, situated in Sulphur Spring Valley; is the centre of trade for quite an extensive stock-raising and mining region. Altitude, 4,164 feet. The Dos Cabezas peaks, where some rich mines of gold and silver are being developed, are twelve miles southeast from this station. Camp

STATE CAPITOL OF CAL.

Bowie, 20 miles.

The valley, in which Willcox is situated, extends north and south about 50 miles each way, and lies between the mountain ranges of Sierra Bonita and Chiricahua, on the east, and the Galinro and Dragoon on the west. The lower portion is called Sulphur Spring Valley, and the upper, Arivaypa Valley. There is no stream of note in these valleys, but along the base of the ranges of mountains and in the foot-hills are many fine springs and some brooks. The grass in and around these valleys is very rich and abundant; and it is recognised by stock men as one of the best stock ranges in the Territory. At several points in the valley sulphur springs have been discovered, and at one place deposits of salt cover several square miles. At Willcox, and in fact throughout the valley, an abundance of good water can be obtained by digging wells from ten to fifteen feet in depth.

Stages leave here, every other day, for Fort Grant, 24 miles; Camp Thomas, 64 miles; San Carlos, 99 miles, and Globe, 132 miles. Fares, about 15 cents per mile.

From Willcox to RAILROAD PASS, 8.3 miles, we ascend 230 feet, reaching an elevation of 4,394 feet, the highest point reached by the Southern Pacific Railroad on its whole line. From this "pass," we descend 635 feet in the next 15.4 miles and arrive at

Bowie—situated in the San Simon Valley, and at this time, prospects to soon become a place of much importance. It is a regular dining station; at the Campbell house, in front of which all through passenger trains stop, the accommodations for guests are first-class, and the meals served the *best* on the road. Water, for use at the station, is obtained from a well 300 feet in depth, but in many places in the valley it can be obtained from 25 to 75 feet. Thus, it will be readily understood that the San Simon Valley is not adapted to agriculture, and to only a limited extent for stock raising, wholly on account of the scarcity of water, as the soil is rich, and the rainfall at certain seasons, just sufficient to cover the whole face of the land with a coating of nutritious grasses. This valley opens in New Mexico and extends in a north-western direction for near 100 miles to a junction with the Gila, affording a natural road-way from this station to the valleys and mining region in the northern part of the Territory.

We understand a railroad is projected down the San Simon Valley, with the coal fields of the San Carlos Indian Reservation, as an objective point 100 miles distant, and that a stage line is soon to be put on this route which will reach Camp Grant in 28 miles; Camp Thomas, 75 miles; and Globe in 135 miles; already a large amount of freight is forwarded from Bowie for the towns, Gov. camps, and mines of this region.

Fort Bowie, is 15 miles south, but we understand, it will be moved to near this station, at an early day.

The Bowie Milling and Mining Co., who own 70 gold and silver claims, ranging from four to fifteen miles south, are about erecting at this station a 40 stamp mill to be run by electricity.

From Bowie it is 15.7 miles to the small station of

San Simon—from which a stage runs daily to Gayleyville, 22 miles. Fare, $4.00.

The territorial line is crossed 10.9 miles east from San Simon Station and 3.8 miles further we are at

Steins Pass—altitude, 4,351. It is reported, there are some good mineral prospects near. From this station eastward to the Rio Grande River there are few objects of interest to the traveler. The face of the land is covered with a rich growth of grass, but devoid of water, except an occasional little lake or sink strongly impregnated with alkali.

Pyramid — a small station is reached in 15.1 miles, from which it is 4.4 miles to

Lordsburg—This is the shipping point for Clifton, a celebrated copper mining town, 80 miles to the northwest. Where are located large smelting furnaces, turning out daily over a carload of bullion. Hydraulic works are also being erected to work placer claims in the vicinity. Stages leave Lordsburg Mondays and Thursdays for Clifton. Fare, $10.

South from Lordsburg, two miles, is the little mining camp of Shakespeare, where is located a smelting furnace. The ores, gold and silver, are said to be rich.

The altitude of Lordsburg is 4,245 feet. It is situated on a broad plain, and being devoid of water the Railroad Company was compelled to dig for it. At a depth of 100 feet their boreing intrument entered a ledge of mineral, and followed it 500 feet. The assays, made in San Francisco, run from $50 to $3,700 per ton. After these facts became known to a few of the officials of the road, the necessary papers were filed to secure the *find*; water was ignored as a useless commodity and the "smile" of satisfaction pervading the countenances of these lucky—embroyo—nabobs—was of that brilliant hue, which dispelled all thought of water for the future. This prospect was sold recently, to the "Wall St. Gold and Silver Mining Co.," for $2,000,000, who are now developing the property and erecting a stamp mill and smelting works near the station.

Leaving Lordsburg, we pass the following small stations: LISBON, 10.7 miles; SEPAR, 9 miles; WILNA, 11.6 miles; GAGE, 8.8 miles; TUNIS, 11.1 miles, and 8.4 miles further, we are at

Deming—The junction of Atchison, Topeka & Santa Fe Railroad, where a connection was made with the Southern Pacific, March 8th, 1881.

Deming, for the first year after the roads reached it, was cursed by swarms of the most vile and dangerous classes of humanity, resulting in many desperate and bloody encounters. This *scum*, has now, nearly all floated away, leaving only a few stores and saloons, a few hundred yards to the south from the station, which appear to be ekeing out a miserable existence on a very limited patronage.

The Railroad Companies have erected at this "junction" a large hotel and depot building, in which are located the usual waiting rooms, ticket and telegraph offices, etc., with a large freight ware house, a short distance to the westward. The hotel—Deming House—contains 25 rooms, with baths, hot and cold water, and is a regular eating station for all passenger trains.

Six horse stages leave Deming daily for Silver City and intermediate points, carrying passengers, mails, and Wells, Fargo & Co's and Adam's & Co's express, through in eight hours, distance 52 miles. At Silver City connections are made with stages for Fort Bayard, 9 miles; Santa Rita Copper Mines, 21 miles; Georgetown, 28 miles; Mogollon Mines, 80 miles, and Clifton, 100 miles. Fare, about 15 cents per mile.

Distances from Deming: Tucson, 220 miles; Yuma, 467 miles; Los Angeles, 711 miles; San Francisco, 1198 miles; Ogden, Utah, via San Francisco, 2,080 miles; Omaha, Neb., via Utah, 3,112 miles; Albuquerque, N. M., 231 miles; Santa Fe, N. M., 316 miles; El Paso, Tex., 88 miles; New Orleans, via T. & P, from El Paso, 1172 miles; Denver, Col., via La Junta, 761 miles; via Espanola, 711 miles; Kansas City, 1149 miles.

With Deming, we conclude our descriptions, for this volume of the "Overland," and refer our readers to the Time Tables of the several diverging Railroad lines from Deming and El Paso.

For Time Tables Southern Pacific, eastward, see pages 271-2.

For Time Tables Atchison, Topeka & Santa Fe, north and east, see page 273.

For Time Tables Texas & Pacific, east and north, see page 274.

For many items of general interest, see Annex, commencing on opposite page.

For information in regard to Arizona, see Annex, No. 64.

For a complete ENCYCLOPEDIA OF COLORADO, Buy **CROFUTT'S GRIP-SACK GUIDE.**
☞SOLD ON THE TRAINS.☜

ANNEX.

In order not to encumber the body of this work with matters that do not directly pertain to the main points at issue, the author has originated an "annex," wherein the reader will find a mass of information which has been prepared with great care, and embraces condensed descriptions and statistical information gathered from the best sources. To these points the reader is frequently referred, throughout the work, by a number to correspond with the annex sought. The numbers at the bottom of the large illustrations, which begin at the first of the book, will be found to correspond with those in the annex, giving a description of the same, and, vice versa.

No. 1 ANNEX. American Progress.—This beautiful picture, which is No. 1 of our large views, is purely national in design, and represents the United States' portion of the American Continent; the beauty and variety, from the Atlantic to the Pacific Ocean, illustrating at a glance the grand drama of Progress in the civilization, settlement, and history of this country.

In the foreground, the central and principal figure, a beautiful and charming female, is floating westward through the air, bearing on her forehead the "Star of Empire." She has left the cities of the East far behind, crossed the Alleghanies and the "Father of Waters," and still her course is westward. In her right hand she carries a book—common school—the emblem of education and the testimonial of our national enlightenment, while with the left hand she unfolds and stretches the slender wires of the telegraph, that are to flash intelligence throughout the land. On the right of the picture, is a city, steamships, manufactories, schools and churches, over which beams of light are streaming and filling the air—indicative of civilization. The general tone of the picture on the left, declares darkness, waste and confusion. From the city proceed the three great continental lines of railway, passing the frontier settler's rude cabin and tending toward the Western Ocean. Next to these are the transportation wagons, overland stage, hunters, gold-seekers, pony express, the pioneer emigrant, and the war-dance of the "noble red man." Fleeing from "Progress," and toward the blue waters of the Pacific, which shows itself on the left of the picture, beyond the snow-capped summits of the Sierra Nevadas, are the Indians, buffalo, wild horses, bears, and other game, moving westward—ever westward. The Indians, with their squaws, pappooses, and "pony-lodges," turn their despairing faces toward the setting sun, as they flee from the presence of the wondrous vision. The "Star" is *too much for them*. What American man, woman or child, does not feel a heart-throb of exultation as they think of the glorious achievements of PROGRESS since the landing of the Pilgrim Fathers, on staunch old Plymouth Rock!

This picture was the design of the author of the TOURIST—is NATIONAL, and illustrates, in the most artistic manner, all those gigantic results of American brains and hands, which have caused the mighty wilderness to blossom like the rose.

No. 2 ANNEX. Passage Ticket Memoranda.

No. 3 ANNEX. Baggage Check Memoranda.

No. 4. ANNEX.—RATES OF FARE.

	1st Class	2d Class	Emigrant
New York to San Francisco, California	$137 35	$102 25	$75 00
Philadelphia to " " "	134 85	100 75	73 50
Baltimore to " " "	131 85	99 50	69 50
Boston to " " "	137 75	104 25	76 00
Cincinnati to " " "	114 25	87 50	60 00
Indianapolis to " " "	113 85	88 00	58 00
Chicago to " " "	113 50	89 00	58 00
St. Louis to " " "	105 85	82 00	52 00
Omaha to Grand Island, Nebraska	6 20		
" North Platte, "	11 65		
" Sidney, "	16 85		
" Denver, Colorado	25 00	22 00	
" Colorado Springs, Colorado	25 00	22 00	
" Pueblo, Colorado	25 00	22 00	
" Cheyenne, Wyoming	22 00	20 00	
" Deadwood, Black Hills, via Stage from Sidney	40 50	35 50	
" Laramie, Wyoming	26 20		24 20
" Ogden, Utah	60 00	50 00	40 00
" Salt Lake City, Utah	62 00	52 00	42 00
" Virginia City, Montana via Stage from Dillon	65 10	50 10	42 00
" Deer Lodge, " " " "	68 10	53 10	45 00
" Helena, " " " "	68 10	53 10	45 00
" Corinne, Utah	61 75	51 35	41 75
" Boise City, Idaho, via Stage from Shoshone	92 80	71 60	53 50
" Silver City, " " " " "	90 80	78 00	60 50
" Baker City, Oregon, " " " "	106 75	85 00	65 45
" Walla Walla, Wash'n, " " " "	106 75	90 75	60 00
" Umatilla, Oregon, " " " "	108 60	85 00	65 45
" Dalles, Oregon, " " " "	114 05	85 00	66 95
" Portland, via Stage from Shoshone	120 00	85 00	66 95
" " " Redding	139 25	114 25	84 25
" " Steamer from San Francisco	120 00	85 00	55 00
" Elko, Nevada	80 50	70 10	43 00
" Battle Mountain, Nevada	86 75	75 00	45 00
" Reno, "	98 00	75 00	45 00
" Virginia City, " via V. & T. R. R. from Reno	101 00	78 00	48 00
" Truckee, California	99 00	75 00	45 00
" Marysville, "	100 00	75 00	45 00
" Sacramento, "	100 00	75 00	45 00
" Stockton, "	100 00	75 00	45 00
" Los Angeles, " all Rail via Lathrop	100 00	78 00	47 50
" Los Angeles, " via Steamer from San Francisco	115 00	85 00	55 00
" Santa Barbara, " " " " " "	110 00	82 50	52 50
" San Diego, " " " " " "	115 00	85 00	55 00
" San Jose "	100 00	75 00	45 00
" San Francisco, "	100 00	75 00	45 00

Children under five years of age, *free*; under twelve years, half-fare.

Cars can be chartered for carrying passengers; each person must be provided with a Ticket.

No. 5.—ANNEX.

OUR WESTERN COUNTRY

Past and Present—This country can no longer be spoken of as the "Far West," as that land is generally conceded to lie nearer sundown, or, at least, beyond the Rocky Mountains. Nebraska, which we enter on crossing the river, so lately opened up to the world, and so lately considered one portion of the "Wild West," forms now one of our central States. It possesses a genial climate, good water, and a fair supply of timber, and the broad prairies of the eastern portion of the State are dotted with well-cultivated and well-stocked farms, that greet the eye of the traveler in every direction, while on all sides may be seen the evidences of thrift and comfort found only in a farming region. Wheat, oats and corn yield luxuriant returns, and all kinds of fruits and garden vegetables, incidental to this latitude, can be grown in profusion. Rarely will the traveler find a more magnificent scene, and more suggestive of real wealth and prosperity, than can be seen on these broad prairies, when the fields of yellow grain or waving corn are waiting for the harvesters. Miles and miles away stretch the undulating plains, far—aye, farther than the eye can see.

In rapid succession we pass the better residence of the "old settler," with his immense fields of grain and herds of stock, on beyond the boundaries of earlier settlements; and now we reach the rude cabin of the hardy settler who has located still "farther west," and here, within a few years, will arise a home as attractive as those we have left behind, surrounded with orchards, gardens and flocks. Here, too, will the snug school-house be found, and the white church with its tapering spire, pointing the people to the abode of Him who hath so richly blessed his children. There is beauty on every hand. The wild prairie flowers, of a thousand different hues and varieties, greet the eye at every step; and the tiniest foot that ever trod Broadway could scarce reach the ground without crushing the life from out some of these emblems of purity. And when the cooling showers have moistened the thirsty earth, or when the morning dew is spangling flower, vine and tree, there is more of quiet, graceful beauty—more of that spirit floating around us which renders man more human, and woman nearer what we desire her to be, than can be found within the walls of any city. Long will the memory of these scenes remain impressed on the mind of the traveler who admires nature in all her phases.

For a long time, Iowa, Indiana, Michigan and Ohio were supposed to contain the wheat-growing soil of the Union, and they became known as the "Granaries of the States." But those "granaries" have pushed themselves a little "farther west," if we may be allowed to use the expression. Nebraska has retained a portion of the name; California and Oregon took the remainder. Nebraska annually produces a large surplus of wheat and corn, which finds its way eastward. With the advantages possessed by this State; with a water-front of several hundred miles on a stream navigable the greater portion of the year; with the grandest railroad on the continent traversing her entire breadth; with all the resources of commerce at her command; with unlimited water power for manufactures, it will be strange, indeed, if Nebraska does not sustain her high rank in the great family of States.

From our present stand-point the quotation, "WESTWARD THE STAR OF EMPIRE TAKES ITS WAY," *must* apply to

The Far West—How often that sentence has been quoted, those who are the *most* familiar with the growth of our western possessions can best remember. So often has it been uttered, that it has passed into a household word, and endowed its innocent and unsuspecting author with an earthly immortality. From the boyhood days of that reliable and highly respectable Individual, the "Oldest Inhabitant" of any special locality in the "Eastern States," it has borne the heading—in large or small caps—of nearly every newspaper notice which chronicled the fact that some family had packed their household goods and gods (mostly goods) and left their native land of woods, rocks, churches and school-houses, to seek a home among the then mythical prairies of the "Far West." But oh! in later years, how that quotation ran across the double columns of these same papers in all conceivable forms of type, when the fact was chronicled that one of our Western Territories was admitted as a State into the Union.

Well, but where was your "Far West" then, where people went when they had "Westward, ho!" on the brain? asks one, who speaks of the West as that part of our country which lies between the summit of the Rocky Mountains and the waters of the Pacific Ocean? Well, the "Far West" of that time, that almost mythical region, was what now constitutes those vast and fertile prairies which lie south and west of the great lakes, and east of and bordering on the Mississippi River. All west of that was a blank; the home of the savage, the wild beast, and all unclean things—at least so said the "Oldest Inhabitant."

But our hardy pioneers passed the Rubicon, and the West receded before their advance. Missouri was peopled, and the Father of Waters became the great natural highway of a mighty commerce, sustained in equal parts by the populous and newly made States lying on both its banks, which had been carved out of the "Far West" by the hands of the hardy pioneers.

Ohio, Indiana, Illinois, Michigan, Minnesota, Missouri and Iowa, had joined the sisterhood, and yet the tide of emigration stayed not. It traversed the trackless desert, scaled the Rocky Mountains, and secured a foothold in Oregon. But it passed not by unheeding the rich valleys and broad prairies of Nebraska, which retained what became, with subsequent additions, a permanent and thriving population. Then the yellow gold, which had been found in California, drew the tide of emigration thitherward, and in a few years our golden-haired sister was added to the number comprising the States of the Union.

Oregon and Nevada on the western slope, Kansas and Nebraska on the east, followed, and, later, Colorado, and still we have Dakota, Idaho, Montana, Washington, Utah, Arizona, and New Mexico Territories, to say nothing of Alaska, waiting the time when they too shall be competent to add their names to the roll of honor and enter the Union on an equality with the others. Thus we see that the "Far West" of to-day has become far removed from the West of thirty—or even ten—years ago, and what is now the central portion of our commonwealth was then the *Far, Far West.*

All is Changed—To-day the foam-crested waves of the Pacific Ocean bear on their bosoms a mighty and steadily increasing commerce. China, Japan, Australia, the Sandwich Islands, South America, and the Orient are at our doors.

A rich, powerful, populous section, comprising three States, has arisen, where but a few years since the Jesuit missions among the savages were the only marks of civilization. And all over the once unknown waste, amid the cosy valleys and on the broad plains, are the scattered homes of the hardy and brave pioneer husbandmen; while the bleak mountains—once the home of the savage and wild beast, the deep gulches and gloomy canyons, are illuminated with the perpetual fires of the "smelting furnaces," the ring of pick, shovel and drill, the clatter of stamps and booming of blasts, all tell of the presence of the miner, and the streams of wealth which are daily flowing into our national coffers are rapidly increasing; for, just in proportion as the individual becomes enriched, to does his country partake of his fortune.

Condensed History—It is only a score of years ago since the Government of the United States, in order to better protect her citizens that had spread themselves over the wild expanse of country between the Missouri River and the Pacific Ocean, and from the Mexican on the south and the British possessions on the north, established a system of military forts and posts, extending north and south, east and west, over this Territory. Though productive of much good, they were not sufficient to meet the requirements of the times, and in many places settlers and miners were murdered with impunity by the Indians. Wise men regarded rapid emigration as the only safe plan of security, and this could not be accomplished without swifter, surer, and cheaper means of transporting the poor, who would gladly avail themselves of the opportunity to possess a free farm, or reach the gold fields of the West. The railroad and telegraph—twin sisters of civilization—were talked of, but old fogies shook their heads in the plentitude of their wisdom, piously crossed themselves, and clasped with a firmer grasp their money bags, when Young America dared broach the subject, "No, sir, no; the thing is totally absurd; impracticable, sir; don't talk any more of such nonsense to me," they would reply, as they turned away to go to their church or to their stock gambling in Wall street—probably the latter occupation. But Young America did not give up to this theory or accept the dictum of Moneybags; and as the counties of the West grew and expanded under the mighty tide of immigration, they clamored for a safe and speedy transit between them and their "Fatherland." Government with its usual red-tape delays and scientific way of how *not to do it*, heeded not the appeal, until the red hand of War—of Rebellion—pointed out to it the stern necessity of securing, by iron bands, the fair dominions of the West from foreign or domestic foe.

Notwithstanding that Benton, Clark, and others had long urged the necessity and practicability of the scheme, the wealth and power which would accrue to the country from its realization, the idea found favor with but few of our wise legislators until they awoke to the knowledge that even the loyal State of California was in danger of being abandoned by those in command, and turned over to the insurgents; that a rebel force was forming in Texas with the Pacific coast as its objective point; that foreign and domestic machinations threatened the dismemberment of the Union into three divisions; not until all this stared them in the face could our national Solons see the practicability of the scheme so earnestly and ably advocated by Sargent of California and his able coadjutors in the noble work. To this threatened invasion of our Western possessions, what had Government to offer for successful defense? Nothing but a few half-finished and illy-manned forts around the bay, and the untaught militia of the Pacific coast. Under this pressure was the charter granted; and it may truly be said that *the road was inaugurated by the grandest carnival of blood the world has ever known;* for, without the pressure of the rebellion, the road would probably be in embryo to-day. Although the American people had been keenly alive to the importance of a speedy transit between the two extremes of the Continent ever since the discovery of gold on the Pacific slope, up to this time the old, vague rumors of barren deserts, dark, deep, and gloomy gorges, tremendous, rugged, snow-clad mountains, and the wild savage, made the idea seem preposterous. Even the reports of the emigrants could not convince them to the contrary; nor yet the reports of the Mormons who marked and mapped a feasible route to Salt Lake City. And it is worthy of remark, that, for over 700 miles the road follows very closely their survey.

Practical, earnest men, disabused the minds of the people regarding the impracticability of the scheme, after the road had became a national necessity—a question of life and unity of the Republic. The great work has been accomplished, and to-day the locomotive whirls its long train, filled with emigrants or pleasure seekers, through that region which, only a few years ago, was but a dim, undefined, mythical land, composed of chaos, and the last faint efforts of nature to render that chaotic State still more inhospitable and uninviting. How great the change from the ideal to the real! For three hundred miles after leaving Omaha, that vague "Great American Desert" proves to be as beautiful and fertile a succession of valleys as can be found elsewhere, under like geographical positions. Great is the change indeed; still greater the changes through which our country has passed during the period from the commencement to the ending of our proudest national civil record, save one. We live in a fast age; the gentle breeze of to-day was the tornado of fifty years ago.

In noting the history of the Continental railroad we must speak of the attempts in that direction which had been made by other parties. Missouri, through her able and liberal legislature, was the first State to move in the construction of a national or continental railroad. The Legislature of that State granted a charter, under which was incorporated the Missouri and Pacific Railroad Co., who were to build a road, diverging at Franklin, southwest, via Rollo, Springfield, Neosho (the Galena district), and along the line of the thirty-sixth parallel to Santa Fe, New Mexico. From Santa Fe, to San Francisco preliminary surveys were made, and had it not been for the rebellion, this road would undoubtedly have been completed long ere this; good authorities placing the limit at 1864. The cause which *compelled* the construction of the Union and Central roads, *destroyed* the Southern. Passing, as it did, mostly through Southern, hostile territory, Government could not aid or protect it in its construction, and consequently the work was suspended. The States of Arkansas and Tennessee, by their legislatures, proposed to assist the work, by constructing a railroad from Little Rock, to connect with the M. & P., somewhere between the ninety-eighth and one hundred and second degree of longitude, and for that purpose a charter was granted.

Organization of the Pacific Railroad—The evident, and we might add, the imperative necessity of connecting the East and West, and the intervening Territories, encouraged the corporators of the great trans-continental line to apply to the Government for aid. Many measures were devised and laid before the people, but the supposed impregnability of the Rocky Mountains, and other natural obstacles to be encountered, caused a hesitancy even then on the part of our energetic people to commence the great work. To attempt to lay the iron rail through vast tracts of unknown country, inhabited by wandering, hostile tribes of savage nomads; to scale the snow-clad peaks of the Rocky Mountains with the fiery locomotive, seemed an undertaking too vast for even the American people to accomplish. But the *absolute* IMPORTANCE, the *urgent* NECESSITY of such a work, overcame all objections to the scheme, and in 1862 Congress passed an act, which was approved by President Lincoln on the first day of July of that year, by which the Government sanctioned the undertaking, and promised the use of its credit to aid in its speedy completion. The act was entitled "An act to aid in the construction of a railroad and telegraph line from the Missouri River to the Pacific Ocean, and to secure to the Government the use of the same for postal, military, and other purposes."

Land Grant—The Government grant of lands to the great national highway, as amended, was, every alternate section of land for 20 miles on each side of the road, or 20 sections, equaling 12,800 acres for each mile of the road. By the Company's table, the road, as completed, is 1,776 18-100 miles long from Omaha to Sacramento. This would give the companies 22,735,104 acres, divided as follows: Union Pacific, 13,295,104; Central Pacific, 9,440,000.

By mutual agreement between the Union and Central companies, made several years ago, Ogden, in Utah, has been decided upon as the "junction" of the two roads.

In addition to the grant of lands and right of way, Government agreed to issue its thirty year six per cent. bonds in aid of the work, graduated as follows: for the plains portion of the road, $16,000 per mile; for the next most difficult portion, $32,000 per mile; for the mountainous portion, $48,000 per mile.

The Union Pacific Railroad Co. built 525 78-100 miles, for which they received $16,000 per mile; 363 602-1000 miles at $32,000 per mile; 150 miles at $48,000 per mile, making a total of $25,236,512.

The Central Pacific Railroad Co. built 7 18-100 miles at $16,000 per mile; 580 32,100 miles at $32,000 per mile; 150 miles at $48,000 per mile, making a total of $25,885,120.

The total subsidies for both roads amount to $52,121,632. Government also guaranteed the interest on the companies' first mortgage bonds to an equal amount.

Cost of construction, material, etc.—In the construction of the whole line, there were used about 500,000 tons of iron rails, 1,700,000 fish plates, 6,800,000 bolts, 6,126,375 cross-ties, 23,505,500 spikes.

Besides this, there was used an incalculable amount of sawed lumber boards for building, timber for trestles, bridges, etc. Estimating the cost of the road with equipments complete by that of other first-class roads ($105,000), per mile and we have the sum of $186,498,900 as the approximate cost of the work.

We have not had much to say heretofore in regard to the

Importance of the Road—to the American people, the Government, or the world at large, simply from the fact that it seemed to us, anything we might say would be *entirely superfluous*, as the incalculable advantages to all could admit of *no possible doubt*. We contented ourselves in annually calling attention to the vast extent of rich mineral, agricultural and grazing country opened up—a vast country which had heretofore been considered *worthless*. We have pointed out, step by step, the most important features, productions, and advantages of each section traversed by the road; stated that the East and West were now connected by a *short* and *quick* route, over which the vast trade of China, Japan, and the Orient could flow in its transit eastward; and, finally, that its importance to the miner, agriculturalist, stock-raiser, the Government, and the world at large, *few, if any*, could estimate.

To those who are continually grumbling about the Pacific railroad, and forget the history of the past, professing to think that these railroad companies are great debtors to the Government, we would most respectfully submit

Facts in Brief.—On the 18th day of March, 1862, before the charter for the Pacific railroad was granted, while the country was in the midst of a civil war, at a time, too, when foreign war was most imminent—the Trent affair showed *how imminent*—and the country was straining every nerve for national existence, and capital, *unusually cautious*, Mr. Campbell, of Penn., Chairman of the House Committee on the "Pacific Railroad" (See *Congressional Globe*, page 1712, session 2d, 37th Congress), said:

"The road is a necessity to the Government. It is the Government that is asking individual capitalists to build the road. Gentlemen are under the impression that it is a very great benefit to these stockholders to aid them to an extent of about half the capital required. I beg leave to call the attention of gentlemen to the fact that it is the Government which is under the necessity to construct the road. If the capitalists of the country are willing to come forward and advance half the amount necessary for this great enterprise, the Government is doing little in aiding the Company to the extent of the other half by way of a loan."

Again, (page 1,911)—"It is not supposed that in the first instance the Company will reimburse the interest to the Government; it will reimburse it in transportation." Mr. White said: "I undertake to say that not a cent of these advances will ever be repaid, nor do I think it desirable that they should be, as this road is to be the highway of the nation."

In the Senate (see *Congressional Globe*, page 2.257, 3d vol., 2d session, 37th Congress) Hon. Henry Wilson, from Mass., said:

"I give no grudging vote in giving away either money or land. I would sink $100,000,000 to build the road, and do it most cheerfully, and think I had done a great thing for my country. What are $75,000,000 or $100,000,000 in opening a railroad across the central regions of this Continent, that shall connect the people of the Atlantic and Pacific, and bind us together? Nothing. As to the lands, I don't grudge them."

Nine years later—after the road had been completed nearly two years—Senator Stewart, from the Committee on the Pacific railroad, said in his report to the U. S. Senate:

"The cost of the overland service for the whole period—from the acquisition of our Pacific coast possessions down to the completion of the Pacific railroad—was over $8,000,000 per annum, and this cost was constantly increasing.

"The cost, since the completion of the road, is the annual interest"—[which includes all the branches—Ed.]—$3,897,129—to which must be added one-half the charges for services performed by the company, about $1,163,138, er annum, making a total expenditure of about $5,000,000, and showing a saving of at least $3,000,000 per annum.

'This calculation is upon the basis that none of the interest will ever be repaid to the United States, except what is paid by the services, and that the excess of interest advanced over freights is a total loss.

"In this statement no account is made of the constant destruction of life and private property by Indians; of the large amounts of money paid by the Secretary of the Treasury as indemnity for damages by Indians to property in the Government service on the plains, under the act of March 3, 1849; of the increased mail facilities, of the prevention of Indian wars, of the increased value of public lands, of the development of the coal and iron mines of Wyoming, and the gold and silver mines of Nevada and Utah; of the value of the road in a commercial point of view in utilizing the interior of the continent, and in facilitating trade and commerce with the Pacific coast and Asia; and, above all, in cementing the Union and furnishing security in the event of foreign wars."

Remember that the Government by charter exacted that these companies should complete their line by 1876; but, by almost superhuman exertion, it was completed May 10, 1869—and the Government has had the benefit of the road *seven years* before the company were compelled by law to finish it.

Now, if we take *no account* of the millions the Government saved during the building of the road—and at *their own* figures—the *saving* during the seven years previous to 1876 has netted the Government $21,000,000, *besides paying the interest on the whole amount of bonds.*

Again, if it cost the Government, before the completion of the Pacific railroad, according to Mr. Stewart, " over $8.000,000 per annum, and this cost was *constantly increasing*" how fast was this increase? Could it be less than six per cent. per annum? Should the figures be made on the basis of six per cent., the Government must have saved, previous to 1876, in the seven years that the line was completed—before the companies were compelled to complete it—over THIRTY MILLIONS OF DOLLARS. This, too, after the Government deducts every dollar of interest on *their own* bonds issued to the companies to *aid* the construction of the road.

The above are some few of the advantages of the Pacific railroad to the Government, and, consequently, to the country at large.

The States and Territories on the line of the Union and Central Pacific railroads, or immediately tributary to it, contained a population, in 1860, of only 554,301, with 232 miles of telegraph line and 32 miles of railway. This same cope of country contained a population, according to the census of 1870, of 1,011,971, and was encompassed by over 13,000 miles of telegraph lines and 4,191 miles of railroads, *completed*, and many more in progress, in which was invested the enormous capital of $363,750,000. Add to the above the immense amount of capital invested—in quartz mills, smelting furnaces, development of mines, and other resources of the country, within the same ten years—then should we bring all the figures down to the present times, the grand total would be comparatively an astonishing romance.

Where, but a few years ago, the buffalo and other game roamed in countless thousands, and the savages skulked in the canyons, and secret hiding-places, where they could pounce out *unawares* upon the emigrant; the hardy pioneers who have made the wilderness *if not* "to blossom like the rose," a *safe* pathway for the present generation, by laying down their lives in the cause of advancing civilization, *now* are to be seen hundreds of thousands of hardy emigrants, with their horses, cattle, sheep, and domestic animals; and the savages are among the things that have " moved on."

Grumblers—The great hue and cry that are made at times by the people and press of the country, in regard to " giving away the lands," "squandering the public domain," etc., which censure the Government for giving, and the railroad company for receiving grants of land in aid of this road, are very surprising in view of the foregoing facts. We would like to know what the lands on the line of these railroads would be worth *without* the road?

Did the Government ever sell any? Could the Government ever sell them? NEVER. It could not realize as much from a million of acres as it would cost their surveyors and land-agents for cigars while surveying and looking after them. When the Pacific road commenced, there was not a land office in Colorado, Wyoming, Montana, Utah, or Nevada, and only one or two in each of the other States or Territories. On the other hand, by the building of the road, many millions of dollars have already found their way into the Government treasury, and at *just double the usual price per acre*. These grumblers would place the Government in the position of the boy who wanted to *eat* his apple, *sell* it, and then get credit for *giving it away*. O! how generous.

No. 6 ANNEX. **The High School at Omaha**—An illustration of which we present on page 29, stands on the site of the old State House of Nebraska, and is known as "Capitol Hill." It was completed in 1876, and cost $280,000. It is 176 feet long and 80 feet wide. The main spire rises 185 feet from the ground.

The building is constructed in the most substantial manner, which, for convenience, beauty in design, and finish throughout, has but few, if any, superiors in the western country,

No. 7 ANNEX. **First Steam Train**—See illustration and description on page 56.

No. 8 ANNEX. **The Madrone Tree**—See description, page 164.

No. 10 ANNEX. **Jack Slade**—Virginia Dale was originally a stage station on the old Denver, Salt Lake and California road, and was laid out and kept by the notorious Jack Slade, who was division superintendent for the old C. O. C. Stage Co., from 1860 to 1863. It was supposed that Slade was the head of a gang of desperadoes who infested the country, running off stock from the emigrants, and appropriating the same. At any rate he was a noted desperado, having, it is said, killed thirteen men. The last of his exploits, east of the mountains, was the wanton and cruel murder of Jules Burg, the person who gave his name to Julesburg. Slade had a quarrel with Jules in 1861, which ended in a shooting scrape, wherein Slade was beaten—or, as their class would say, "forced to take water." In 1863 some of the drivers on the line, friends and employes of Slade's, decoyed Jules to the Cold Spring ranche, on the North Platte River, kept at the time by old Antoine Runnels, commonly known as "the Devil's left bower." He was a great friend of Slade's, who appears to have rightfully earned the title of "right bower" to that same warm-natured individual. The place where this tragedy occurred is 50 miles north of Cheyenne, and 25 miles below Fort Laramie, whither Slade repaired from Cottonwood Springs (opposite McPherson station) in an extra coach as soon as he was notified of the capture of his old enemy. He drove night and day, arriving at Cold Spring ranche early in the morning. On alighting from the coach he found Jules tied to a post in a coral, in such a position as to render him perfectly helpless. Slade shot him twenty-three times, taking care not to kill him, cursing all the time in a most fearful manner, returning to the ranche for a " drink " between shots. While firing the first twenty-two shots, he would tell Jules just where he was going to hit him, adding that he did not intend to kill him immediately; that he intended to torture him to death. During this brutal scene, seven of Slade's friends stood by and witnessed the proceedings. Unable to provoke a cry of pain or a sign of fear from the unfortunate Jules, he thrust the pistol into his mouth, and at the twenty-third shot blew his head to pieces. Slade then cut the ears from his victim, and put them in his pocket.

In the saloons of Denver City, and other places, he would take Jules' ears out of his pocket, throw them down on the bar, and openly boasting of the act, would demand the drinks on his bloody pledges, which were never refused him. Shortly after this exploit, it became too hot for him in Colorado, and he was forced to flee. From thence he went to Virginia City, Montana, where he continued to prey upon society. The people in that country had no love or use for his kind of people, and after his conduct had become insupportable, the Vigilantes hung him.

His wife arrived at the scene of execution just in time to behold his dead body. She had ridden on horseback, 15 miles, for the avowed purpose of shooting Slade, to save the disgrace of having him hung, and she arrived at the scene with revolver in hand, only a few minutes too late to execute her scheme—Jack Slade, the desperado, was dead—and he died—"with his boots on."

No. 14 ANNEX. Snow Difficulties—The Central Pacific company commenced the erection of snow-sheds at the same time with their track-laying over the Sierra Nevada Mountains, and the result has been their trains have never been delayed as often or as long as on many roads in the Eastern States. The depths of snow-fall and the necessities for snow-sheds over the Sierras were *known*, and could be guarded against, but further to the eastward, over the Rocky Mountains, on the route of the Union Pacific, no such necessity for protection against snow was thought to exist; hence the blockade of February and March, 1869.

The Union Pacific Company immediately took, as was thought by everybody at the time—ample precautions to protect their cuts from the drifting snow, by the erection of snow-fences and snow-sheds at every exposed point, but the winter of 1871-2 proved to be one of unusual—unheard-of severity. The snow caused annoying delays to passenger and freight traffic, as well as costing the company a large amount of money to keep the road open. But the lesson taught was a good one in enabling the company to take such measures as were necessary to protect their road against all possible contingencies in the future, which they *have done*, by raising their tracks and building additional snow-sheds and fences.

On the "Central" there are nearly 50 miles of snow-sheds; one continuous of 28 miles in length. On the "Union" there are about 20 miles, and innumerable snow-fences.

No. 15 ANNEX. State Capitol of California—See page 173.

No. 16 ANNEX. Castellated Rocks at Green River—As the subject of the large illustration, No. 2, is described on page 72; it will be unnecessary to repeat it here.

No. 17 ANNEX. Memories of Fort Bridger—which were handed to us by one of our friends, who was with the first party of soldiers who arrived at the place where the fort now stands:

"Early in the winter of 1857, on the 23d of November, the winds were blowing cold and bleak over the snow-covered ridges surrounding Bridger—a town with a significant name, but nothing but a name except an old stone building with the appellation of fort attached to it, built by the Mormons, and surrounded by a small redoubt and *chevaux de frise* pierced for three six-pound mountain howitzers."

"The U. S. forces, comprising the fifth, seventh and tenth Infantry, second dragoons, and four companies of the fourth artillery, the whole under command of Brigadier-General Albert Sidney Johnson, were on their way to Salt Lake City. The fifth, under Major Rugg'es; the seventh, under Colonel Morrison; the second dragoons, under Colonel Howe; the fourth artillery, under Major Williams, entered Bridger on the 23d of November, and established a camp; while a part of the supply train accompanying the expedition, numbering at least 160 wagons, was behind, delayed by heavy snows, entirely separated from the command, and forced to encamp about one mile from each other on the Big and Little Sandy Rivers." [NOTE—These streams are tributaries of Green River on the east, rising near South Pass, about 160 miles north of Bridger.]

"While encamped there, a party of Mormons, under command of Orson Pratt, the generalissimo of the so-called Mormon Legion, assisted by one Fowler Wells, another formidable leader of the Mormon church militant, dashed in and surrounded the trains in the dark hours of the night, completely surprising the entire party, not one escaping to give the alarm. After taking the arms and equipments from the men, they gave them a very limited amount of provisions to last them through to Leavenworth, Kansas, allowing them at the rate of five head of cattle for twenty men, and then started them off in the wilderness to reach that place—about 1,000 miles distant—with no weapons other than their pocket knives with which to protect themselves against the Indians, or to procure game when their limited supply of provisions should become exhausted. After accomplishing this soldierly, humane and Christian act, the Mormons set fire to the train, burning up everything which they could not carry away, and retreated, driving the stock with them, while those left to starve turned their faces eastward. There were 230 souls in that despoiled party, only *eight* of whom ever reached the border settlements; the knife of the savage, and starvation, finishing the cruel work begun by the *merciful* Mormons. The survivors reached Leavenworth in June, 1858, bringing the sad intelligence of the fate of their comrades.

"The loss of these trains necessarily cut short the supplies in Bridger. The troops were put on short rations, and, to add to their horror, the beef cattle accompanying the expedition had nearly all frozen to death, leaving but a few head in camp.

"At Black Fork, the command lost over 300 head in one night; the horses and mules dying in about an equal ratio. Before reaching Bridger, the dragoons were compelled to leave their saddles, which they buried in the snow, the horses being unable to carry them. The animals were compelled to subsist on sage-brush, for two-thirds of the time, and then, to obtain this fibrous shrub, they were compelled to remove snow several feet deep. The men had no other fuel; no water only as they melted snow, for three weeks before reaching Bridger.

"When the news arrived at the camp that the trains were destroyed, the troops immediately began to forage for anything that was palatable, well knowing that no supplies could reach them before late in the spring. The snow was then, on an average, from six to seven feet deep, and the game had mostly left the hills. The rations were immediately reduced to one-half, but even this pittance failed on the 28th day of February, when one-quarter ration per man was issued, being the last of all their stores. Two 100-pound sacks of flour were secured by Maj. E. R. S. Canby, who gave for them $300 in gold. They were placed in his tent, which stood where the old flagstaff now stands, and he supposed his treasure secure.

"But that night a party of men belonging to Company I, 10th Infantry, commanded by Lieut. Marshall, made a *coup d'etat* on the tent, pulling out the pins and throwing the tent over the astonished Major, but securing the flour, with which they escaped in the darkness, and succeeded in hiding it about a mile from camp, in the sage-brush. All was confusion. The long-roll was beaten; the troops turned out and answered to their names, no one being absent. So the matter ended for the time. The next day, at guard mount, the Major commenced a personal search among the tents for his flour. He found—what? In one tent, two men were cooking a piece of mule meat; in another, he found five men cutting up the frozen skin of an ox, preparatory to making soup of it, the only other ingredient to the savory mess being a little flour. Overcome by the sight of so much wretchedness, the Major sat down and cried at his inability to assist them. He asked the men if they could obtain nothing better to eat, and was answered in the negative.

"The severity of the suffering endured by the men nearly demoralized them, still they went out foraging, dragging their wasted forms through the snow with great difficulty. Some would meet with success in their hunts at times; others would not. The mules and horses were either killed and eaten by the men, or died of cold and hunger, which left them without the means of supplying their camp with wood, only as they hauled it themselves. But the men did not murmur. Twenty or thirty would take a wagon and haul it five or six miles to the timber, and after loading it with wood, haul it to camp. Each regiment hauled its own wood, thus securing a daily supply. Some days a stray creature would be slain by the hunters, and there would be rejoicing in the camp once more.

"Early in the spring of 1858 most of the men departed for Salt Lake City, leaving companies B, D and K, of the 10 Infantry, and company F, 7th Infantry. Twenty-seven men from each company were detailed to go to the pineries, 25 miles away, to cut timber with which to erect quarters. On arriving in the pinery, they found an old saw mill and race, which had been used by the Mormons, and everything convenient but the necessary machinery. Luckily the quarter-master's department had the required machinery, and soon they had a saw mill in good running order. By the 15th of of September, 1858, the quarters were up and ready for use. They were large enough for five companies, including a chapel, hospital, sutler's store, guard house, etc.

"The Fourth of July, 1858, was duly observed and honored. The flag-staff was raised in the center of the parade ground, the flag hoisted by Major Canby, and prayers said by Major Gatlin.

"On the 23d of September, 1858, a large train of supplies arrived, causing great joy among the troops. Two days later three long trains of supplies filed through the place on the way to Salt Lake City.

No. 18 ANNEX. **Hanging Rock, Utah**—See description on page 97 of this book."

No. 19 ANNEX. **Steamboat Rock**—The large illustration, No. 6, is one of many beautiful vtews to be seen while passing through Echo and Weber canyons, Utah. From our point of view the appearance of Steamboat Rock is exceedingly perfect. The lines (seams in the rocks) run gracefully up for 300 or 400 feet, and in the heen of the moon the sage-brush, dwarf cedars, and other shrubs, growing along the upper crevices can easily be conjectured into a load of passengers worthy of the mighty vessel, but she stands in stone, and the ship carpenters—the elements—are steadily taking her timbers apart.

No. 20 ANNEX. **Paddy Miles' Ride**—Mr. Miles, or "Paddy," as he was familiarly called, was foreman to the Casement Brothers, who laid the track of the Union Pacific railroad. One morning, Paddy started down Echo Cañon with a long train of flat cars, sixteen in number, loaded with ties and iron rails for the road below Echo City, where were then, as now, the station, switches, etc. The reader will remember that from the divide to the mouth of Echo Cañon is a heavy grade, no level place on which cars would slack their speed.

The train had proceeded but a few miles down the cañon, going at a lively rate, when the engineer discovered that the train had parted, and four loaded cars had been left behind. Where the train parted the grade was easy, hence that portion attached to the locomotive had gained about half-a-mile on the stray cars. But when discovered they were on heavy grade and coming down on the train with lightning speed. What was to be done? The leading train could not stop to pick them up, for at the rate of speed at which they were approaching, a collision would shiver both trains, destroying them and the lives of those on board.

There were two men, Dutchmen, on the loose cars, who might put on the brakes, and stop the runaway. The whistle was sounded, but they heard it not; they were fast asleep behind the pile of ties. On came the cars, fairly bounding from the track in their unguided speed, and away shot the locomotive and train. Away they flew, on, around curves and over bridges, past rocky points and bold headlands; on with the speed of the wind, but no faster than came the cars behind him.

"Let on the steam," cried Paddy, and with the throttle chock open, with wild, terrible screams of the whistle, the locomotive plunged through the gorge, the mighty rocks sending back the screams in a thousand ringing echoes.

"Off with the ties," shouted Paddy, once more, as the whistle shouted its warning to the stationmen ahead to keep the track straight and free, for there was no time to pause—that terrible train was close on to them, and if they collided the canyon would have a fearful item added to its history. On went the train past the sidetracks, the almost frantic men throwing off the ties, in hopes that some of them would remain on the track, throw off the runaways, and thus save the forward train. Down the gorge they plunged, the terror keeping close by them, leaping along—almost flying, said one, who told us the tale—while the locomotive strained every iron nerve to gain on its dreaded follower. Again the wild scream of the locomotive of "switches open," rung out on the air and was heard and understood in Echo City. The trouble was surmised, not known, but the switches were ready, and if the leading train had but the distance it could pass on and the following cars be switched off the track, and allowed to spend their force against the mountain side. On shot the locomotive, like an arrow from the bow, the men throwing over the ties until the train was well-nigh unloaded, when just as they were close to the curve by which the train arrives at the station, they saw the dreaded train strike a tie, or something equally of service, and with a desperate plunge, rush down the embankment, into the little valley and creek below. "Down brakes," screamed the engine, and in a moment more the cars entered Echo City, and were quietly waiting on the sidetrack for further developments. The excited crowd, alarmed by the repeated whistling, was soon informed of the cause of these screams, and immediately went up the track to the scene of the disaster, to bring in the dead bodies. When they arrived they found the poor unfortunates sitting on the bank, smoking their pipes and unharmed, having just woke up. The first they knew of the trouble was when they were pitched away from the broken cars on the soft greensward. The debris of car frames, wheels, and ties, gave them the first intimation they had received that something was the matter.

No. 21 Annex. **Great Salt Lake**—Behind the station at Promontory the hills rise into the dignity of mountains. To the top of the left hand point we strolled one bright, spring morning. After an hour's toilsome walking through sage-brush and bunch-grass, then among sage-brush and rocks, until we had attained a height to which that persistent shrub could not attain, then among rocks, stunted cedars, tiny, delicate flowers and blooming mosses, until we stood on the summit of the peak, on a narrow ridge of granite, not over four feet wide, and there, almost at our feet (so steep was the mountain) lay the Great Salt Lake, spread out like a vast mirror before us, its placid bosom glittering in the morning sun, like a field of burnished silver. Mile after mile it stretched away, placid and motionless, as though no life had ever caused a vibration of its currents, or given one restless impulse to its briny bosom.

By the aid of the glass, Church or Antelope and other mountain islands could be distinctly seen, rearing their towering crests far above the silver border at their base, their sloping sides enrobed in the greenest of all green coverings. Standing there as lone sentinels in the midst of this waste of waters, they possess a wondrous beauty as a recompense for their utter isolation. Away beyond these islands rise the white-crested Wasatch Mountains, and we believe that we can pick out the curve in their brown sides where nestles Salt Lake City, secure and beautiful in her mountain fastness. Far away to the southward the range blends with the sky and water, and the dim, indistinct lines of green, brown and silver blend in one, while above them the clear blue of the mighty dome seems to float and quiver for a space, and then sweeps down to join them, blending with them in one waving mass of vanishing color, which slowly recedes in the dim distance, until the eye can follow its course no farther. Turn now to the left, and there, sweeping up far behind Promontory Point, is the northwestern arm of the lake—Monument Bay. That long, green line is Monument Point, throwing its long ridge far out into the bosom of the lake, as though it would span the waters with a carpet of green. Away to the west Pilot Knob rears its crest of rocks from out the center of the great American desert. Do not look longer in that direction, all is desolation; only a barren plain, and hard, gray rocks, and glinting beds of alkali meet the vision.

One more view to the north, one look at the lines of green hills and greener slopes which sweep down toward the sandy, sage-clad plateau on which stands the station; another and last look at the placid lake, and now, cooled and refreshed by the mountain breeze, we pluck a tiny moss bell from the cleft in the rock, and then descend the rugged mountain. We have seen Salt Lake from the most commanding point of view from the north, and now we are better able to understand its shape and comprehend its dimensions, which are 126 miles in length by 45 in width. The principal islands are Antelope (15 miles long), Sheep's, Hot, Stansbury, Carrington and Egg. They possess many charming summer retreats, many natural bathing places. The water is so buoyant that it is difficult for the bather to sink.

The lake has no outlet for the waters continually pouring into it from Bear, Jordan, Weber and other rivers. Evaporation absorbs the vast volume, but it is a noticeable fact, and one worthy of consideration, that since the settlements have been made in the Territory, and the bosom of the earth has been turned with the plow, rendering the barren waste blooming and productive, that the waters of the lake have risen steadily, and are now 12 feet higher than they were 20 years ago. Fences, which once enclosed fine meadow lands, are now just peering above the flood, marking its encroachment on the fertile bottom lands. The grand old mountains bear unmistakable evidence of the water's presence far up their rocky sides. At what time the floods reached that altitude, or whether those mountains were lifted from the present level of the lake by volcanic action, and carried those water lines with them, are questions no one can answer. Savans may give learned theories regarding things they know nothing of; they may demonstrate that Salt Lake is held in its present position by immutable laws, but they cannot destroy the ocular evidence that it has been rising slowly and steadily for 20 years.

No. 23 Annex. **California**—Page 226.

No. 24 Annex. **Hauling Ores in Hides**—See page 146.

No. 25 Annex. **Brigham Young**—Late President and Prophet of the Mormon "Church of the Latter Day Saints," was born in Whittingham, Vermont, on the first day of June, 1801. His father, John Young, was a Revolutionary veteran, and served three campaigns under Washington. The family consisted of six daughters and five sons, of whom Brigham was the fourth. In early life he was connected with the Methodists, and at this time he followed the occupation of carpenter, joiner and glazier.

Young was *first* married in 1824, and in the spring of 1830 first saw "The Book of Mormon," which was in the possession of one of his brothers, and made a great impression upon him, and of which he afterwards became so firm a believer and prominent supporter. In April, 1832, he was baptized a member of the Mormon Church.

Before becoming a Mormon, Brigham Young made himself thoroughly acquainted with their principles, and then clung to his belief in the teachings of the "Book of Mormon" with great tenacity to the close of his eventful life. It was characteristic of the man that he was deliberate in arriving at an opinion, but when it was once formed he was steadfast to his convictions. While Joseph Smith was alive, by whom he was baptized, he was his friend and firm supporter, and from the time when the church of his choice was composed of but a persecuted and incipient handful, fleeing from place to place, until the day of his death, his was the master-spirit that controlled all their deliberations and ruled in all their prominent councils. Brigham Young was the great organizer and master spirit that enabled them, by practical councils and directions, to cross the wide and unknown desert plains of America in the year 1847, when possessed of the scantiest resources, and establish among the far-off mountains of Utah Territory, a prosperous and thriving community.

He was equal to the grand occasion of his life in rescuing the church from disorganization at

Nauvoo, in 1844, where he stepped to the front and took the helm. The good of the Church was always his first and foremost consideration; he laid plans for its prosperity, and in their successful execution, he made vast sums of money for himself. Like all new organizations, especially those of an ecclesiastical character, there were many schisms and rivals to be put down, and in doing away with these, he was frequently forced to take measures that drew down upon his head the odium of the outside world. With the same opportunities for becoming a tyrant and despot, with a large, ignorant element among his subjects, few men with the same tenacity of will, and force of character, would have been less of an oppressor than the late Prophet priest and Revelator of the Mormon Church.

President Young has taken a prominent part in all public improvements, in every plan calculated to facilitate communication between the Territory and the Eastern States; materially assisting in forming several express companies and stage lines. He built several hundred miles of the Western Union Telegraph, graded 150 miles of the Union Pacific railroad, and has ever offered his assistance to every enterprise of the kind which had a material bearing on the interests of Utah.

He died regretted and respected by his followers, and admired by the world at large, Wednesday, Aug. 29, 1877, in the seventy-sixth year of his age. His funeral took place on Sunday, September 2, 1877, amid a great popular demonstration, the body being viewed by over 20,000 people.

The following characteristic document, prepared by the diseased about four years previous to his death, contains his instructions for the conduct of the funeral obsequies. The paper was read by George Q. Cannon before the assembled multitude on the day of the funeral, and the instructions therein contained were carried out to the letter.

"I, Brigham Young, wish my funeral services to be conducted after the following manner:

When I breathe my last I wish my friends to put my body in as clean and wholesome state as can conveniently be done, and preserve the same for one, two, three or four days, or as long as my body can be preserved in a good condition. I want my coffin made of plump 1¼ inch redwood boards, not scrimped in length, but two inches longer than I would measure, and from two to three inches wider than is commonly made for a person of my breadth and size, and deep enough to place me on a little comfortable cotton bed with a good suitable pillow for size and quality; my body dressed in my Temple clothing and laid nicely into my coffin, and the coffin to have the appearance that if I wanted to turn a little to the right or to the left I should have plenty of room to do so; the lid can be made crowning.

At my interment I wish all of my family present that can be conveniently, and the male members wear no crape on their hats or their coats; the females to buy no black bonnets, nor black dresses, nor black veils; but if they have them, they are at liberty to wear them. The services may be permitted, as singing and a prayer offered, and if any of my friends wish to say a few words, and really desire, do so; and when they have closed their service, take my remains on a bier and repair to the little burying ground which I have reserved on my lot east of the White House on the hill, and in the southeast corner of this lot have a vault built of mason work, large enough to receive my coffin, and that may be placed in a box, if they choose, made of the same material as the coffin - redwood. Then place flat rocks over the vault, sufficiently large to cover it, that the earth may be placed over it—nice, fine, dry earth—to cover it until the walls of the little cemetery are reared, which will leave me in the southeast corner. This vault ought to be roofed over with some kind of a temporary roof. There let my earthly house or tabernacle rest in peace and have a good sleep until the morning of the first resurrection; no crying, nor mourning with any one that I have done my work faithfully and in good faith.

I wish this to be read at the funeral, providing that if I should die anywhere in the mountains, I desire the above directions respecting my place of burial to be observed; but if I should live to go back with the Church, to Jackson County, I wish to be buried there.

BRIGHAM YOUNG,
President of the Church of Jesus Christ of Latter-day Saints.
SUNDAY, November 9th, 1873.
Salt Lake City, Utah Ter.

Brigham Young will ever stand prominently forward on the pages of the world's history, as one of the most remarkable men of the nineteenth century, respected by his followers and admired by the world at large, whose vices and virtues will go hand in hand adown the stream of time.

No. 26 ANNEX **National Park—**The explorations of Dr. Hayden, United States Geologist, have demonstrated that *this, our own country*, contains natural wonders, which, in extent, grandeur, and wondrous beauty, far surpass those of any other portion of the known world. The result has been, a bill has passed Congress setting apart a tract of country 55 by 65 miles in extent as a great NATIONAL PARK, or mammoth pleasure-ground, for the benefit and enjoyment of the people. The entire area within the limits of the reservation is over 6,000 feet in altitude. Almost in the centre of this tract is located the Yellowstone Lake, a body of water 15 by 22 miles in extent, with an elevation of 7,427 feet. The ranges of mountains that hem the numerous valleys on every side rise to the height of from 10,000 to 12,000 feet, and are covered with perpetual snow.

This country presents the most wonderful volcanic appearance of any portion of this continent. The great number of hot springs and the geysers represent the last stages—the vent or escape pipes—of these remarkable volcanic manifestations of the internal forces. All these springs are adorned with decorations more beautiful than human mind ever conceived, and which have required thousands of years for the cunning hand of nature to form. The most remarkable of these geysers throws a column of boiling hot water 15 feet in diameter to a *measured* altitude of 150 feet. This display is continued for hours together, and so immense is the quantity of water discharged, that during the eruption, the volume of water in the river is doubled. Another throws a column of hot water 200 feet in height, and over a foot in diameter. It is said the geysers of Iceland, which have been the objects of interest for scientists and travelers of the entire world for years, sink into insignificance in comparison with the Hot Springs of the Yellowstone and Fire-hole Basins.

The *most wonderful* story about this remarkable region is told by Langford, one of the first discoverers. He says: "At a certain point on the Yellowstone River, the water runs down a steep and perfect grade over a surface of slate-rock, which has become so smooth from the velocity of the rushing torrent, that, at a distance of twenty miles, the *friction* becomes so great that the water is *boiling* hot." We do not vouch for the truth of this story, and we are not certain that Langford will swear to it.

The mountain rim of the Yellowstone Lake rises from 1,500 to 4,000 feet above its surface, and, except in two directions, is unbroken. To the west and southwest are breaks in the chain, through one of which appear the outlines of a conspicuous conical peak, 10,500 feet in height. In the mountain system which surrounds the lake are born the tributaries, almost the principal sources, of three of the largest rivers on the continent. Four of the most important tributaries of the Missouri—namely, the Big Horn, the Yellowstone, the Madison and the Gallatin, have their springs here. Flowing first north, then east, they strike the Missouri, which, in its turn, flows southeasterly to the Mississippi Valley, where its waters are blended with the stately stream that empties its tides at least 3,500 miles below into the Gulf of Mexico. The Snake River, whose sources are actually interlaced with those of the Madison and the Yellowstone, turns westward, and traverses nearly a thousand miles of territory before it joins the Columbia on its way to the Pacific Ocean. Again, the Green River, rising but a few miles from the sources of the others, seeks the Colorado of the South, which, after innumerable windings through deserts, and a roaring passage of hundreds of miles in the abysses of canyons surpassing even those of the Yellowstone in grandeur, depth, and gloom, reaches the gulf of California. Penetrating to the lofty recesses where these springs rise, the explorer stands, as it were, astride of the grandest water-shed in the world. A pebble dropped into one spring touches a water-nerve of the Pacific; a pebble cast into another touches a similar nerve of the Atlantic Ocean. It is a thought to cause the wings of the spirit of a man in such a place to expand like an eagle's. (See large illustrations, Nos. 35 and 36.)

No. 27 ANNEX. **Ocean Steamships.**— The steamers of the Occidental and Oriental Line, between San Francisco and Yokohama, leave San Francisco about the 15th of each month. The passage rates are:

Payable in U. S. Gold Coin.	*First Class or Cabin.	European Steerage.	Chinee Steerage.	Distance fr'm San Franci'o.
SAN FRANCISCO TO				Miles.
Yokohama, Japan,	$250 00	$ 85 00	$53 00	4,764
Hiogo, "	270 00	100 00	.....	5,104
Nagasaki, "	290 00	100 00	62 00	5,444
Shanghae, China,	300 00	100 00	65 00	5,964
Hongkong, "	300 00	100 00	53 00	6,384
Calcutta, India,	450 00	.....	.....	9,385

Children under 12 years of age, one-half rates; under five years, one-quarter rates: under one year, free.

Family Servants, (European) eating and sleeping in cabin, full cabin rates.

250 lbs. baggage allowed each adult, first-class or cabin passenger; 150 lbs. each, European steerage; 100 lbs. each, Chinese steerage: proportionate to children.

☞ **Round Trip Tickets**, good for twelve months, will be sold at a reduction of 12½ per cent. from regular rates.

An allowance of 20 per cent. on return passage will be made to passengers who paid full fare to Japan or China, or *vice versa*, re-embarking within six months from date of landing, and an allowance of 10 per cent. to those who return within twelve months.

Families whose fare amounts to FOUR FULL PASSAGES will be allowed 7 per cent. reduction.

Exclusive use of staterooms can be secured by the payment of half-rate for extra births.

The Pacific Mail steamships leave San Francisco about the 1st of every month, for Yokohama and Honkorg, and for Sidney and Aukland via Honolulu, at about the same time—1st of each month—and for New York. via Panama, about the 1st and 15th of each month. For Victoria, B. C., Port Townsend, Seattle and Tacoma, the 10th, 20th and 30th of each month.

The Oregon Steamship Co. send steamers to Portland from San Francisco every five days.

Other steamers for up and down the coast, leave at changeable intervals; about weekly, however.

No. 28 ANNEX. **Col. Hudnut's Survey.** —On the west side of Promontory Point, the line known as Colonel Hudnut's survey of the Idaho and Oregon branch of the U. P. R. R., passes north to Pilot Springs; thence down Clear Creek or Raft River to Snake River, and along the southern bank of this stream to Old's Ferry; thence across the country to Umatilla, on the Columbia River. For the entire distance between Promontory and Raft River, the country is uninviting, though not barren. From thence the route passes through a country abounding in fertile valleys and bold mountains—the latter well-wooded. There is plenty of wood and other materials for building the proposed road along the whole length of the line. To the mouth of Raft River from Promontory is about 100 miles. The scenery along the line is varied, from smiling, fertile valleys to lofty, snow-clad mountains. We will speak only of the general characteristics of the route and of one or two points of remarkable interest. The main feature of the Snake or Shoshone River is its majestic cataracts. The stream, sometimes called Lewis River, is the South Fork of the Columbia, and was discovered by Lewis and Clark, who ventured westward of the Rocky Mountains in 1804. It rises in the Rocky Mountains, near Fremont's Peak, in the Wind River Range, which divides Idaho and Wyoming Territories. The head waters of the stream are Gros Ventre, John Craig's and Salt Creeks on the south, with the outlets of Lyon's and Barret's lakes on the north. The general course of the river from its source to Big Bend is northwest. At this point Henry's Fork, a large stream flowing from the the north, empties its waters into the main river. Thence the course is southwesterly until the first falls are reached— about 400 miles from the river's source. These are called the AMERICAN FALLS and are very fine, but do not present so sublime an appearance as will be seen about 100 miles further down the river, where the waters leave the elevated plains of Idaho by a series of cascades, known as the SHOSHONE FALLS, from 30 to 60 feet high, closing the scene in one grand leap of 210 feet perpendicular. The width of the river at the point of taking the last leap is about 700 feet. The form of the fall is circular—somewhat like those of the Niagara. Before the river reaches the cascades it runs between lofty walls, which close in around it until but a narrow gorge is left for the passage of the water

1,000 feet below the tops of the bluffs. The most complete view of the falls is obtained from Lookout Point, a narrow spit of rocks which projects from the main bluffs a short distance down the stream from the falls. From this point Eagle Rock rises before us in the midst of the rapids, and almost overhanging the falls, fully 200 feet high; its pillar-like top surmounted by an eagle's nest, where, year after year, the monarch of the air has reared its young. Near the center of the river are several islands covered with cedar, the largest one being called Ballard's Island. Two rocky points, one on either side of the falls, are called the Two Sentinels. Excepting in point of the volume of water, the falls will compare favorably with Niagara.

From this point the river runs nearly west until it reaches War Eagle Mountains, about 80 miles from its source, when it turns due north, following that course for 150 miles, then bending again to the west it unites with Clark's River, forming the Columbia. After leaving the last falls the country is less broken, and the work of building the road would be comparatively light for most of the way.

No. 29 ANNEX. Western Stock Raising.

DURBIN, ORR & Co.—Cattle branded \B; also some of them \B, and horses the same.
Post-office, Cheyenne, W. T. Range, Bear Creek.

CREIGHTON & Co.—Horses branded quarter circle open block, on left shoulder. Also, part cattle branded half-circle on shoulder.
Post-office Pine Bluffs, W. T. Range, Horse and Pumpkin creeks.

Stock raising is an important industry. We have often expressed our belief that, ultimately, it would be found there was not one foot of valueless land on the line of the Pacific railroad. The Bitter Creek country, previous to 1868, for 80 miles was universally admitted by all who knew anything about that section of country, to be utterly valueless. Coal, in immense quantities, was discovered all along the creek—great veins—and it is now the most valuable section of the Union Pacific railroad. Portions of the Humboldt and Nevada Desert were also set down as valueless; now, see what irrigation has done for a portion of it, where the people have had the enterprise to adopt a system of irrigation, as at Humboldt Station. We contend that *all* the lands on the line of this road are valuable, some as mineral, some as agricultural, but the greater portion is the finest grazing land in the world. This fact, of late years, is becoming thoroughly understood, as in 1868 there would not exceed twenty thousand head of cattle on the whole line of the Pacific railroad, across the continent; *now* there are over 700,000 head of cattle, 30,000 head of horses, and full 450,000 head of sheep.

The range is enormous, taking in broad plains, grass-covered mountains, and thousands of as beautiful little foot-hills and mountain valleys as there are in the world. This section commences about 250 miles west of the Missouri River, and extends to the eastern base of the Sierra Nevada Mountains, all of which, with only a few miles intervening, is the stock-raiser's paradise. The absence of water is the only drawback in this intervening section, and in time wells will be sunk and that obstacle overcome. The valley bluffs, low hills and mountain sides of this whole section are covered with a luxuriant growth of gramma or "bunch" grass, one of the most nutritious grasses grown, together with white sage and grease-wood, upon which all kinds of stock thrive all the season, without care, excepting what is necessary to prevent them from straying beyond reach. Old work-oxen that had traveled 2,500 miles ahead of the freight wagon during the season, have been turned out to winter by their owners, and by the following July they were "rolling fat"—fit for beef. We know this to be a fact from actual experience.

This country is the great pasture land of the continent. There is room for millions of cattle in this unsettled country, and then have grazing land enough to spare to feed half the stock in the Union.

In the foot-hills and mountainous portion of this great grazing range, and along the line of the great water courses, there is no trouble from lack of water, for the mountain valleys are each supplied with creeks and rivers. Springs abound in various sections, so that no very large tract of land is devoid of natural watering places. The grass grows from nine to twelve inches high, and is peculiarly nutritious. It is always green near the roots, summer and winter. During the summer the dry atmosphere cures the standing grass as effectually as though cut and prepared for hay. The nutritive qualities of the grass remain uninjured, and stock thrive equally well on the dry feed. In the winter what snow falls is very dry, unlike that which falls in more humid climates. It may cover the grass to the depth of a few inches, but the cattle readily remove it, reaching the grass without trouble.

Again, the snow does not stick to the sides of the cattle and melt there, chilling them through, but its dryness causes it to roll from their backs, leaving their hair dry. The cost of keeping stock in this country is just what it will cost to employ herders—no more. The contrast between raising stock here and in the East must be evident. Again, the stocking of this country with sheep, is adding an untold wealth to the country. The mountain streams afford ample water power for manufactories, and wool enough could be grown here with which to clothe all the people of the Union, when manufactured into cloth. With the railroad to transport the cattle and sheep to the

Eastern and Western markets, immense fortunes are now being made, and the business is comparatively new—in its infancy.

No drouths which have been experienced in this great range have ever seriously affected the pasturage, owing to the peculiar qualities of the grasses indigenous to the country. So with storms; it has seldom happened that any storms are experienced which cause loss, and none ever need to, and none ever do, when the stock is properly attended to and herded.

On these ranges it is common for stock of many owners to range together, and a system of brands has been adopted, and recorded with the county clerk in the section of country where the herds belong. The recording of the brands is a protection against theft and loss by straying, as each cattle man knows the brands in use in his range, and each endeavors to protect the other's interest.

The illustrations that we present, show two of the brands in use, and the method adopted by all cattle men to make known their brand, and the particular range, or *home range* of the cattle. [These are actual names, brands, range and addresses.]

The ANNUAL "ROUND-UP."—One of the most important and interesting features of the stock-raising business is the cattle "round-up." In the "free and easy" manner of raising cattle on the broad, western plain, where the owner may not see one-half of his herd for six months at a time, it may be imagined that the restless Texans scatter almost from Dan to Beersheba, and that extra effort is necessary when they are finally collected by the regular spring "round up." Companies of herders are organized to scour certain sections of country, and bring every animal to a grand focal point, no matter who that animal may belong to or what its condition may be. The old-fashioned "husking bee," "'possum hunt" or "training day" is vastly outdone by this wild revelry of the herders. Mounted upon their fleetest ponies, the cow-boys scatter out in all directions, gather in "everything that wears horns," and at night may have the property of half-a-dozen owners in one immense, excited herd. Then, while a cordon of herders hold the animals together, representatives of the different "brands" ride into the herd, single out their animals, one by one, and drive them off to be branded or marketed. Moving along, day after day, the scene is repeated, until the whole plains country has been visited, and every breeder has had an opportunity to take an inventory of his stock. Of course the participants "camp out" wagons, following the herd, with blankets and provisions, the "round-up" season, being one of mirth and frolic, as well as of work, from beginning to end.

No. 30 ANNEX. The Great Cave—of Eastern Nevada, lies about forty-five miles to the southwest of Eureka. It is situated in one of the low foot-hills of the Shell Creek Range, which extends for about two miles into a branch of Steptoe Valley. The ridge is low, not over 60 or 65 feet high, and presents no indications which would lead one to suspect that it guarded the entrance to an immense cavern. The entrance to the cave would hardly be noticed by travelers, it being very low and partly obscured. A rock archway, small and dark, admits the explorer, who must pass along a low passage for about 20 feet, when it gradually widens out, with a corresponding elevation of roof. Many of the chambers discovered are of great size; one, called the "dancing hall," being about seventy by ninety feet. The roof is about forty feet from the floor, which is covered with fine gray sand. Opening into this chamber are several smaller ones, and near by, a clear, cold spring of excellent water gushes forth from the rock. Further on are more chambers, the walls of which are covered with stalactites of varied styles of beauty. Stalagmites are found on the floors in great numbers. It is not known how far this cave extends, but it has been explored over 4,000 feet, when a deep chasm prevented further exploration.

INDIAN LEGEND—The Indians in this vicinity have a curious fear of this place, and cannot be tempted to venture any distance within its haunted recesses. They have a legend that "heap" Indians went in once for a long way and none ever returned. But one who ventured in many moons ago, was lucky enough to escape, with the loss of those who accompanied him, and he is now styled, "Cave Indian." According to the legend, he ventured in with some of his tribe and traveled until he came to a beautiful stream of water, where dwelt a great many Indians, who had small ponies and beautiful squaws. Though urged to stay with his people, "Cave" preferred to return to sunlight. Watching his chances, when all were asleep, he stole away, and, after great suffering, succeeded in reaching the mouth of the cave, but his people still live in the bowels of the earth. The Indians thoroughly believe the story, and will not venture within the darkness. Another story is current among the people who live near by, which is, that the Mormons were once possessors of this cave, and at the time when they had the rupture with the United States Government, used it as a hiding place for the plate and treasures of the Church and the valuables of the Mormon elders. The existence of the cave was not known to the whites, unless the Mormons knew of it, until 1866.

A LITTLE HISTORY—In the latter part of the summer of 1858, a party of prospectors from Mariposa, in California, crossed the Sierra Nevada Mountains *via* Yo-Semite to Mono Lake, then in Utah, but now in that part of the country set off to form Nevada. For three years the party worked placer mines and other gold along the various canyons and gulches extending eastward from the Sierras, which led others to continue prospecting further north, and who discovered Comstock Ledge. Other prospectors followed, and the discovery of rich veins in Lander, Esmeralda, Nye and Humboldt counties, and in the adjoining Territory of Idaho, was the result. The great "unexplored desert," on the map, was avoided until 1865 and 1866, when parties began to branch out and discover the rich argentiferous quartz and fine timber land, extending along a series of parallel valleys, from the Humboldt to the Colorado River. Several New York companies became interested in these discoveries, and erected a 20-stamp mill at Newark, 22 miles north of where Treasure City now stands, to work veins in the Diamond Range. Across the valley, opposite Newark, White Pine Mountain rises 10,285 feet. Here the "Monte Christo" mill was erected, at which a Shoshone Indian came one day with a specimen of better "nappias" than had yet been discovered, and, by his guidance, the rich mines discovered at Treasure Hill and the "Hidden Treasure" mine were located and recorded on the 14th of September, 1867. But, aside from the production of mineral, along these mountain ranges, another source of wealth exists in the valleys extending through Nevada and Utah. We refer to that branch of business which has been gradually increasing—one which will bring a large revenue to the settlers along these valleys in stock-raising. Bunch grass grows in abundance, and cattle are easily wintered and fattened, finding a ready market in the mining districts and westward to Sacramento and San Francisco.

No. 31 ANNEX. Nevada Falls.—209.

No. 32 ANNEX. Pioneer Mail Enterprises.—(See page 218.)

No. 33 ANNEX. The Donner Party.—(Illustration page 71). Around this beautiful sheet of water—nestled so closely in the embrace of these mighty mountains, smiling and joyous in its matchless beauty, as though no dark sorrow had ever occurred on its shores, or its clear waters reflected back the wan and haggard face of starvation—is clustered the saddest of memories—a memory perpetuated by the name of the lake.

In the fall of '46, a party of emigrants, mostly from Illinois, arrived at Truckee River, worn and wasted from their long and arduous journey. Among that party was a Mr. Donner, who, with his family, were seeking the rich bottom lands of the California rivers, the fame of which had reached them in their Eastern home. At that time a few hardy pioneers had settled near Sutter's Fort, brought there by the returning trappers, who, with wondrous tales of the fertility of the soil and the genial climate of California, had induced some of their friends to return with them and settle in this beautiful land. The Donner party, as it is generally called, was one of those parties, and under the guidance of a trapper, was journeying to this then almost unknown land. Arriving at the Truckee, the guide, who knew the danger threatening them, hurried them forward, that they might cross the dreaded Sierras ere the snows of winter should encompass them. Part of the train hurried forward, but Mr. Donner, who had a large lot of cattle, would not hurry. Despite all warnings, he loitered along until, at last, he reached the foot of Donner Lake, and encamped there for the night. The weather was growing cold, and the black and threatening sky betokened the coming storm. At Donner Lake, the road turned to the left in those days, following up Coldstream, and crossing the Summit, near Summit Meadows, a very difficult and dangerous route in fair weather. The party who encamped at the lake that night numbered 16 souls, among whom were Mrs. Donner and her four children. During the night, the threatened storm burst over them in all its fury. The old pines swayed and bent before the blast which swept over the lake, bearing destruction and death on its snow-laden wings. The snow fell heavily and fast, as it *can* fall in those mountains. Most of the frightened cattle, despite the herder's vigilance, "went off with the storm."

In the morning the terror-stricken emigrants beheld one vast expanse of snow, and the large white flakes falling thick and fast. Still there was hope. Some of the cattle and their horses remained. They could leave wagons, and with the horses they might possibly cross the mountains. But here arose another difficulty, Mr. Donner was unwell, and could not go—or preferred to wait until the storm subsided; and Mrs. Donner, like a true woman, refused to leave her husband.

The balance of the party, with the exception of one, a German, who decided to stay with the family, placed the children on the horses, and bade Mr. and Mrs. Donner a last good-by; and after a long and perilous battle with the storm, they succeeded in crossing the mountains and reaching the valleys, where the danger was at an end. The storm continued, almost without intermission, for several weeks, and those who had crossed the Summit knew that an attempt to reach the imprisoned party would be futile—worse than folly, until the spring sun should melt away the icy barrier.

Of the long and dreary winter passed by these three persons, who shall tell? The tall stumps standing near where stood the cabin, attest the depth of snow. Some of them are twenty feet in height.

Early in the spring a party of brave men, led by Claude Cheney, started from the valley to bring out the prisoners, expecting to find them alive and well, for it was supposed that they had provisions enough to last them through the winter, but it seems they were mistaken.

After a desperate effort, which required weeks of toil and exposure, the party succeeded in scaling the mountains, and came to the camp of the Donners. What a sight met the first glance! In a rudely constructed cabin, before the fire, sat the Dutchman, holding in a vice-like grasp a roasted arm and hand, which he was greedily eating. With a wild and frightened look he sprang to his feet and confronted the new comers, holding on to the arm as though he feared they would deprive him of his repast. The remains of the arm were taken from him by main force, and the maniac secured. The remains of Mr. Donner were found, and, with those of his faithful wife, given such burial as the circumstances would permit, and taking the survivor with them, they returned to the valley.

The German recovered, and still lives. His story is, that soon after the party left, Mr. Donner died, and was buried in the snow. The last of the cattle escaped, leaving but little food; and when that was exhausted, Mrs. Donner died. Many dark suspicions of foul play on the part of the only survivor have been circulated, but whether they are correct will never be known, until the final unraveling of time's dark mysteries.

No. 34 ANNEX. "Roll I'm Through."—Oct. 17th, 1872, as an excursion train, loaded with passengers, most of whom were women and children, rounded the curve close below the tunnel, and with No. 6 train thundering along close behind, the timbering in the tunnel was discovered by the fireman to be on fire. The engineer, Johnny Bartholomew, comprehended the position at a glance, made one of the most brilliant dashes, under the circumstances, on record. The train past through the tunnel safely, when to have stopped short would have been sure death. G. H. Jennings, Esq., of Brooklyn, N. Y., has put the following words in the mouth of the brave engineer:

I ain't very much on the fancy,
   And all that sort of stuff,
For an engineer on a railroad
   Is apt to be more "on the rough;"
He don't "go much" on "his handsome,"
   I freely "acknowledge the corn,"
But he has got to "git up" on his "wide-awake,"
   That's "just as sure's you're born."

Now, I'll tell you a little story,
   'Bout "a run" we had for our necks,
When we thought "old Gabe" had called us,
   To "ante up our checks."
We came 'round the curve by the tunnel,
   Just beyond the American Flat,
When my fireman sings out, "Johnny!
   Look ahead! My God, what's that?"

You bet, I warn't long in sightin'—
   There was plenty for me to see,
With a train full of kids an' wimmen,
   And their lives all hargin' on me—
For the tunnel was roarin' and blazin',
   All ragin' with fire an' smoke,
And "Number Six" close behind us—
   "Quick, sonny! shove in the coke."

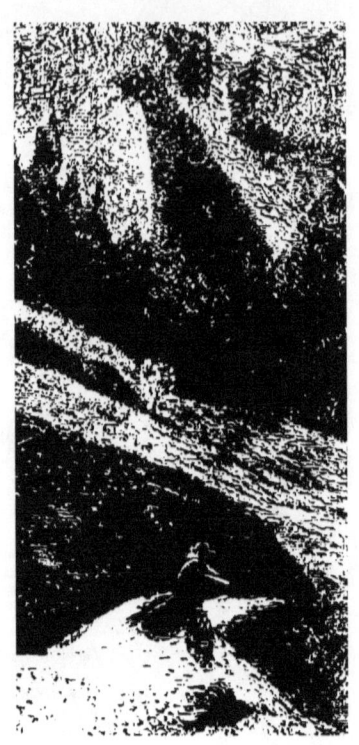

"Whistle 'down brakes,'" I first thought;
  Then, think's I, "old boy, 't won't do;"
And with hand on throttle an' lever,
  I knew I *must roll 'em through!*
Through the grim mouth of the tunnel—
  Through smoke an' flame, as well—
Right into the "gateway of death," boys;
  Right smack through the "jaws of hell!"

The staunch "old gal" felt the pressure
  Of steam through her iron joints;
She acted just like she was human—
  Just like she "knew all the points;"
She glided along the tramway,
  With speed of a lightning flash,
With a howl assuring us safety,
  Regardless of wreck or crash.

I 'spose I might have "jumped the train,
  In hope to save sinew and bone,
And left them wimmen and children
  To take that ride alone;
But I tho't of a day of reck'nin';
  And whatever "Old John" done here,
No Lord ain't going to say to him then,
  "*You went back* as an engineer!"

No. 35 ANNEX. **The Valley of the Yellowstone**—No. 7, of the large illustrations, is engraved from a photograph taken by Prof. Hayden, the great explorer of the West. It presents a view of one of the finest and most picturesque portions of the valley. It is looking southward, above the first or lower canyon, and directly on the Snowy Range, whose white-capped summits may be seen on the left of the picture, extending up the river. Below is the first canyon, between the high, narrow, limestone walls of which, the Yellowstone flows, about three miles, and then makes its exit from the mountain region proper. The valley is about 20 miles in length, and from four to five miles in width, and is one of the most delightful portions of Montana. (See ANNEX No. 26 and 36.)

No. 36 ANNEX. **The Falls of the Yellowstone**—as illustrated in No 8 of our series of large views is one of the most perfect pictures ever made. It is from a photograph taken by Prof. Hayden, and engraved by Bross, of New York. It represents the Lower Falls of the Yellowstone, where the waters make a leap into the canyon, a distance of 350 feet. Prof. Hayden, in his report, says: "After the waters of the Yellowstone roll over the upper falls, (140 feet,) they flow with great rapidity over an apparently flat, rocky bottom, which spreads out to nearly double its width above the falls, and continues thus until near the Lower Fall, when the channel again contracts, and the waters seem, as it were, to gather themselves into one compact mass, and plunge over the precipice in detached drops of foam, as white as snow, some of the huge globules of water shooting down through the sunlight, like the white fire contents of an exploded rocket. It is a spectacle infinitely more beautiful than the grandest picture ever presented of the famous Falls of Niagara. In the immediate vicinity of the Lower Falls, and in the grand canyon, the scene is indescribably beautiful. A heavy mist arises from the water at the foot of the falls, so dense that one cannot approach within from two to three hundred feet of them, and even then the clothes will be drenched in a few moments. Upon the glowing, yellow, nearly vertical walls of the west side, the mist mostly falls, and for 300 feet from the bottom, the wall is covered with a thick matting of mosses, sedges, grasses, and other vegetation of the most vivid green, which have sent their small roots into the softened rocks and are nourished by the ever-ascending spray. (See ANNEX No. 26 and 35.)

No. 37 ANNEX. **Falls of the Willamette River.** (See page 215.)

No. 38 ANNEX. **Cape Horn.** (See page 232.)

No. 39 ANNEX. **Wood Hauling in Nevada.** (See page 232.)

No. 40 ANNEX. **Mirror Lake Yo-Semite Valley.** (See page 209.)

No. 41 ANNEX. **The Pony Express.** (See page 151.)

No. 42 ANNEX. **Sierra Nevada Mountains.** (See page 138.)

No. 43 ANNEX. **Mount Shasta.** (See page 214.)

No. 44 ANNEX. **Woodward Gardens**—These Gardens were laid out in 1860 by R. B. Woodward, Esq., a gentleman of enterprise and refined taste, to surround, adorn and beautify his private residence, situated near the center of the grounds. To this end the continents of both America and Europe were searched to procure every variety of ornamental trees, exotics, indigenous plants, or articles of *rare virtue* and *value*. For us to attempt to describe these beautiful grounds, and do justice to the subject, were we able, would take a larger book than the TOURIST. They must be seen to be appreciated. You will find in the "Art Gallery" rare paintings and statuary; in the "Zoological department a great variety of different kinds of wild animals, including the California lion, and a mammoth grizzly bear, weighing 1.600 pounds; also a great variety of California birds.

In these grounds are towering evergreen trees and crystal lakes, oriental arbors and beautiful statuary, delightful nooks and shady retreats, with creeping vines, fragrant flowers, sparkling fountains, sweet music, and, above all, the glorious California sky. Possessed of all these luxuriant surroundings, and with ample income, could any person be surprised that Mr. Woodward should persistently decline to open them to the curious public? But the time came at last. It was when the soldiers and sailors of this country lay bleeding in the hospitals, on the ships, in the camps, and on the battle-fields, with widows, orphans, suffering, and death on every side. The sanitary fund was low. *Money must be had!* Then it was that his noble heart leaped to the rescue. The grounds were then thrown open to the public *in aid of the Sanitary Fund*. The receipts were princely; and no one can say how many lives were saved, or the sufferings of the last moments of life alleviated, by the aid of the generous proprietor of the Woodward Gardens? These gardens were opened permanently to the public in May, 1866. They occupy five acres of ground, four of which are bounded by Market, Mission, 13th and 14th streets, with one acre to the south of 14th street, connected by tunnel under that street from the main garden.

No. 45 ANNEX.—**The Geysers**—Page 184.

No. 46 ANNEX—**The large view of San Francisco**—See page 214.

No. 47 ANNEX. [From page 234.] **The Ancient Ruins** of Arizona are very extensive, and are scattered over a large portion of the Territory. These ruins consist, in part, of towns, cities, and scattered hamlets; castles, temples and great buildings; fortifications, huge walled enclosures and causeways, together with great canals, cisterns and reservoirs of immense feet in width; one building is 350 feet long by 150 wide; an immense castle is situated on the apex of a mountain, 3,000 feet above the valley. Potteryware and stone implements in great variety are scattered about the ruins, while rude hieroglyphics and pictures of men, animals, birds, turtles and reptiles are painted on or cut deep into the rocks, at different places in the country. Burial, or cremation urns are often found, some containing ashes and partly burned human bones.

Casa Grande, (signifying "Big House,") one of these ruins, is situated a few miles south of the Gila river, on a great plain, about 14 miles north of the station of Casa Grande. This building is one of the best preserved, but unde. the medium size; is 63 feet long by 45 feet wide; the walls are of concrete, five feet thick, and are now standing about 40 feet in height, or a little over four stories.

These ruins are pre-historic; the builders have long since passed away, leaving no records of who they were; when they lived; whence they came, or whither they went.

The first account in history of these ruins date back over 300 years—to the Sixteenth Century—at which time the Jesuit Fathers explored and described the country; its ruins and people; their reports differ in no essential points from those of a later period, nor could the Fathers procure any information from the Indians then inhabiting the country, that would aid them to establish the identity of the people who had built the cities, towns and other improvements, and who undoubtedly possessed much civilization, and lived by cultivating the soil.

No. 48. ANNEX. **Painted Rocks.**—Mr. Hodge, in his work, "*Arizona as it is,*" says: "This mass of rock rises from the surface of the plain to a height of perhaps 50 feet, the uppermost being a broken ledge from which masses have fallen off, and the whole covering less than an acre of ground. On the standing ledge and, on the broken masses at its base, are carved deep in the surface rude representations of men, animals, birds and reptiles, and of numerous objects, real or imaginary, some of which represent checker boards, some camels and dromedaries, insects, snakes, turtles, etc., etc.; and on the other broken rocks at the base of the ledge, are found on all sides like sculptured figures, some of which are deeply imbedded in the sand. These pictured rocks present much of interest to the thinking mind, and when examined by some one versed in hieroglyphical reading, may be found to give some clue to the time of making and the people who made them."

The earliest account we have of these "Painted Rocks," as well as the "Ancient Ruins" of Arizona, comes from the exploration of the Jesuit Fathers, who traversed the country in the Sixteenth Century. In comparing their reports with the appearance of the "Rocks" and "Ruins" of the present time, very little, if any, change is noticeable. The Indians, in answer to all questions of the Fathers, as to who carved the rocks, or built the towers and cities then in ruins, received the same answer as the Pima Indians give at the present time, which was, "Moc-te-zu-ma."

No. 49. ANNEX. **Viewing Progress**—During the construction of the Pacific Railroad—and, in fact, for many years afterwards—the locomotive, cars, and all persons connected with the road, were viewed with great curiosity by the Indians in the country through which it was built. The engines — "fire wagons" — and the long train of cars — "heap wagon no hoss" — received the Indians' special attention; and they would gather around on the hills at first, and then cautiously approach and watch every movement—lying around for days and months at a time. From the commencement of the building of the road through the Indian country until its completion, the Indians had a wholesome fear of the "fire wagons." They would often attack small parties of graders, or stragglers from the camps; but only in two or three instances did they attempt to disturb the passing trains, and in those cases they were punished *so severely,* that ever afterward they declared "fire-wagon" bad medicine.

The illustration on page 233 represents a small party of Apache Mohaves, headed by their famous war chief, Mi-ra-ha, in 1868, who, having heard about the terrible "fire-wagons," left their country in northern Arizona, and made a pilgrimage to the northward, many hundred miles to view the great curiosity. We learned from a half-breed, on a recent visit to Arizona, that Mi-ra-ha, after his return to Arizona, resolved to gather his forces and capture one of these Pacific railroad 'fire-wagons.' But as Gen. Crook made it *very lively* for him at home for many years after his return, and as Capt. Porter sent him to his "happy hunting ground," in 1876, he has probably got all he can attend to.

No. 50. ANNEX. **Palace Hotel at San Francisco.**—This monster hotel of the world, is situated in the city of San Francisco occupying one entire block of ground, 344 by 265 feet, bounded by New Montgomery, Market, Annie and Jessie streets. It is seven stories high (115 feet), the foundation walls are twelve feet thick, while the exterior and interior walls range from 1½ feet to 4½ feet in thickness.

The foundation walls at their base, are built with inverted arches. All exterior, interior and partition walls, at every five feet, commencing from the bottom of the foundation, are banded together with bars of iron, forming, as it were, a perfect iron basket-work filled in with brick. The quantity of iron so used, increases in every story towards the roof, and in the upper story the iron bands are only two feet apart.

The roof is of tin, the partitions of brick and the cornice of zinc and iron. The building has three courts, the center one having an iron-framed glass covering, and is 144 by 84 feet, with a drive-way and sidewalk opening on New Montgomery street, forty-feet wide. The two outer courts, from the basement level, are each 22 by 135 feet, with two drive-ways, 20 feet wide, one from Market and Annie streets, and one from Annie and Jessie streets. These are connected by two brick-arched passage-ways, ten feet in width, allowing ample space for a four-in-hand team to pass under and through them.

Besides the city water-works, a supply of water comes from four artesian wells of a ten-inch bore, which have a capacity of 28,000 gallons per hour. A reservoir is located under the centre court, capable of containing 630,000 gallons. On the roof are seven tanks, which will contain 128,000 gallons.

The hotel is supplied with two steam force pumps for water, two additional for fire, five elevators, together with all the modern improvements, and built throughout in the most substantial manner. (See illustration page 180.)

No. 52 ANNEX. **"Prickey," the Horned Toad.** See page 126.

No. 53 ANNEX. **Yo-Semite** and **Big Trees.** See page 184.

No. 54 ANNEX. **Our Artists.**—The greater portion of the engraving in this work was executed by R. S. Bross, of New York, and C. W. Chandler, cor. Madison and Dearborn Sts., Chicago. Nearly all the large views, including "Utah's Best Crop," "Women of the Period," and "Brigham Young," and most of the large views were engraved from photographs, by Mr. Bross, while the "Orange Orchard," "The Loop," "Crossing the Sangre de Christo Mountains," "Yucca Palm," "Hanging Rock of Utah," etc., were engraved by Mr. Chandler.

The photographs were by Savage, of Salt Lake City, and Watkins and Houseworth, of San Francisco. All of these artists we take pleasure in recommending.

No. 55 ANNEX. **The "Boss" Cactus of the World**—on page 233, a simple reference has been made to this wonderful *Sprout* as being peculiar to the Gila Desert. It is possible they may grow in other portions of the Territory, but certain it is, these are the *first* on our route.

These Cacti are of different shades of green and yellow, and rise from the ground in the shape of a *huge* cone, many to the height of 60 feet, with a diameter of three feet near the ground. Some of these great cones have from one to five smaller cones that branch out from the main trunk at different heights, and shoot up parallel with it to various heights, all presenting the same general appearance.

All these cones are grooved from top to bottom, the grooves being from one to three inches in depth and as many inches apart; the whole surface is covered with thorns of various sizes,—some three inches in length; and all very sharp.

These cacti have a tough flaxen shell or exterior, but a soft, pithy inside, and produce one blossom annually—on the top—and yield a kind of fruit much prized by the natives.

On the Gila Desert, no tree or shrub grows more than a few feet from the ground, and rattlesnakes, lizards, owls, and woodpeckers are about the only living things noticeable.

How the lizard and owl manage to raise the young, and keep them from being devoured by the snake, is a problem which the woodpecker alone has solved by pecking a hole in the tall cactus near its top, making its nest, and raising its young secure from the snake and all its creeping enemies. As we ride along and see Mr. Woodpecker peeking out from his fortress in the tall cactus, we cannot help but admire the sagacity of the little fellow, while condemning his judgment for attempting to live and bring up a respectable family in such a "God-forsaken country" as the Gila Desert.

No. 58 ANNEX. **New Sacramento Depot.** See page 173.

No. 63 ANNEX. **The Mammoth Snow Plow.** See page 126.

No. 64 ANNEX. **Arizona** is a Territory of 122,000 square miles, more than double the size of the State of Pennsylvania, and, if reports are true, contains a wealth of minerals *far* exceeding any portion of the United States. Indian difficulties have had much to do in retarding the settlement of the Territory, but happily they are now at an end, and the proximity of the "iron horse" has had a tendency to direct attention to this heretofore almost inaccessible region, the result of which will soon enable the land of *Arisunna*—"The Beautiful of the Sun"—to come forward and demand admission into the Union of States as one more star in the bright constellation. Spanish Adventurers penetrated Arizona as early as 1540, but no permanent settlement was made until 1560, when the Jesuit Fathers settled with their followers at Tucson. In 1725 there were thirty missions within the present limits of Arizona, besides seventy-one Indian villages in charge of the Jesuit Missionaries. At that time these missions were in the height of their prosperity, and from which time they commenced to decline, owing principally to Indian difficulties. Many of the Missions were burned and the priests murdered.

The mineral deposits of Arizona are very extensive and very rich; principally gold, silver, copper and coal, but we have no space to particularize. We will simply present a few "items," and chronological events: In 1540, the Spanish viceroy—Mendoza -ordered an exploration of Arizona, at which time there were 200 silver mines being worked. The miners and people were despoiled and immense wealth carried away to Madrid, Spain.

In March, 1539, Padre Marco de Niza, and Senor Estivanico—a man of color—left Caliacana—New Spain—and reached the Gila River, and discovered the Pima Indian villages, at the same point in the valley that their descendents are now found.

In 1560, the first settlement was made at Tucson. An Indian outbreak in 1802, and again in 1827, made the tenure of the Spanish rule precarious and practically ended it; yet, soldiers remained in the country in small numbers until 1840. In 1824, Sylvester Pattie and his son James—formerly of Bardstown, Ky.—visited Arizona on a trapping expedition, failing in their efforts to find game, they were arrested by order of the Mexican commander of San Diego and imprisoned, where the father died. Sometime afterwards James was released, and joined the famous Walker expedition and was killed in battle.

In 1849, the "Southern Route"—through Arizona—was much frequented by emigrants enroute for the gold fields of California, which resulted in much suffering and loss of life.

Arizona and New Mexico were ceded to the United States by Mexico, February 2d, 1848—or that portion laying north of the Gila and Masilla valleys. The portion to the south, was not acquired until December 30, 1853, and was known as the "Gadsden's Purchase." This last acquisition formed a part of Sonora. The Boundary Commission commenced its work in the summer of 1849 and completed it in 1855.

Cap. John Moss—is said to be the first explorer of the Grand Cañon of the Colorado, in 1850. Major Heintzelman located Fort Yuma in 1851. Lieut. Ives, explored the Colorado River and its lower cañons, in 1854.

Lieut. A. B. Gray, in 1854, made a survey for a railroad from Marshall, Texas, to El Paso and thence westward to Tubac, from which point branch surveys were made to Post Labos, on the Gulf of California, and via Yuma to San Diego.—This line was known as the Hon. Robert J. Walker route, for which bonds were issued in 1852.

Lieut. Park, in 1854-5, made a survey from San Diego, Cal., via Yuma and Tucson to El Paso, Texas.

Yuma—first called Arizona City—was laid out in 1854. In August, 1856, a party left San Antonio, Texas, to prospect for mineral in Arizona. Their route was via Apache Pass—and a perilous one. Upon their arrival at Tubac—the party was divided and a hunt for silver mines commenced through the mountains of Santa Rita, Arivaca and the Cerro Colorado. Many mines were discovered and several companies formed for working them, but the Apaches got away with most of the mineral and all the scalps.

The Crabb expedition—of 1,000 men—raised in California, in 1856-7, for colonizing Sonora, reached Sonoita, Arizona, in March, 1857. By invitation of the Sonora authorities, an advance of 100 men with their commander, Henry A. Crabb, entered Sonora soon after, and were met at Coborca, by Pesqueira, Governor of the State of Sonora, and every person killed. Crabbs' head was cut off and sent by the Governor to the City of Mexico, as an evidence of his loyalty to his government. This act checked immigration to Sonora.

In 1857, Senator Guinn of California, endeavored to secure a territorial organization for Arizona, but failed.

In August, 1857, J. C. Woods, established a semi-monthly stage line, between San Antonio, Texas, and San Diego, Cal. In 1858, the Butterfield semi-weekly stage line was established, between St. Louis, Mo., and San Francisco, Cal., with a subsidy from the Government of $600,000 a year.—*Time 22 days.* His service was faithfully performed—without a single failure—until 1861, the outbreak of our civil war, when the route was changed, leaving St. Joseph, Mo., and going via Salt Lake City, Utah.

In 1861, the Government troops were withdrawn from Arizona and the Indians and desperados took possession.

In 1860, Senator Green of Missouri, endeavored to have a bill passed for a temporary Government for Arizona, but failed.

In 1860, a filibustering expedition was fitted out in California to invade Sonora, but without result.

The 27th of February, 1862, Cap. Hunter of the Southern Confederacy with 100 men, took possession of Tucson. The advance of Gen'l Carlton, with the California column of Federal troops, met Capt. Hunter's forces, near Pecacho Peak—40 miles west of Tucson—resulting in a victory for Gen'l Carlton, and the retreat of all Confederate troops into Texas.

February 24th, 1863, Congress passed the Organic Act, establishing Arizona as a Territory, with John N. Goodwin, of Maine, as Governor. Arizona, until February 24th, 1863, was a part of New Mexico.

The Territory was formally organized on Dec. 24th, 1863, at Navajo Springs, 40 miles northwest from the noted Zuni Pueblo.

When the Territory was organized, it was said to contain 32,400 Indians and only 580 Whites.

An old Arizonian said, when he saw the first locomotive coming: "I felt just as though I must go and hug it."

In 1880, the production of precious metals in Arizona was $4,472,471. In 1881, the same was $8,198,766—an increase of $3,726,295. Should we venture to prophesy—for 1882—our figures would exceed $10,000,000, and — for 1890 — $25,000,000. This Territory—in a mineral point of view—is greatly *underrated*.

The Indian troubles of last year have tended to make, not only every soldier, but every teamster, wood-chopper, burro puncher, mule-skinner, bull-whacker and all other men — traveling arsenals; with a belt about the waist loaded with cartridges, a pair of six-shooters, a formidable knife and a rifle for long range.

The soil in the valleys of Arizona is a rich gravelly alluvium, and with sufficient water, would produce abundant crops; in some portions, two crops a year.

The rains come in July, August and September, and the sand storms cloud bursts and "blizzards" occasionally.

During our civil war, Arizona was one field of carnage. Indian depredations, nearly devastated the country.

Over 40,000 square miles of coal formation can be traced in the Territory; one of the most important is the San Carlos Indian Reservation.

Just east of Old Maricopa Wells stage station, at the base of the Estrella range, may be seen a remarkable formation, representing distinctly the perfect face of a man reclining, with his eyes closed, as though in sleep. The Indians in this country have a legend concerning this face. They believe it is Montezuma's face—and that he will awaken from his long sleep some day, will gather al. the brave and the faithful around him, uplift his down-trodden people, expell the invaders on his rights, and restore to his kingdom all the great power and glory, as it was before the white man visited it.—It is reported, that in some localities, watch-fires are kept constantly burning in anticipation of Montezuma's early coming.

No. 65 ANNEX. **EMIGRANT SLEEPING CARS**—Of all the improvements adopted by Railroad companies in this country, for the comfort and convenience of their passengers, the Sleeping Car, is the most important, and we might add, the most expensive to the passengers. Indeed, the charges for berths in Sleeping Cars, on many roads—together with onerous perquisites, virtually exclude the greater portion of the traveling public. Most men emigrating with their wives and families to the El Dorado of the West, start with small means, depending upon pluck, energy and hard work, for the future outcome; these parties *cannot* pay three or more dollars a day for sleeping accommodations, however anxious they may be, to alleviate the hardships incidental to a long journey in the emigrant cars.

The honest, sober, industrious, economical and enterprising emigrant, is the *germ of life* in our trans-Missouri country. To cherish, aid, and ameliorate the condition of the emigrant, is to hasten the settlement of the vast amount of unoccupied land, and the developement of the enormous mineral and other resources of the great West, the results of which, are not only of great interest to *all* good citizens, but of *paramount importance* to the great Railroads of the country, the basis of whose existence are founded upon the very class of emigrants named, without which they would *never* have been built.

Families emigrating, should have the *special* care, attention, and *protection*, of the Railroad companies' agents, over whose road they are traveling, t gether with all the comforts and conveniences possible, and at the *lowest* rates of fare.

The Pacific Railroad companies' agents have always been noted for the fatherly attention shown their emigrant passengers, and the Sleeping Cars now in use by this line, especially for their emigrants, are for comfort and convenience, far in advance of any car heretofore used on emigrant trains. These cars are 44 feet long, 9 feet 4 inches wide, with raised roof, patent air brake couplers, and all modern Sleeping Car improvements,—excepting only upholstery,—and will accommodate 48 persons. The seat frames, are of iron, the back and seats and upper berths are wood slats. The seats let down, and the upper berths fold up, the same as those in the Palace Sleepers now in use on the first-class trains. The woodwork about the seats and upper berths is ash, polished and varnished—without paint.

As most emigrants are provided with blankets, and more or less bedding of their own, they are enabled to get along very comfortably, and as no extra charge is made for the Sleepers—economically.

No. 66 Annex. [From page 32.] **THE OMAHA, NIOBRARA & BLACK HILLS R. R.** This road was commenced in 1879, at Jackson, Neb., on the line of the Union Pacific, seven miles west of Columbus, and 99 west of Omaha, and is now completed and running to Norfolk, 46 miles north, at which place it connects with the railroad running up the Elkhorn Valley from Fremont. [See page 29.] This new road runs through a rich agricultural and well-settled section of country, and, as its name implies, its objective point is the gold regions of the Black Hills, towards which it is being pushed with the usual energy displayed by the Union Pacific management, by whom it is controlled. The line of this road has recently been changed and now runs from Columbus.

ITEMS—Gold—It is reported that Sir Francis Drake was the first discoverer of gold on the Pacific Coast. He landed on the coast a few miles north of the Bay of San Francisco, in the summer of 1578, and reported to Queen Elizabeth: "There is no part of earth here to be taken up wherein there is not a reasonable quantity of gold and silver." Yet the discovery was not followed up.

ITEMS.—The Southern Pacific Railroad reached El Paso, Texas, May 13th, 1881. The Texas & Pacific connects with the Southern at Sierra Blanca, 91½ miles east from El Paso, and runs into El Paso on the Southern's track. The first train of the Texas & Pacific rolled into El Paso, December 31, 1881. January 16th, the Southern—or the Galveston, Harrisburg & San Antonio Railway—was completed to Marfa, 195 miles east of El Paso. The Mexican Central, had February 1st, 1882, over 50 miles of track laid from El Paso, south, into Old Mexico. The elevation of El Paso, is 3,500 feet above sea level. Its population is about 3,000. El Paso Del Norte, Mexico, has a Mexican population of about 4,000.

CROFUTT'S
GRIP-SACK GUIDE OF COLORADO.
A COMPLETE ENCYCLOPEDIA OF THE STATE.

☞ Will you please to read a few of the opinions of this book from Coloradoans—those most competent to judge of its merits?

Indorsement by the Denver Board of Trade.

GEORGE A. CROFUTT, Esq.—The copy of your "Grip Sack Guide of Colorado," presented to this Board came duly to hand. I have to report to you that the Board of Directors, by resolution, passed you a vote of thanks and expressed the opinion that it is the most complete, concise and truthful book ever published on the resources of our State, and give it their *unqualified indorsement*. To this I would add my own approval, assuring the tourist, emigrant and the public generally who desire a most complete encyclopedia of Colorado that they will find it full of interest from beginning to end. Yours respectfully, J. T. CORNFORTH, President Board of Trade.

——" The most complete, most thorough and reliable guide that has ever been offered to the community. The book, which has nearly 200 pages and near 100 illustrations, contains in a condensed form all that the traveler or the tenderfoot need to know about the wonderful land that surrounds us."—*Rocky Mountain News*, Denver, Colo.

——" The subject matter is carefully prepared, and a large amount of excellent matter is condensed in its columns. San Juan is treated in a fair manner."—*Silver World*, Lake City, Colo.

——" To the tourist and traveler the work is indispensable; in fact, none traveling over any portion of Colorado, can afford to do without it."—*Tribune*, Greeley, Colo.

——" It is the most complete, authentic and concise work over written on the Centennial State." —*Elk Mountain Pilot*, Irwin, Colo.

——" The Grip Sack is a model for all books of its kind. It is elegantly and copiously illustrated, and furnishes the most valuable information in the most convenient form."—*Gazette*, Colorado Springs, Colo.

——" The most complete work ever published on this State, as the author has made personal tours through the State and knows whereof he speaks, and is also a man capable of seeing and describing."—*Independent*, Alamosa, Colo.

——" It is free from gloss or 'taffy' so often found in works of this kind."—*San Juan Herald*.

——" Without hesitation we pronounce it the most perfect, complete and convenient work of the kind that ever came under our notice. It is a work invaluable to those visiting the State, and of great interest and importance to everybody."—*Mountaineer*, Colorado Springs, Colo.

——" It is a perfect gem of typographical art and is chuck full of information, which makes it a most desirable book for tourists and others coming into the State, as it tells, to use its own language, what is worth seeing, where to see it, how to go, where to stop and what it costs."—*Sierra Journal*, Rosita, Colo.

☞ Crofutt's Grip-Sack Guide is published by The Overland Publishing Co., of Omaha, Neb., printed on fine tinted paper, magnificently illustrated, bound in full cloth and gold, for the library, and a Railway Edition in flexible cloth. It contains a complete map (colored) of the State, and is for sale by news agents on all regular passenger trains in the Western country.

☞ *Don't fail to buy the Grip-Sack if you want to know all about the great Centennial State; the State that produces annually the largest amount of precious metals in the world*—OVER *$26,000,000.*

OVERLAND TIME TABLE.

UNION PACIFIC............................EASTERN DIV.
† Meals.   P. J. NICHOLS, *Division Supt.*, OMAHA, NEB.   * Telegraph.

WEST BOUND.			Dist. from Omaha.	OMAHA TIME. STATIONS.	Elevation	EAST BOUND.		
Denver Express.	Daily Emigr'nt.	Daily Ex. 1st & 2d Class.				Daily Ex. 1st & 2d Class.	Daily Emigr'nt.	Denver Express.
7.00† P M	5.20 P M	11.20 A M		Lv........Transfer........Ar		4.00 P M	6.10 A M	8.00† A M
7.35	5.47	12.15 P M†		Lv....*Omaha Depot.....Ar	966	3.25 P M†	5.20	7.35
7.45	6.28	12.25	4	......Summit Siding.......	1142	3.10	5.00	7.22
8.00	7.00	12.40	10	..........*Gilmore..........	976	2.55	4.30	7.07
8.13	7.25	12.52	15	.........*Papillion.........	972	2.42	4.05	6.55
8.30	7.55	1.09	21	..........*Millard..........	1047	2.25	3.35	6.40
8.50	8.32	1.30	29	..........*Elkhorn..........	1150	2.06	2.55	6.21
8.57	8.42	1.37	31	..........Waterloo..........	1140	2.00	2.45	6.15
9.10	9.00	1.50	35	...........*Valley...........	1147	1.50	2.25	6.05
9.25	9.40	2.05	42	...........Mercer...........	1120	1.34	1.50	5.50
9.37	10.03 †	2.15 †	47	..........*Fremont..........	1176	1.20 †	1.25 †	5.38
9.54	10.35	2.53	54	............Ames............	1270	12.38	12.50	5.21
10.13	11.10	3.12	62	........*.North Bend........	1259	12.18	12.13 A M	5.02
10.30	11.40	3.30	69	...........Rogers...........	1359	12 00 noon	11.45	4.45
10.47	12.17 A M	3.48	76	..........*Schuyler.........	1335	12.43	10.47	4.27
11.05	12.55	4.08	84	..........*Benton...........	1440	11.25	10.08	4.08
11.23	1.35	4.20	92	.........*Columbus..........	1432	11.05	9.35	3.49
11.40 P M	2.15	4.47	99	..........*Duncan...........	1470	10.47	9.00	3.31
12.04	3.07	5.12	109	.........*Silver Creek.......	1534	10.21	8.15	3.07
12.18	3.33	5.26	115	..........Havens............		10.08	7.50	2.45
12.31	4.00	5.40	121	..........*Clark's..........	1610	9.54	7.50	2.41
12.45	4.27	5.54	126	.........Thummel's.........		9.40	7.00	2 26
12.58	4.53	6.08	132	.......*Central City........	1686	9.25	6.33	2.15
1.07	5.03	6.16	135	..........Paddock...........		9.16	6.16	2.07
1.24	5.40	6.33	142	.........*Chapman's........	1760	9.00	5.46	1.50
1.37	6.05	6.46	148	..........Lockwood..........	1800	8.46	5.25	1.37
1.55	6.35 †	7 00 †	154	.......*Grand Island.......	1850	8.30 †	5.00 †	1.20
2.23	7.50	7.42	162	...........Alda............	1907	7.50	4.03	12.52
2.42	8.30	8.05	170	.........*Wood River........	1974	7.30	3.28	12.33
3.00	9.05	8.25	178	...........Shelton...........	2010	7.10	2.55	12.15 P M
3.15	9.35	8.42	183	...........*Gibbon..........	2046	6.54	2.25	12.00 night
3.35	10.12	9.05	191	............*Buda...........	2106	6.32	1.45	11.40
3.45	10.30	9.16	196	.....*Kearney Junction,.....	2150	6.21	1.25	11.30
4.00	11.00	9.33	201	..........Stevenson.........	2170	6.05	12.55	11.16
4.00	11.20	9.44	205	............Odessa..........		5.53	12.35	11.06
4.25	11.50	10.08	212	.........*Elm Creek.........	2241	5.35	12.05 P M	10.51
4.48	12.45 P M	10.30	221	..........*Overton..........	2305	5.12	11.23	10.30
5.00	1.05	10.42	225	..........Josselyn..........	2330	5.00	11.03	10.17
5.15	1.35 †	10.59	230	.........*Plum Creek........	2370	4.45	10.34	10.03
5.33	2.10	11.20	239	...........Cayotte...........	2440	4.25	9.59	9.46
5.48	2.37	11.35	245	............Cozad...........	2480	4.10	9.33	9.32
6.00	3.00	11.48	250	........*Willow Island.......	2511	3.58	9.10	9.21
6.24	3.45	12.15 A M	260	...........Warren...........	2570	3.33	8.25	8.58
6.43	4.20	12.37	268	........*Brady Island........	2637	3.14	7.46	8.40
6.55	4.39	12.50	278	...........Hindrey..........	2695	2 02	7.25	8.30
7.05	5.00	1.02	282	...........Maxwell...........		2 52	7.05	8.21
7.23	5.30	1.23	285	...........Gannett...........	2752	2.35	6.27	8.05
7.40	6.00 †	1.40	291	Ar.....*North Platte.....Lv	2789	2.20 A M	6.00 A M	7.50 P M

MOUNTAIN DIVISION.
ROBERT LAW, *Div. Supt.*..NORTH PLATTE, NEB.

8.00	6.30 P M	2.00 A M	291	Lv......North Platte......Ar	2789	2.00	5.35	7.30
8.18	7.10	2.25	299	...........Nichols...........	2892	1.38	5.05	7.10
8.25	7.50	2.45	308	.........*O'Fallon's........	2976	1.13	4.30	6.50
8.55	8 38	3.00	315	............*Dexter..........	3000	12.53	4.05	6.35
9.13	9.03	3 20	322	............*Alkali...........	3038	12.33	3.20	6.18
9.36	9.47	3.45	332	...........Roscoe...........	3105	12.10 A M	2.22	5.56
9.57	10.33	4.07	342	..........*Ogalalla..........	3190	11.45	1.27	5.36
10.02	10.40	4.12	344	............Boslor...........		11.40	1.20	5.32
10 21	11.20	4.30	351	.............Brule...........	3266	11.20	12.42 A M	5.15
10.44	12.00 night	4.55	361	..........*Big Spring.........	3325	10.53	12.00 night	4.53
11.02	12.37 A M	4.20	369	............Barton...........	3431	10.32	11.18	4.37
11.10 A M	12.55	5.32 A M	371	.......*Denver Junction.......	3130	10.22	11.00	4.30 P M

NOTE.—See third page ahead for continuation of Overland Time.

UNION PACIFIC............KANSAS PACIFIC DIV.

D. E. CORNELL, *Gen'l Agt. Pass'r and Ticket Departments*......KANSAS CITY, MO.

WEST BOUND TRAINS.			Dis. from Kan. City	STATIONS.	Elevati'n	EAST BOUND TRAINS.		
Emigr'nt	Col. Ex.	Pac. Ex.		* Tel. † Meals.		Atlantic Ex.	Eastern Ex.	Emigr'nt
3.00 P M	6.00 P M	3.00 P M	....	Lv........Boston........Ar	....	2.40 P M	6.25 A M	2.40 P M
5.55	5.55	5.55	....	........New York........	....	10.35 A M	9.30	10.36
6.00	12.30	6.00	....	........Chicago........	....	6.20	2.25 P M	6.20 A M
8.52 A M	8.32	8.52 A M	....	........St. Louis........	....	5.52 P M	6.22 A M	6.05
10.45 P M	9.45 A M	10.10 P M	....	Lv....*Kansas City.....Ar	766	6.00 A M	5.35	3.10
	9.50		....	....*Leavenworth....	783		5.55	....
11.00	9.50	10.15	....	....*State Line	763	5.55	5.30	3.05
11.05	9.55	10.20	1	....*Armstrong....	773	5.50	5.25	2.55
11.55	10.26	10.54	12	....*Edwardville....	801	5.18	4.55	2.05
12.12 A M	10.36	11.04	16	........Tiblow........	811	5.08	4.46	1.49
12.25	10.46	11.14	19	........Loring........	799	4.59	4.39	1.36
12.37	10.54	11.22	22	........Lenape........	799	4.51	4.33	1.23
12.56	11.07	11.35	27	........*Linwood........	807	4.38	4.22	1.04
1.36	11.32	12.01 A M	26	....L. & L. Junction....	831	4.13	4.00	12.25
1.40	11.35	12.04	37	....Bismarck Grove....	....	4.10	3.58	12.20
2.00	11.38	12.07	38	........*Lawrence........	845	4.07	3.55	12.15 P M
2.40	12.05 P M	12.35	47	........Williamston........	869	3.40	3.31	10.38
2.54	12.15	12.44	51	........Perryville........	870	3.31	3.24	10.21
3.00	12.19	12.48	52	........Medina........	871	3.27	3.20	10.14
3.20	12.26	12.55	55	........Newman........	879	3.20	3.13	10.01
3.40	12.41	1.10	60	........*Grantville........	895	3.07	3.01	9.35
4.08	12.58	1.29	66	...A., T. & S. F. Crossing...	....	2.50	2.47	9.03
4.20	1.00 †	1.31	67	........*Topeka........	904	2.48	2.25	9 00
4.39	1.31	1.45	71	........Menoken........	....	2.35	2.14	8.24
5.03	1.46	2.02	77	........*Silver Lake........	933	2.18	2.01	7.56
5.28	2.01	2.19	83	........*Rossville........	951	2.02	1.46	7.30
6.00	2.20	2.41	90	........*St. Marys........	973	1.42	1.30	6.55
6.26	2.36	3.00	96	........Belvue........	....	1.24	1.16	6.26
7.20	2.55	3.30	103	........*Wamego........	1018	1.05	12.53	5.45
7.50	3.23	3.48	110	........St. George........	1018	12.42	12.35	4.45
8.30	3.45	4.10	118	........*Manhattan........	1042	12.22 A M	12.19 P M	4.10
9.24	4.17	4.41	129	........*Ogdensburg........	1078	11.53	11.51	3.00
9.55	4.32	5.00	134	........Ft. Riley........	1090	11.35	11.36	2.25
10.08	4.43	5.08	138	......*Junction City......	1100	11.20	11.30	2.10
11.18	5.17	5.45	151	........*Hazleton........	1132	10.54	10.58	1.08
11.46	6.01	6.01	157	........*Detroit........	1153	10.36	10.48	12.30
12.20 P M	5.50 †	6.35 †	162	........*Abilene........	1173	10 22	10.30	12.08 A M
12.57	6.33	7.02	171	........*Solomon........	1193	9.50	10.10	11.15
1.31	6.53	7.26	179	......New Cambria......	....	9.38	9.50	10.25
1.56	7.08	7.45	185	........*Salina........	1243	9.21	9.35	10.00
2.30	7.30	8.10	193	........*Bavaria........	1280	8.59	9.15	9.10
3.15	7.45	8.40	200	........*Brookville........	1366	8.40 †	8.40 †	8.45
3.35	8.10	8.52	204	........Arcola........	1459	8.08	8.30	7.10
4.00	8.27	9.12	210	........Alum Creek........	1586	7.50	8.15	6.30
4.35	8.12	9.27	215	........Mount Zion........	1672	7.35	8.01	6.00
4.46	8.49	9.35	218	........Fort Harker........	1600	7.28	7.53	5.45
5.12	9.03	9.50	223	........*Ellsworth........	1556	7.13	7.42	5.12
5.40	9.25	10.10	230	........Black Wolf........	1583	6.55	7.17	4.37
6.32	9.48	10.85*	239	........*Wilson........	1702	6.32	6.50	3.53
7.40	10.22	11.16	252	........*Bunker Hill........	1882	5.53	6.25	2.45
8 31	10.50	11.46	262	........*Russell........	....	5.27	6.08	1.55
9.20	11.11	12.15 P M	271	........*Gorham........	....	5.00	5.37	1.10
9.33	11.19	12.22	274	........Walker........	1962	4.52	5.30	12.55
9.54	11.30	12.35	278	........*Victoria........	....	4.42	5.20	12.35 P M
10.50 A M	11.56	1.05	289	........*Hays........	2009	4.10	4.55	11.25
12.10	12.27 A M	2.10 †	302	........*Ellis........	2135	3.35	4.17 †	10.15
1.05	12.57	2.40	312	........*Ogallah........	2385	2.40	3.54	8.35
2.20	1.17	3.05	321	........*Wa Keeney........	....	2.10	3.32	7.55
3.50	1.35	3.45	335	........*Co-lo-no........	....	1.35	3.14	6.40
5.10	2.25 †	4.25	350	........Buffalo Park........	2773	1.00	2.25	5.10
5.35	2.40	4.40	355	........Grainfield........	....	12.45	2.06	4.44
6.10	3.02 *	5.04	364	........*Grinnell........	2923	12.22 P M	1.43 A M	3.45 A M

Jefferson City time—30 minutes faster than Denver.

UNION PACIFIC..................KANSAS PACIFIC DIV.

CONTINUED.

TRAINS BOUND WEST.			Dist. fr Kan Cy	STATIONS.		Elevation.	TRAINS BOUND EAST.		
Emigr't.	Col. Exp.	Pacific Express.		†Meals.	*Tel.		Atlantic Express.	Eastern Express.	Emigr't.
7.05	3.34	5.40	376	........Cleveland........		3064	11.52	1.16 A M	2.45
8 00	4.01	6 10	387	........*Monument........		3199	11.21	12 49	1.45
9.20	4.44	7.00	405	........*Sheridan........		3121	10.40	12.06 A M	12.20 A M
10.16 A M	5.20 †	7.10 †	420	........*Wallace........		3319	10.00	10.35	11.00
1.05	6.02	8.55	452	........Arapaho........		4024	7.25	9.25	7.05
1.55	6.30	9.30	461	........*Cheyenne Wells........		4295	6.55	9.05	6.25
2.55	7.00	10.00	472	........First View........		4595	6.30	8.41	5.37
4.55	7.39	10.45	487	........*Kit Carson........		4307	5.50	8.00	3.57
6.10	8.22	11.35 P M	509	........*Aroya........		4666	5.00	7.15	2.20
7.30	9.05	12.30	523	........Mirage........		4859	4.14	6.34	12.45
8.20	9.25 †	1.10	534	........*Hugo........		5068	3.50	6.15	12.05 P M
9.50	10.44	2.10	555	........*River Bend........		5511	2.45	5.04	9.10
11.50 P M	11.55	3.40	583	........*Deer Trail........		5203	1.25	3.52	6.50
12.50	12.28	4.15	595	........*Byers........			12.50 A M	3.25	5.50
2.30	1.31	5.25	617	........*Box Elder........		5546	11.46	2.33	4.05
4.00	2.30 P M	6.35	639	Ar........Denver........Lv		5197	10.45	1.30 P M	2.00
4.10	....	0.00	639	Lv........Denver........Ar		5197	6.50		1.40 A M
7.45 A M	....	11.26 A M	690	........Greeley........		4479	4.32		10.05
11.50 P M	....	1.50 P M	741	Ar........Cheyenne........Lv		6041	2 10		6.00 P M
....	....	7.00 A M	639	Lv........Denver........Ar		5197	10 00		
....	....	7.30	655	........Golden........		5387	9.15		
....	....	9.25	684	........Boulder........		5184	7.37		
....	....	11.25	727	........Fort Collins........		4966	5 45		
....	....	1.25 P M	760	Ar...Colorado Junction...Lv		6325	4 00 P M		

UNION PACIFIC............St. JOSEPH & WESTERN DIV.

J. HANSEN, *General Agent*..ST. JOSEPH, MO.

GOING WEST.			Dist. fr St Jos'h	STATIONS.		GOING EAST.		
	No. 1.	No. 3. Pac. Ex.		†Meals.		No. 4.	No. 2.	
....	....	6.30 A M		Lv........St. Joseph........Ar		9.00 P M		....
....	....	6.42	1	........Elwood........		8.48		....
....	....	6.54	6	........Wathena........		8.35		....
....	....	7.21	14	........Troy........		8.10		....
....	....	7.25	15	........A. & N. Junction........		8.06		....
....	....	7.53	25	........Severance........		7.37		....
....	....	8.05	29	........Leona........		7.25		....
....	....	8.22	34	........Robinson........		7.08		....
....	....	8.47	43	........Hiawatha........		6.43		....
....	....	9.10	50	........Hamlin........		6.21		....
....	....	9.21	54	........Morrill........		6.09		....
....	....	9.40	61	........Sabetha........		5.50 †		....
....	....	10.05	69	........Oneida........		5.06		....
....	....	10.31	77	........Seneca........		4.42		....
....	....	11.04	89	........Axtell........		4.08		....
....	....	11.33	99	........Beattie........		3.40		....
....	....	12.30† P M	112	........Marysville........		3.00		....
....	....	12.45	118	........Merkimer........		2.44		....
....	....	1.10	128	........Hanover........		2.20		....
....	....	1.45	137	........Hollenberg........		1.45		....
....	....	1.57	142	........Steele City........		1.32		....
....	....	2.26	152	........Fairbury........		1.05 †		....
....	....	3.05	167	........Alexandria........		12.07 P M		....
....	....	3.30	175	........Belvidere........		11.44		....
....	....	*3.50	183	........Carleton........		11.25		....
....	....	4.10	191	........Davenport........		11.07		....
....	....	4.34	200	........Edgar........		10.42		....
....	....	4.56	209	........Fairfield........		10.21		....
....	6.35 A M	5.45	227	........Hastings........		9.35	8.47 A M	....
....	7.15	6.15	240	........Doniphan........		9.05	8.07	....
....	7.50	6.47† P M	252	Ar........Grand Island........Lv		8.35 A M	7.30	....

NOTE.—Trains run on St. Joseph time between St. Joseph and Grand Island.

This Train connects with Express on Main Line for Omaha, and all points East

This Train connects closely with Regular Express from Omaha.

OVERLAND TIME TABLE. 265

UNION PACIFIC..MOUNTAIN DIV.
CONTINUED.

† Meals.   * Telegraph.

WEST FROM OMAHA.		Dis. from Omaha.	OMAHA TIME.	Elevati'n	EAST FROM CALIFORNIA.	
Daily Emigr'nt.	Daily Express 1st&2d c's		STATIONS.		Daily Express 1st&2d c's	Daily Emigr'nt.
....	12.55	5.32	371	......Denver Junction......	3430	10.22 P M	11.50 P M	....
....	1.2	5.48	377	..............Weir...........	3500	10.08	10.30	....
....	2.15	6.20	387	............Chappel.........	3702	9.40	9.40	....
....	3.00	6.48	397	..........*Lodge Pole.......	3800	9.15	8.55	....
....	3.45	7.18	407	.............Colton..........	4022	8.50	8.15	....
....	4.30 †	7.45 †	414	............*Sidney.........	4073	8.05 †	7.05	....
....	5.30	8.35	423	...........Brownson........	4200	7.40	6.27	....
....	6.20	9.0	433	............*Potter..........	4370	7.15	5.45	....
....	7.00	9.3	442	...............Dix............	4580	6.50	5.07	....
....	7.45	10.00	451	............*Antelope........	4712	6.27	4.28	....
....	8.12	10.18	457	..............Adams..........	4784	6.12	4.05	....
....	8.40	10.35	463	............Bushnell.........	4860	5.50	3.37	....
....	9.30	11.05	473	........*Pine Bluffs, W. T...	5026	5.30	2.52	....
....	10.00	11.22	479	..............Tracy...........	5149	5.15	2.27	....
....	10.32	11.40	484	.............Egbert...........	5272	5.00	2.00	....
....	11.10	12.00 noon	490	..............Burns...........	5428	4.45	1.32	....
....	11.45	12.17 P M	496	............*Hillsdale........	5591	4.30	1.05	....
....	12.20 P M	12.37	502	..............Atkins..........	5900	4.12	12.20 P M	....
....	12.55	12.55	508	.............*Archer.........	6000	3.57	11.57	....
....	1.40 †	1.20 †	516	............*Cheyenne.......	6041	3.15 †	10.50	....
....	2.55	2.00	522	.......Colorado Junction.....	6325	2.55	10.15	....
....	3.17	2.15	526	...............Borie..........	6469	2.43	9.50	....
....	3.40	2.30	531	...............Otto...........	6724	2.30	9.25	....
....	4.05	2.50	536	.........*Granite Canon.....	7298	2.13	8.40	....
....	4.40	3.15	543	.............*Buford.........	7780	1.50	8.00	....
....	5.20	3.43	549	............*Sherman........	8242	1.23	7.00	....
....	6.20	4.08	555	...........Tie Siding.........	7085	1.00	6.25	....
....	6.50	4.25	559	.............Harney..........	7857	12.45	6.00	....
....	7.20	4.43	564	...........*Red Buttes.......	7336	12.30	5.35	....
....	8.05	5.10	570	.........*Fort Sanders......	7163	12.12	4.57	....
....	8.20 P M	5.20 P M	573	............*Laramie........	7123	12.05 P M	4.45 A M	....

LARAMIE DIVISION.

E. DICKINSON, Div. Supt. ....LARAMIE, W. T.

....	9.00 P M	5.00 P M	573 Lv	......*Laramie......Av	7123	11.20	3.15	....
....	9.40	5.22	581	..............Howell..........	7090	11.00	2.38	....
....	10.09	5.37	588	...........*Wyoming........	7068	10.46	2.10	....
....	10.45	5.57	595	..............Buttons.........	7048	10.30	1.42	....
....	11.20	6.15	602	..........*Cooper's Lake.....	7044	10.14	1.15	....
....	11.47	6.30	606	............*Lookout.........	7109	10.00	6.45	....
....	12.15 A M	6.46	611	............Harper's..........	7140	9.42	12.15 A M	....
....	12.40	7.05	614	..............*Miser...........	6810	9.20	11.40	....
....	1.00	7.20 †	623	...........*Rock Creek......	6690	9.05 †	11.13	....
....	1.30	8.05	630	..............Wilcox..........	7033	8.22	10.35	....
....	2.19	8.30	638	..............Aurora..........	6080	8.00	9.55	....
....	2.55	8.50	645	..........*Medicine Bow.....	6550	7.42	9.00	....
....	3.33	9.24	652	..........Niles Junction......	6540	7.22	8.25	....
....	3.50	9.35	656	..............*Carbon.........	6750	7.14	8.10	....
....	4.25	10.00	662	..............Simpson.........	6898	6.55	7.42	....
....	4.50	10.13	669	..............*Percy..........	6950	6.40	7.22	....
....	5.19	10.29	675	...............Dana...........	6875	6.23	6.55	....
....	6.00	10.48	680	..............*Edson..........	6751	6.00	6.20	....
....	6.34	11.08	688	..............Wolcott.........	6800	5.37	5.45	....
....	7.00	11.24	694	............Fort Steele........	6840	5.20	5.19	....
....	7.42	11.45	702	............Grenville.........	6560	4.59	4.43	....
....	8.20	12.05 A M	709	............*Rawlins.........	6732	4.40	4.10	....
....	10.00	12.50	716	...............Solon...........	6821	4.05	2.55	....
....	10.30	1.10	723	...........*Separation.......	6900	3.35	2.15	....
....	11.10	1.36	730	.............Fillmore.........	6885	3.10	1.35	....
....	11.45	1.55	737	.............*Creston.........	7030	2.50	1.05	....
....	12.20 P M	2.20	744	..............Latham.........	6900	2.20	12.20 P M	....
....	12.55	2.40	752	..........*Wash-a-kie.......	6697	1.50	11.50	....
....	1.40	3.10	761	...........*Red Desert.......	6710	1.20	10.57	....

UNION PACIFIC .... **LARAMIE DIV.**
CONTINUED.

† Meals. * Telegraph.

WEST FROM OMAHA		Dis. from Omaha	LARAMIE TIME. STATIONS.	Elevati'n	EAST FROM CALIFORNIA.	
Daily Emigr'nt	Daily Express 1st&2d c's				Daily Express 1st&2d c's	Daily Emigr'nt
....	2.15	3.35	768	........Tipton..........	6600	1.0′	10.30	....
....	2.50	3.57	775	......*Table Rock........	6890	12.3	9.50	....
....	3.15	4.12	780	........Monell..........	6785	12.10	9.24	....
....	3.35	4.26	785	......*Bitter Creek......	6685	12.01 A M	8.40	....
....	4.18	4.53	791	......*Black Buttes......	6600	11.35	7.55	....
....	4.40	5.08	798	........Hallville........	6590	11.20	7.30	....
....	5.08	5.25	805	...*Point of Rocks.......	6490	11.00	7.00	....
....	5.32	5.40	810	........Thayer..........	6425	10.45	6.32	....
....	6.00	6.00	817	......*Salt Wells........	6360	10.25	6 00	....
....	6.30	6.20	825	........Baxter..........	6300	10.02	5.28	....
....	7.00	6.40	831	......*Rock Springs......	6280	9.45	4.55	....
....	7 40	7.05	839	........Wilkins..........	6200	9.18	4.15	....
....	8.10P M†	7.25 †	845	Ar......Green River......Lv	6140	9.00 †	3.45	....

WESTERN DIVISION.

W. B. DODDRIDGE, *Div. Supt.* .... EVANSTON.

....	8.40 P M	7.45 PM†	845	Lv....*Green River....Ar	6140	8.40 †	3.00 A M	....
....	9.25	8.12	853	.........Peru..........	....	8.15	2.15	....
....	9.55	8.30	858	........*Bryan.........	6340	7.50	1.50	....
....	10.25	8.50	866	........Marston........	6245	7.35	1.20	....
....	11.10	9.15	876	......*Granger.........	6270	7.10	12.40 A M	....
....	11.53	9.45	887	...*Church Buttes......	6317	6.45	11.55	....
....	12.25 A M	10.05	896	........Hampton........	6500	6.25	11.25	....
....	1.10	10.35	904	........*Carter.........	6550	5.55	10.40	....
....	1.55	11.00	913	........*Bridger........	6780	5.30	10.00	....
....	2.15	11.15	918	..........Leroy.........	7123	5.15	9.40	....
....	3.00	11.45	928	.......*Piedmont,.......	6540	4.40	8.55	....
....	4.00	12.25 P M	937	........*Aspen..........	7835	4.15	7.52	....
....	4.30	12.42	942	.......*Hilliard........	7310	4.02	7.20	....
....	4.50	12.55	945	.........Millis.........	6790	3 50	6.50	....
....	5.40	1.25 †	955	.......*Evanston........	6870	3.00 †	5.20	....
....	6.18	1.53	956	....Almy Junction.......	6872	2.53	5.10	....
....	6.47	2.10	961	.........Midway.........	6876	2.37	4.50	....
....	7.10	2.25	966	.......*Wasatch........	6879	2.25	4.30	....
....	8.00	2.58	975	......*Castle Rock......	6290	1.40	3.15	....
....	8.35 †	3.20	982	.........*Emory.........	5974	1.20	3.20	....
....	9.35	3.50	991	..........*Echo.........	5315	12.35	2.30	....
....	10.55	4.20	999	........*Croydon.......	5250	12.10 P M	1.50	....
....	11.40	4.45	1007	.........*Weber.........	5130	11.50	1.12	....
....	12.35 P M	5.05	1015	.......*Peterson........	4903	11.28	12.35	....
....	1.00	5.18	1019	......Devil's Gate......	4870	11.15	12.15 P M	....
....	1.30	5.40	1024	........*Ulutah.........	4500	11.00	11.30	....
....	2.10 †	6.00 †	1032	Ar......*Ogden......Lv	4340	10.40 AM†	11.15 A M	....

Utah & Northern Branch U. P. R'Y.

NORTH-WARD. Pass'ngr	Miles	STATIONS.	SOUTH-WARD. Pass'ngr
7.00 P M	....	Lv.†Ogden.Ar	7.45 A M
7.30	9	.Hot Springs.	7.00
8.15	21	†Brigham..	6.30
10.55	58	....Logan.....	3.40
12.40	80	...Franklin...	2.20
5.45 A M	158	...Pocatello...	9.08
6.55	181	..†Blackfoot..	7.30
8.45	206	..Eagle Rock..	6.10 P M
10.50	245	....Camas.....	4.00
12.40 P M	274	†Beaver Canon	2.25
5.20	350	...†Dillon....	9.25
7.20	380	...Melrose...	7.05
10.30 P M	416	Ar..Butte..Lv	4.20 A M

UTAH CENTRAL R. R.

SOUTHWARD - DAILY.		Miles	STATIONS.	NORTHWARD DAILY.	
Pass'ngr	Pass'ngr			Pass'ngr	Pass'ngr
6.20 P M	9.40 A M	..	Lv. Ogden..Ar	9.00 A M	5.40 P M
7.10	10.31	16	..Kaysville..	8.12	4.52
7.31	10.52	21	.,Farmington.	7.50	4.33
7.44	11.04	26	..Centreville.	7.33	4.16
7.53	11.13	28	Wood's Cros'g	7.25	4.08
8.20 P M	11.40 A M	37	Ar.S'lt L'k.Lv	7.00 A M	3.40 P M

☞ At Salt Lake City connection is made with the Utah Southern Railroad for 'Frisco, Silver Reef (or Leeds) and all points in Southern Utah, Southeastern Nevada and Northern Arizona.

OVERLAND TIME TABLE.

CENTRAL PACIFIC .... SALT LAKE DIV.

A. G. Fell, *Division Supt.*, Ogden, Utah.

† Day Telegraph.   ‡ Day and Night Telegraph.   * Meals.

WEST FROM OMAHA.		Dist. from Omaha.	SACRAMENTO TIME.	Elevation.	EAST FROM CALIFORNIA.	
Daily Emigr'nt	Daily Ex 1st & 2d Class.		STATIONS.		Daily Ex 1st & 2d Class.	Daily Emigr'nt
2.00* P M	6.00* P M	1032	Lv........‡Ogden........Av	4340	8.30* A M	8.00 A M
2.40	6.23	1041	........Bonneville........	4251	8.07	7.20
2.05	6.42	1048	........Brigham........	4240	7.50	6.55
3.40	7.03	1056	........†Corinne........	4220	7.30	6.10
4.10	7.23	1064	........Quarry........	4271	7.12	5.40
4.55	7.45	1075	........†Blue Creek........	4379	6.50	5.00
6.00	8.30	1084	........‡Promontory........	4905	6.15	3.40
7.10	8.59	1092	........Rozel........	4588	5.45	2.55
7.55	9.25	1101	........Lake........	4223	5.15	2.05
8.25	9.42	1108	........Monument Point........	4226	5.01	1.25
9.00	9.58	1110	........Seco........	4224	4.45	12.45
9.30	10.15	1123	........‡Kelton........	4222	4.30	12.05 A M
10.40	11.00	1135	........Ombey........	4310	4.05	10.40
11.30	11.30	1145	........Matlin........	4630	3.40	9.55
12.25 A M	12.01 A M	1150	........‡Terrace........	4619	3.00	8.30
2.30	12.35	1166	........Bovine........	4346	2.30	7.35
3.25	1.00	1179	........Lucin........	4404	2.05	6.55
4.25	1.35	1188	........†Tecoma........	4812	1.35	6.10
5.15	2.10	1198	........Montello........	4990	1.08	5.25
6.20	2.55	1207	........Loray........	5555	12.40	4.40
7.15	3.30	1214	........‡Toano........	5070	12.15 A M	4.05
8.05	4.03	1224	........Pequop........	6183	11.50	3.20
8.35	4.22	1230	........†Otego........	6153	11.33	2.55
9.00	4.37	1235	........Independence........	6004	11.15	2.25
9.30	4.55	1241	........Moors........	6118	10.55	1.50
9.45	5.02	1244	........Cedar........	5978	10.45	1.35
10.15	5.20 A M	1250	Av........‡Wells........Lv	5628	10.20 P M	1.00 P M

HUMBOLDT DIVISION.

G. W. Coddington, *Division Supt* .... Carlin, Nev.

12.30 P M	5.30 A M	1250	Lv........‡Wells........Ar	20.8	10.10 P M	12.3? P M
1.05	5.45	1258	........Tulasco........	2483	9.50	11.50
1.30	5.55	1263	........Bishop's........	5400	9.40	11.25
2.10	6.11	1270	........Deeth........	5340	9.25	10.45
2.32	6.20	1276	........Natchez........	....	9.16	10.25
3.10	6.38	1283	........†Halleck........	5227	8.57	9.50
3.25	6.47	1287	........Peko........	5204	8.47	9.35
4.15	7.14	1298	........Osina........	5135	8.20	8.40
4.25	7.34	1307	........*Elko........	5065	7.40	7.54
5.55	8.18	1319	........Moleen........	4981	7.16	7.00
6.50	8.46	1330	........†Carlin........	4903	6.40	5.20
8.10	9.20	1339	........Palisade........	4840	6.15	4.40
8.50	9.41	1349	........Cluro........	4766	5.53	3.45
9.25	10.00	1358	........†Be-o-wa-we........	4690	5.36	3.05
10.05	10.21	1368	........Shoshone........	4636	5.16	2.20
10.55	10.45	1379	........Argenta........	4548	4.54	1.30
11.55	11.11	1390	........†Battle Mountain........	4508	4.30	12.35
12.15 A M	11.22	1395	........Piute........	4506	4.21	12.15 A M
12.55	11.37	1403	........Coin........	4505	4.08	11.45
1.30	11.52	1410	........Stone House........	4505	3.55	11.15
1.33	12.20 P M	1423	........Iron Point........	4421	3.30	10.25
3.30	12.45	1436	........Golconda........	4375	3.03	9.35
4.13	1.08	1445	........Tule........	4387	2.40	8.55
4.40 A M	1.20	1451	Ar........‡Winnemucca........Lv	4315	2.30 P M	8.30 P M

TRUCKEE DIVISION.

Frank Free, *Division Supt.* .... Wadsworth, Nev.

5.30 A M	1.30 P M	1451	Lv........Winnemucca........Ar	4315	2.20 P M	7.35
6.15	1.53	1461	........Rose Creek........	4331	1.55	6.50
6.55	2.18	1471	........Raspberry........	4322	1.33	6.08
7.25	2.32	1478	........†Mill Creek........	43.7	1.16	5.35
8.15	2.55	1491	........†Humboldt........	4228	12.50	4.45
9.00	3.42	1502	........†Rye Patch........	4239	12.05	3.42
10.10	4.03	1513	........†Oreana........	4256	11.40 P M	2.55 P M

CENTRAL PACIFIC............................TRUCKEE DIV.
CONTINUED.

† Day telegraph.   ‡ Day and night telegraph.   * Meals.

WEST FROM OMAHA.		Mls. from Omaha.	SACRAMENTO TIME.	Elevati'n	EAST FROM CALIFORNIA.	
Daily Emigr'nt & Freight	Daily Express 1st&2d c's		STATIONS.		Daily Express 1st&2d c's	Daily Emigr'nt & Freight
.... 11.10 A M	4.28 P M	1524	........†Lovelocks........	3977	11.10	1.50 ....
.... 11.55	4.45	1533	......Granite Point......	3017	10.50	1.10 ....
.... 12.35 P M	5.03	1540	..........†Brown's........	3925	10.30	12.35 P M ....
.... 1.35	5.30	1552	........White Plains......	3893	10.00	11.35 ....
.... 2.10	5.50	1559	..........Mirage..........	4199	9.43	11.00 ....
.... 2.40	6.05	1567	........†Hot Springs......	4070	9.27	10.20 ....
.... 3.35	6.30	1577	..........Desert..........	4017	9.00	9.25 ....
.... 4.30	6.44	1584	.......Two Mile Station...	4155	8.46	8.55 ....
.... 5.30	7.00	1586	........†Wadsworth.......	4077	8.30	8.00 ....
.... 6.00	7.17	1593	..........Salvia..........	4130	8.10	7.30 ....
.... 6.30	7.35	1601	..........Clark's.........	4263	7.51	7.00 ....
.... 7.25	8.02	1613	...........Vista..........	4403	7.20	6.10 ....
.... 8.50	8.50	1622	.........‡Reno...........	4507	6.30	4.55 ....
.... 9.40	9.20	1631	.........†Verdi..........	4927	6.00	4.15 ....
.... 9.55	9.25	1632	..........Essex..........	5010	5.56	4.10 ....
.... 10.30	9.48	1639	..........Mystic.........	5216	5.40	3.45 ....
.... 10.50	10.08	1641	.........†Bronco.........	5340	5.30	3.30 ....
.... 11.20	10.25	1647	..........‡Boca..........	5533	5.13	3.05 ....
.... 11.30	10.30	1649	.......†Prosser Creek.....	5010	5.09	2.58 ....
.... 11.55	10.47	1652	........Proctors.........	5720	4.57	2.42 ....
.... 12.10 A M	10.55 P M	1655	Ar......‡Truckee......Lv	5845	4.50 A M	2.30 ....

SACRAMENTO DIVISION

J. B. WRIGHT, *Div. Supt*..SACRAMENTO.

.... 1.55 A M	11.05 P M	1655	........‡Truckee.........	5845	4.40	1.40 ....
.... 2.45	11.30	1662	......Strong's Canyon.....	6780	4.15	1.00 ....
.... 3.45	12.08 A M	1671	..........‡Summit........	7017	3.45	12.05 A M ....
.... 4.00	12.16	1673	.......Soda Springs.......	....	3.35	11.40 ....
.... 4.20	12.27	1675	.........Cascade.........	6519	3.23	11.25 ....
.... 4.40	12.41	1679	.........Tamarack........	6401	3.05	11.05 ....
.... 5.00	12.53	1683	.........‡Cisco..........	5939	2.50	10.45 ....
.... 5.45	1.25	1691	......‡Emigrant Gap......	5229	2.12	9.55 ....
.... 6.30	2.00	1697	.......‡Blue Canyon......	4677	1.25	8.25 ....
.... 7.00	2.20	1702	.........Sandy Run.......	4154	1.02	7.35 ....
.... 7.35	2.40	1706	..........†Alta..........	3612	12.40	6.25 ....
.... 7.45	2.50	1708	........Dutch Flat........	3403	12.30	5.55 ....
.... 8.00	3.00	1710	........†Gold Run........	3206	12.20 A M	5.35 ....
.... 8.30	3.25	1617	.........C. H. Mills......	2691	11.57	4.55 ....
.... 9.00	3.45	1721	.........†Colfax.........	2421	11.15 *	4.20 ....
.... 9.25	3.57	1724	.........Lauder..........	....	11.05	3.30 ....
.... 9.35	4.05	162.	.........N. E. Mills.......	2280	10.57	3.10 ....
.... 9.55	4.18	1728	.........Applegate.......	2000	10.47	2.45 ....
.... 10.10	4.30	1732	........Clipper Gap.......	1759	10.38	2.25 ....
.... 10.55	4.55	17 9	.........†Auburn.........	1392	10.15	1.40 ....
.... 11.25	5.15	1744	........†New Castle.......	969	9.55	1.00 ....
.... 11.45	5.27	1747	..........Penryn..........	....	9.37	12.25 ....
.... 12.05	5.37	1750	...........Pino..........	403	9.25	12.05 P M ....
.... 12.35 P M	5.55	1752	.........‡Rocklin.........	248	8.55	11.00 ....
.... 1.00	6.05	1757	.........‡Junction........	163	8.42	10.30 ....
.... 1.20	6.15	1760	.........Antelope.........	154	8.30	10.05 ....
.... 1.50	6.33	1767	..........Arcade..........	55	8.15	9.35 ....
.... 2.10	6.45	1771	.......‡A. M. Bridge......	52	8.04	9.15 ....
.... 2.30 P M	6.55 A M	1775	Ar......Sacramento.....Lv	30	7.55 P M	9.00 A M ....

NOTE.—There are two routes south from Sacramento. Passengers for Stockton, Los Angeles Southern California and intermediate points will take the route described on page 173. See Western Division time table, further on. Those for San Francisco direct, take route described on page 183—New Short Line. Time table on next page.

CENTRAL PACIFIC............SHORT LINE via BENICIA.

J. B. WRIGHT, *Division Supt.,* SACRAMENTO.

* Trains stop on Signal.   § Trains will not stop.   †Telegraph.   ‡ Meals.

WEST FROM OMAHA		Dis. from Omaha	SAN FRANCISCO TIME. STATIONS.	Altitude.	EAST FROM SAN FRAN'SCO	
Daily Emigr'nt & Fre'ght	Daily Ex 1st & 2d Class.				Daily Ex 1st & 2d Class.	Daily Emigr'nt & Freight
4.00 P M	7.20 A M	1776	Lv....†Sacramento......Ar	30	7.30 P M	6.40 A M
4.30	7.36 §	1784	........Webster......	28	7.12 §	6.00
4.40	7.45	1789	........† Davis........	25	7.00	5.30
5.05	7.55 *	1793	........Tremont......	24	6.50	5.15
5.20	8.05	1797	........†Dixon........	25	6.40	4.55
5.35	8.13	1800	........†Batavia......	26	6.31	4.40
6.20	8.25	1805	........Elmira........	26	6.20	4.18
6.45	8.35 *	1809	........Cannon.......	24	6.11	4.00
7.15	8 50	1816	........†Suisun.......	24	5.55	3.30
7.45	9.03 *	1821	........Teal..........	24	5.44	3.10
8.05	9.16 *	1826	........†Goodyears...	26	5.38	2.53
8.20	9.26	1830	........Army Point...	22	5.19 *	2.38
8.25	9.28	1831	........Mail Dock....	10	5.17 §	2.34
8.30 P M	9.30 A M	1832	........†Benicia......	10	5.15 P M	2.30 A M

Steam Ferry-Boat "Solano" Across Straits of Carquinez.

A. D. WILDER, *Division Supt.,* OAKLAND WHARF.

9.40 P M	9.50 A M	1833	........†Port Costa....	9	4.55 §	1 40
9.52	9.56 §	1836	........†Valona.......	18	4.47 §	1.23
9.55	9.58	1837	........†Vallejo Junction..	18	4.45	1.21
10.03	10.04 §	1839	........Tormey.......	18	4.40 §	1.09
10.15	10.11 §	1841	........†Pinole........	17	4.35 §	12.55
10.30	10.20 §	1845	........Sobrante......	16	4.29 §	12.36
10.47	10.30	1848	........†San Pablo....	15	4.23 §	12.18 A M
10.04	10.40 §	1852	........Stege.........	15	4.15 §	11.59
		1853	........Point Isabel...	14		
11.15	10.46 §	1854	........Highland......	14	4.10 §	11.47
11.22	10.49 §	1855	........†Delaware St..	13	4.08 §	11.40
11.30	10.53 §	1857	........†Stock Yards..	13	4.04 §	11.30
11.45	11.00	1859	........†Oakland 16th St.	13	4.00	11.10
11.55		1860	........West Oakland..	11		11.00
11.10		1862	........†Oakland Wharf..	10	3.50	
6.05 A M	11.35 A M	1865	........†San Francisco..		3.30	5.30 P M

Old Overland Route from Sacramento via Stockton....Western Div.

See page 173.

	11.50 A M	1776	Lv....†Sacramento......Ar	30	2.10 P M	
	12.05 P M	1780	........†Brighton......	55	1.55	
	12.13	1784	........Florin........	32	1.46	
	12.20	1791	........†Elk Grove....	53	1.34	
	12.33 *	1794	........McConnell's...	49	1.26 *	
	12.48	1802	........‡Galt.........	49	1.12	
	1.00	1807	........Acampo.......	51	1.00	
	1.06	1810	........†Lodi.........	55	12.50	
	1.20 *	1817	........Castle........	27	12.33 *	
	1.35	1823	........†Stockton.....	23	12.21 P M	
	1.55 ‡	1832	........†Lathrop......	25	11.59	

Connect at Lathrop for the South.   See pages 209 and 270.

	2.02 *	1835	....San Joaquin Bridge...		11.35	
	2.14	1840	........Banta........		11.43 *	
	2.20	1843	........†Tracy........	58	12.51	

Connect at Tracy for San Francisco via Martinez.   See pages 178 and 270.

	2.25 *	1845	........Ellis.........	76	11.25 *	
	2.41	1850	........Medway.......	357	11.09	
	3.02 *	1858	........†Altamont.....	740	10.47	
	3.24	1867	........†Livermore....	485	10.25	
	3.40	1873	........‡.Pleasanton..	351	10.12	
	3.55	1878	........Sunol........	170	10.00	
	4.15 P M	1884	........†Niles........	86	9.40 A M	

Connect at Niles for San Jose and Santa Clara Valleys.   See page 180.

CENTRAL PACIFIC......WESTERN DIVISION.
CONTINUED.

* Trains stop only on Signal. † Telegraph. § Trains will not stop. ‡ Meals.

WEST FROM OMAHA.		Dis. from Omaha.	SAN FRANCISCO TIME.	Altitude.	EAST FROM SAN FRA'CIS'O		
	Daily Express 1st&2d c's		STATIONS.		Daily Express 1st&2d c's		
....	5.27	1887	........Decota........	7	9.33	....	....
....	4.44	1893	........†Haward's........	48	9.18	....	....
....	4.50	1896	........Lorenzo........	32	9.12	....	....
....	4.57	1900	........†San Leandro........	48	9.06	....	....
....	5.08	1903	........†Melrose........	20	8.56	....	....
....	5.15	1905	........†East Oakland........	12	8.50	....	....
....	5.24	1907	........†Oakland........	13	8.40	....	....
....	5.30	1909	........†West Oakland........	12	8.34	....	....
....	5.43	1911	........†Oakland Wharf........	10	8.20	....	....
....	6.05 P M	1914	Ar....†San Francisco....Lv	....	8.00 A M	....	....

TOWARDS SUNRISE—(See page 209.)

CENTRAL PACIFIC......WESTERN DIV.
NORTHERN RAILWAY AND SAN PABLO AND TULARE RAILROAD.

FROM SAN FRANCISCO.			SAN FRANCISCO TIME.	Altitude.	TOWARDS SAN FRA'CISCO.		
	Daily Emigr'nt & Freight	Daily Express 1st&2d c's	Miles. STATIONS.		Daily Express 1st&2d c's	Daily Emigr'nt & Freight	
....	4.30 P M	9.30 A M	0 Lv....†San Francisco....Ar	....	2.35 P M	6.05 A M	....
....	....	9.50	4 ....†Oakland Wharf........	10	2.10	....	....
....	7.15	10.00	7 ....†Oakland (16th St).....	11	2.00	4.45	....
....	7.25	10.06 *	9 ....†Stock Yards........	13	1.54 §	4.34	....
....	7.46	10.20 *	14 ....Stege........	15	1.42 *	4.10	....
....	8.08	10.30	18 ....†San Pablo........	15	1.34 *	3.53	....
....	8.15	10.40 *	21 ....Sobrante........	16	1.26 *	3.38	....
....	8.30	10.49 *	24 ....†Pinole........	17	1.19 *	3.24	....
....	8.55	11.04	29 ....†Vallejo Junction........	18	1.07	3.02	....
....	9.15	11.13	32 ....†Port Costa........	19	1.00	2.45	....
....	9.30	11.22	36 ....†Martinez........	19	12.52	2.20	....
....	9.46	11.30 *	39 ....†Avon........	19	12.44 *	2.13	....
....	10.00	11.38 *	42 ....†Bay Point........	19	12.37 *	1.59	....
....	10.33	11.58	50 ....†Cornwall........	20	12.20 §	1.24	....
....	10.53	12.10 P M	55 ....†Antioch........	20	12.10 P M	1.03	....
....	11.30	12.28	63 ....†Brentwood........	22	11.54	12.23 A M	....
....	11.50	12.38	68 ....†Byron........	28	11.44	11.50	....
....	12.43 A M	12.56	77 ....†Bethany........	39	11.27	11.10	....
....	1.15	1.10	83 ....†Tracy........	58	11.14	10.33	....
....	1.30	1.16 *	86 ....†Banta........	46	11.08	10.15	....
....	2.15	1.35 ‡	94 ....†Lathrop........	25	10.50 A M	9.30	....

CENTRAL PACIFIC......VISALIA DIV.
W. W. PRUGH, Ass. Div. Supt.

....	3.15 A M	2.00 P M	94 Lv....Lathrop....Ar	25	10.35 A M	8.00 P M	....
....	3.41	2.14 *	99 ....Morrano........	....	10.23 *	7.35	....
....	4.02	2.26 *	104 ....Ripon........	....	10.12 *	7.14	....
....	4.16	2.34 *	107 ....Salida........	....	10.05 *	7.00	....
....	4.47	2.52	114 ....†Modesto........	91	9.45	6.28	....
....	5.09	3.02 *	118 ....Ceres........	....	9.38 *	6.06	....
....	5.48	3.24	127 ....†Turlock........	....	9.18	5.25	....
....	6.45	3.49 *	137 ....Chessey........	....	8.54 *	4.38	....
....	7.20	4.06	144 ....Atwater........	....	8.38	4.06	....
....	8.20	4.25	151 ....†Merced........	171	7.55 ‡	3.20	....
....	9.05	4.55	161 ....†Athlone........	....	7.38	2.30	....
....	9.32	5.10 *	168 ....Minturn........	....	7.19 *	2.00	....
....	10.15	5.35	177 ....Berenda........	....	6.58	1.17	....
....	10.46	5.55	185 ....†Madera........	....	6.43	12.45	....
....	11.00	6.27	187 ....†Barden........	....	6.35	12.32 P M	....
....	11.50	6.46 *	196 ....Sycamore........	....	6.16 *	11.50	....
....	12.35 P M	7.07	206 ....†Fresno........	292	5.55	11.07	....
....	1.15	7.28 *	216 ....Fowler........	....	5.35 *	10.24	....
....	2.03	7.51	220 ....†Kingsburg........	....	5.13	9.36	....
....	2.38	8.08	234 ....Cross Creek........	....	4.55	9.00	....
....	3.05	8.22	240 ....†Goshen........	278	4.43	8.33	....
....	3.50 P M	8.45 P M	251 Lv....†Tulare....Ar	282	4.20 A M	7.45 A M	....

SOUTHERN PACIFIC.................................TULARA DIV.

† Telegraph.　　* Trains stop only on signal.　　§ Trains will not stop.　　‡ Meals.

FROM SAN FRANCISCO.		Dis. from San Fran.	SAN FRANCISCO TIME.	Altitude.	TOWARDS SAN FRA'CISCO.	
Daily Emigr'nt & Freight	Daily Express 1st&2d c's		STATIONS.		Daily Express 1st&2d c's	Daily Emigr'nt & Freight
....	4.30 P M	8.50 P M	251	Lv......†Tulare........Ar	282	4.15 A M	7.15 A M	....
....	5.23	9.11	261	......Tipton..........	267	3.52	6.23	....
....	6.23	9.35	273	........Alila.........	....	3.27 *	5.24	....
....	7.05	9.51 *	281	.....†Delano.........	313	3.10	4.43	....
....	8.04	10.14 *	293	.......†Poso.........	....	2.45	3.45	....
....	8.45	10.30	301	.......Lerdo.........	....	2.28 *	3.03	....
....	10.00	10.55	314	.....†Sumner........	415	2.08	2.03	....
....	11.17	11.17	321	......Wade..........	....	1.40	*12.45 A M	....
....	12.01 A M	11.35	329	......Pampa.........	....	1.30 *	11.35	....
....	1.05	11.50	336	....†Caliente........	1290	1.05	10.40	....
....	1.45	12.40 A M	341	.....Bealville........	....	12.40	10.04	....
....	2.39	1.23	349	......†Keene.........	....	12.05 A M	9.13	....
....	3.15	1.52 *	355	......Girard.........	....	11.40 *	8.40	....
....	4.00	2.30	361	†Tehachapia Summit...	3964	11.10	8.00	....
....	4.54	2.57 *	370	.....Cameron........	....	10.39 *	7.05	....
....	6.00	3.30 A M	381	Ar......†Mojava......Lv	2757	10.00	6.00 P M	....

LOS ANGELES DIVISION.

E. E. HEWITT, Asst. Supt..LOS ANGELES.

....	7.00 A M	3.35 A M	381	Lv......†Mojava......Ar	2751	9.55 P M	4.45 P M	....
....	7.50 *	4.05 *	395	.....Sand Creek......	2315	9.12 *	3.45 *	....
....	8.30	4.30	406	.....†Lancaster......	2350	8.45	3.00	....
....	9.30 *	5.10 *	417	......Alpine.........	2823	8.20 *	2.10 *	....
....	10.00 *	5.25 *	421	.....Vincent........	....	8.08 *	1.45 *	....
....	10.30 *	5.40 *	427	......Acton.........	3211	7.50 *	1.05 *	....
....	10.50	5.50	430	.....†Ravena........	2350	7.35	12.45 P M	....
....	11.55	6.10 *	439	.......Lang.........	1681	7.10 *	11.55	....
....	12.50 P M	6.40	452	.....Newhall........	1152	6.40	10.55	....
....	1.15 *	6.55 *	455	...S. F. Tunnel......	1469	6.25 *	10.25	....
....	1.45	7.10	461	†San Fernando......	1066	6.05	10.00	....
....	2.35 *	7.32 *	478	....Sepulveda.......	461	5.35 *	9.15 *	....
....	3.10	7.55 A M	482	Ar...†Los Angeles...Lv	265	5.15 P M	8.45 A M	....
....	5.30	8.25 ‡	482	Lv....Los Angeles....Ar	265	4.45 ‡	4.15 A M	....
....	6.00	8.48	491	.....San Gabriel.....	400	4.23	3.40	....
....	6.10	8.55	493	.....†Savanna.......	....	4.18	3.30	....
....	6.15	9.00	495	.......†Monte.......	260	4.15	3.25	....
....	6.35 *	9.15	501	......Puente........	323	4.00	3.00 *	....
....	7.10	9.37	511	......†Spadra.......	706	3.36	2.20	....
....	7.25	9.47	515	.....Pomona........	856	3.28	2.05	....
....	8.00 *	10.10	524	.....Cucamonga.....	952	3.05	1.30 *	....
....	9.00	10.45	539	......†Colton........	965	2.00 ‡	12.15	....
....	10.20 *	11.10 *	543	....Mound City.....	1055	1.50	12.01 A M*	....
....	11.15	11.45	554	.....El Casco.......	1874	1.17	11.15	....
....	12.01 A M*	12.15 P M	562	....San Gorgonio....	2592	12.55	10.25 *	....
....	12.30 *	12.37	569	......Banning.......	....	12.37	9.50 *	....
....	1.00 *	12.50 *	574	.....Cabazon.......	1779	12.17 *	9.20 *	....
....	1.35 *	1.10 *	583	....White Water.....	1126	11.50 *	8.35 *	....
....	2.05 *	1.28 *	590	....Seven Palms.....	584	11.25 *	8.00 *	....
....	3.00	1.50 *	602	.....Dry Camp......	....	10.55 *	7.15 *	....
....	3.45 *	2.15 *	611	......Indio..........	20	10.33 *	6.35	....
....	4.40	2.45	625	.....Walters........	135	10.00	5.45	....
....	5.55	3.20	642	....†Dos Palmas....	253	9.15	4.25	....
....	6.45 *	3.45	653	...Frink's Spring....	200 (Below Sea.)	8.50 *	3.45	....
....	8.05	4.25 *	671	...Flowering Well....	45	8.05	2.25 *	....
....	8.30 *	4.40 *	677	.....Tortuga........	183	7.50 *	2.00 *	....
....	8.52 *	5.00 *	683	....Mammoth Tank...	257	7.33 *	1.40 *	....
....	9.35 *	5.25 *	694	.....Mosquito......	294	7.05 *	1.00 *	....
....	10.25 *	5.55 *	707	......Cactus........	396	6.30 *	12.02 P M*	....
....	10.50	6.15 *	715	.....Ogilby.........	....	6.10 *	11.40	....
....	11.15	6.30	721	....Pilot Knob......	285	5.55 *	11.15	....
....	11.30	6.40	725	......El Rio.........	....	5.45	11.00	....
....	12.01 P M	7.00 ‡	730	Ar......†Yuma......Lv	123	5.30 A M ‡	10.30 A M	....

SOUTHERN PACIFIC............GILA AND TUCSON DIV.

A. A. BEAN, *Asst. Supt*, TUCSON.

‡ Meals.   * Trains stop on signal.   . Telegraph.

FROM SAN FRANCISCO.		Dis. from San Fran.	SAN FRANCISCO TIME.	Elevat'on	TOWARD SAN FRANCISCO	
Daily Emigr'nt	Daily Express 1st&2d c's		STATIONS.		Daily Express 1st&2d c's	Daily Emigr'nt
....	5.00 PM‡	7.30 PM‡	730	Lv........†Yuma........Ar	.....	5.00‡ A M	5.45 A M
....	6.03 *	8.03 *	744	..........Gila City.........	.....	4.27	4.27
....	7.11 *	8.42 *	760	..........Adonde...........	.....	3.48 *	2.55 *
....	7.55 *	9.05 *	770	...........Tacna...........	.....	3.24 *	1.55 *
....	9.0 *	9.43 *	787	......Mohawk Summit.......	.....	2.45 *	12.22* A M
....	9.31 *	10.00 *	793	.........Texas Hill.........	.....	2.32 *	11.45 *
....	10.30 *	10.30 *	806	...........Aztec............	.....	2.00 *	10.30
....	11.18 *	10.52 *	815	..........Stanwix...........	.....	1.37 *	9.40 *
....	11.42 *	11.03 *	820	..........Sentinel...........	.....	4.27 *	9.17 *
....	12.53 A M	11.37 *	834	........Painted Rock........	.....	12.53	8.05 *
....	2.00 *	12.15 A M	850	.........Gila Bend..........	.....	12.15 A M	6.15
....	3.15 *	12.39 *	859	..........Bosque............	.....	11.50 *	5.30 *
....	3.57 *	1.03 *	869	..........Estrella...........	.....	11 25 *	4.48 *
....	4.40 *	1.25 *	878	.........Montezuma.........	.....	11.00 *	4.05 *
....	5.50 *	1.47 *	887	.........†Maricopa..........	.....	10.35	3.25
....	7.20 *	2.23 *	902	........Sweet Water........	.....	9.55 *	2.10 *
....	8.10 *	2.50 *	913	........†Casa Grande........	.....	9.25	1.15
....	8.50 *	3.13 *	922	..........Toltec............	.....	9 00 *	12.27* P M
....	9.30 *	3.36 *	931	..........Picacho...........	.....	8.35 *	11.40 *
....	10.30 *	4.10 *	945	..........Red Rock..........	.....	7.57 *	10.30
....	11.28 *	4.50 *	961	..........Rillito............	.....	7.15 *	8.50 *
....	12.30 P M	5.30 A M	978	Ar........‡Tucson........Lv	.....	6.30 P M	7.00 A M
....	2.00 PM‡	6.00 AM‡	978	Lv........‡Tucson........Ar	.....	6.00‡ P M	5.00 A M
....	3.25 *	6.46 *	993	..........Papago............	.....	5.15 *	3.57 *
....	4.38 *	7.29 *	1006	.........†Pantano...........	.....	4.38 *	3.00 *
....	5.30 *	7.58 *	1015	..........Mescal............	.....	4.10 *	2.20 *
....	6.15 *	8.25 *	1024	.........†Benson...........	.....	3.45 *	12.15
....	8.20 *	8.55 *	1034	...........Ochoa............	.....	3.16 *	11.27 *
....	9.05 *	9.25 *	1043	.......Dragoon Summit.......	.....	2.48 *	10.44 *
....	9.55 *	9.56 *	1053	..........Cachise...........	.....	2.19 *†	9.55 *
....	10.57 *	10.30 *	1064	.........†Wilcox...........	.....	1.46 ‡	9.03 *
....	11.45 *	10.55 *	1072	.......Railroad Pass........	.....	1.22 *	8.23 *
....	{1.15 A M, 1.45}	{11.45, 12.10 PM}	1088	...........Bowie............	.....	{12.35, 12.10 PM}	{7.05, 6.33}
....	3.00 *	12.52 *	1103	........†San Simon.........	.....	11.25 *	5.18 *
....	4.12 *	1.32 *	1118	.........Steins Pass........	.....	10.43 *	4.05 *
....	5.25 *	2.12 *	1133	..........Pyramid...........	.....	10.00 *	3.00 *
....	6.12 *	2.25 *	1137	........†Lordsburgh.........	.....	9.48 *	2.25 *
....	6.57 *	2.50 *	1148	...........Lisbon...........	.....	9.18 *	1.40 *
....	7.30 *	3.15 *	1157	...........Separ............	.....	8.53 *	1.00 *
....	8.18 *	3.46 *	1169	............Wilna............	.....	8.18 *	12.12* P M
....	8.53 *	4.10 *	1178	............Gage............	.....	7.54 *	11.35 *
....	9.40 *	4.38 *	1189	............Tunis...........	.....	7.22 *	10.48 *
....	10.15 A M	5.00 PM‡	1197	Ar........‡Deming........Lv	.....	7.00‡ A M	10.15 A M

Trains *west* of Deming run on San Francisco time. Those *east* of Deming on Jefferson City (Mo.) time, which is *two hours faster* than San Francisco time.

SOUTHERN PACIFIC...................RIO GRANDE DIV.

JAMES CAMPBELL, *Asst. Supt.*, EL PASO.

....	12.30 PM‡	7.45 PM‡	1197	Lv........†Deming........Ar	232.2	8.00‡ A M	2.00‡ P M
....	1.15	8.08 *	1208	...........Zuni............	221.2	7.35 *	1.15
....	2.15 *	8.42 *	1223	.........Cambray..........	206.2	7.00 *	12.12* P M
....	3.10 *	9.12 *	1236	...........Aden............	192.8	6.30 *	11.15 *
....	4.00 *	9.40 *	1249	...........Afton...........	180.7	6.02 *	10.22 *
....	4.40 *	10.01 *	1258	..........Lanark...........	171.0	5.40 *	9.40 *
....	5.30 *	10.30 *	1271	.........Strauss...........	158.5	5.10 *	8.48 *
....	6.10 *	11.00 *	1281	..........Rogers...........	148.5	4.43 *	8.05 *
....	6.30 P M	11.15 P M	1285	Ar........†El Paso........Lv	144.0	4.30 A M	7.45‡ A M

ATCHISON, TOPEKA & SANTA FE RAILROAD,

DEMING TO KANSAS CITY.		Miles.	STATIONS. † Meals.	Elevation	KANSAS CITY TO DEMING.	
Emigr'nt	Express				Express	Emigr'nt
....	12.30 P M	8.00 P M	1149	Lv......Deming......Ar	....	7.45 AM†	7.00 A M	....
....	3.20	9.34	1110	........Sellers..........	....	5.38	3.45	....
....	5.00	10.05	1097	........Rincon..........	....	4.43	2.30	....
....	11.45	1.15 A M	1021	.......San Marcial......	....	1.15	8.00	....
....	1.32 A M	2.25	994	........Socorro..........	4665	12.10 A M	5.00	....
....	5.12	4.50	938	........Los Lunas........	4914	9.52	12.08 P M	....
....	7.15	5.40	918	......Albuquerque.......	5006	9.05	10.30 A M	....
....	8.24	6.19	902	........Bernalillo.......	5104	8.26	8.24	....
....	10.03	7.10	881	........Wallace..........	5329	7.35	7.10	....
....	1.00 P M	9.00 †	851	.........*Lamy..........	6531	6.05 †	4.47	....
....	2.10	10.07	841	........Glorieta.........	7537	4.50	3.20	....
....	8.25	1.25 P M	786	.......Las Vegas........	6452	1.45 †	8.45	....
....	11.00	2.50	758	.......Shoemaker........	....	12.01 P M	6.05	....
....	12.40 A M	3.50	741	.....Wagon Mound......	6247	11.05 A M	4.40 P M	....
....	7.05	6.42	681	........Otero...........	....	8.32	11.35 A M	....
....	8.00	7.20 †	676	........Raton...........	6688	8.20 †	11.10	....
....	11.50 A M	9.40 P M	652	........Trinidad.........	6031	5.50	8.00	....
....	6.50 P M	2.00 A M	571	........**La Junta.......	....	1.00	12.30 A M	....
....	8.20	2.55	552	....West Las Animas....	....	11.45 A M	10.35	....
....	8.40 P M	3.05	548	......Las Animas........	3959	11.30	10.16	....
....	12.45 A M	5.35	497	........Granada.........	3468	8.45	6.50	....
....	1.45	6.30 †	484	Lv......Coolidge......Ar	3418	8.00 †	5.40	....
....	3.45	7.43	458	........Aubrey..........	....	6.35	3.28	....
....	6.18	9.10	425	........Sherlock.........	2925	5.12	1.00	....
....	7.10	9.25	418	......Garden City.......	....	4.55	12.25 P M	....
....	8.10	9.57	406	......Pierceville........	2800	4.25	11.05 A M	....
....	9.30	10.43	387	........Cimarron.........	2655	3.37	9.30	....
....	11.00 A M	11.40 A M	369	......Dodge City........	2499	2.50	7.40	....
....	2.10 P M	1.15 P M	333	........Kinsley..........	2207	1.15	4.00	....
....	2.40	1.35	325	........Nettleton........	....	1.10	3.22	....
....	3.05	1.52	319	........Garfield..........	....	12.45	2.45	....
....	3.50	2.40 †	308	........Larned..........	2018	12.22 PM†	1.47	....
....	4.28	3.04	299	......Pawnee Rock......	1986	11.40 A M	1.02 A M	....
....	5.22	3.40	286	......Great Bend........	1859	11.10	11.50 P M	....
....	6.05	4.08	276	........Ellinwood........	1738	10.45	11.03	....
....	6.50	4.36	265	........Raymond........	1679	11.23	10.18	....
....	7.40	5.10	253	........Sterling..........	1494	9.56	9.25	....
....	8.10 P M	5.30 P M	245	Ar.....Nickerson......Lv	....	9.40 A M	8.50 P M	....
....	9.15	6.00	234	......Hutchinson........	1182	9.07	7.40	....
....	10.35	6.35	220	........Burrton..........	1410	8.30	6.35	....
....	11.25 P M	6.58	211	........Halstead.........	1320	8.07	5.40	....
....	12.15 A M	7.38	201	........Newton..........	1433	7.40	4.55	....
....	1.00	7.55	194	........Walton...........	1432	7.15	4.10	....
....	2.00	8.20	184	........Peabody.........	1256	6.53	3.20	....
....	3.00	9.10 †	173	........Florence.........	1277	6.30 †	2.30	....
....	5.05	10.12	148	......Cottonwood........	1183	5.05	12.25 P M	....
....	5.48	10.40	137	........Plymouth........	....	4.35	11.40 A M	....
....	7.00	11.10	128	........Emporia.........	1161	4.10	11.00	....
....	8.00	11.59 P M	113	........Reading.........	1074	3.20	9.35	....
....	8.50	12.29 A M	101	......Osage City........	1082	2.53	8.50	....
....	9.30	12.52	93	......Burlingame........	1050	2.33	8.15	....
....	9.55	1.05	88	........Scranton.........	....	2.20	7.53	....
....	10.15	1.15	84	......Carbondale........	1081	2.10	7.37	....
....	11.20 A M	2.00 A M	67	Ar.......Topeka......Lv	904	1.15 A M	6.30 A M	....
....	11.45 A M	2.25 A M	67	Lv.......Topeka......Ar	904	12.55	6.00 A M	....
....	12.47 P M	3.04	51	......Lecompton.........	....	12.17 A M	4.35	....
....	1.30	3.30	40	........Lawrence.........	....	11.50	3.30	....
....	4.38 P M	5.30 A M	0	Ar....Kansas City....Lv	765	10.00 P M	12.05 A M	....
....		2.20 A M	51	Lv.......Topeka......Ar	904	12.50 A M		....
....		5.10 A M	....	Ar......Atchison......Lv	803	10.25 P M		....

* Junction for Santa Fe, 18 miles distant.   ** Junction for Pueblo, Denver and Colorado.

A., T. &S. F. R. R. Trains are run by Jefferson time, being 2 hours faster than San Fran'co time.

TEXAS PACIFIC LINE.

FROM ST. LOUIS.		ST. LOUIS TIME.		FROM SAN FRANCISCO.	
	Express	Miles	STATIONS.	Miles	Express
	9.00 A M	0	Lv......St. Louis......	2645	6.00 P M
	9.26	6	......Carondelet........	2639	5.20
	10.45	42	......Desoto.......	2603	4.00
	12.17 P M	75	......Bismarck.........	2560	2.40
	1.15	89	......Arcadia........	2556	1.37 P M
	4.27	166	......Poplar Bluff......	2479	10.00
	8.29	262	......Newport........	2383	5.37
	12.15 A M	345	......Little Rock.......	2300	1.35 A M
	2.03	388	Ar......Malvern.........	2257	11.49 P M
	4.30 P M	413	Lv......Hot Springs......Ar	2232	3.50 A M
	2.58 A M	410	Lv......Arkadelphia......	2235	10.53 P M
	4.18	442	......Prescott.........	2030	9.32
	6.50 A M	490	Ar......Texarkana.......	2155	7.00 P M
	7.50 A M	490	Lv......Texarkana.......	2155	6.30 P M
	10.52	548	......Jefferson........	2097	3.37
	10.40	564	......Marshall........	2080	2.50
	1.10 P M	587	......Longview........	2057	11.30
	2.43	610	......Big Sandy........	2034	12.13 P M
	3.55	633	......Mineola........	2011	11.05
	5.45	663	......Will's Point......	1981	9.30
	6.33	679	......Terrell........	1965	8.47
	8.40	711	......Dallas........	1933	6.50
	10.10	743	......Ft. Worth........	1901	5.15
	11.55	774	......Weatherford......	1870	3.36
	12.55 A M	798	......Brazos........	1846	2.30 A M
	3.25	848	......Eastland........	1796	11.50
	3.50	858	......Cisco........	1786	11.25
	5.05	883	......Baird........	1761	10.10
	6.12	903	......Abilene........	1741	9.15
	7.58	945	......Sweetwater......	1699	7.25
	9.11	972	......Colorado........	1672	6.13
	11.15	1012	......Big Springs......	1632	4.10
	12.28 P M	1032	......Grelton........	1612	3.05 P M
	5.02	1143	......Pecos........	1501	10.30
	5.50	1163	......Toyah........	1482	9.40
	9.20	1232	......Van Horn........	1412	6.05
	10.50	1267	......Sierra Blanca.....	1377	4.30
	3.30 A M	1359	......El Paso........	1286	12.10 A M
	4.30 A M	1359	......El Paso........	1286	11.15 P M
	5.10	1373	......Strauss........	1272	10.30
	8.00	1447	Ar......Deming........	1198	7.45 P M
	7.00	1447	Lv......Deming........	1198	5.00 P M
	9.48	1507	......Lordsburg........	1138	2.25
	11.25	1541	......San Simon........	1104	12.52 P M
	3.45 P M	1620	......Benson........	1025	8.25
	6.00 P M	1666	Ar......Tucson........	979	6.00 A M
	6.30 P M	1666	Lv......Tucson........	979	5.30 A M
	10.35	1757	......Maricopa........	888	1.47 A M
	5.00 A M	1914	Ar......Yuma........	731	7.30 P M
	5.30 A M	1914	Lv......Yuma........	731	7.00 P M
	9.15	2002	......Dos Palmos......	643	3.20
	12.55 P M	2082	......San Gorgoina.....	563	12.15 P M
	1.50	2101	......Mound City.......	544	11.10
	2.25	2105	......Colton........	540	11.00
	3.36	2133	......Spadra........	512	9.37
	4.23	2153	......San Gabriel......	492	8.48
	4.45	2162	Ar......Los Angeles.....	483	8.25 P M
	5.15 P M	2162	Lv......Los Angeles.....	483	7.55 A M
	10.00	2263	......Mojave........	382	3.35 A M
	2.03 A M	2330	......Sumner........	315	11.00
	4.43	2404	......Goshen........	241	8.22
	6.43	2459	......Madera........	186	6.20
	10.50	2550	......Lathrop........	95	2.00 P M
	2.35 P M	2645	Ar......San Francisco....	0	9.30 A M

www.ingramcontent.com/pod-product-compliance
Lightning Source LLC
Chambersburg PA
CBHW032048220426
43664CB00008B/908